Akaike Information Criterion Statistics

Mathematics and Its Applications (*Japanese Series*)

Y. Sakamoto, M. Ishiguro, and G. Kitagawa

Institute of Statistical Mathematics, Tokyo, Japan

Akaike Information Criterion Statistics

KTK Scientific Publishers / Tokyo

D. Reidel Publishing Company

A MEMBER OF THE KLUWER ACADEMIC PUBLISHERS GROUP

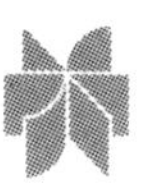

Dordrecht / Boston / Lancaster / Tokyo

Library of Congress Cataloging-in-Publication Data

CIP

Sakamoto, Y. (Yosiyuki), 1943–
Akaike information criterion statistics.

(Mathematics and its applications. Japanese series)
Translation of: Johoryo tokeigaku.
Bibliography: p.
Includes index.
1. Multivariate analysis. 2. Analysis of variance.
3. Distribution (Probability theory) I. Ishiguro, M. (Makio), 1946– . II. Kitagawa, G. (Genshiro), 1948– . III. Title. IV. Title: Information criterion statistics. V. Series: Mathematics and its applications (D. Reidel Publishing Company). Japanese series.
QA278.S2513 1986 519.5′35 86-13113
ISBN 90-277-2253-6 (D. Reidel)

Published by KTK Scientific Publishers (KTK),
307 Shibuyadai-haim, 4-17 Sakuragaoka-cho, Shibuya-ku, Tokyo 150, Japan,
in co-publication with D. Reidel Publishing Company, Dordrecht, Holland

Sold and distributed in the U.S.A. and Canada
by Kluwer Academic Publishers,
190 Old Derby Street, Hingham, MA 02043, U.S.A.,
in Japan by KTK Scientific Publishers (KTK),
307 Shibuyadai-haim, 4-17 Sakuragaoka-cho, Shibuya-ku, Tokyo 150, Japan

In all other countries, sold and distributed
by Kluwer Academic Publishers Group,
P. O. Box 322, 3300 AH Dordrecht, Holland

Printed in Japan

TABLE OF CONTENTS

SERIES EDITOR'S PREFACE

Approach your problems from the right end and begin with the answers. Then one day, perhaps you will find the final question.

'The Hermit Clad in Crane Feathers' in R. van Gulik's *The Chinese Maze Murders*.

It isn't that they can't see the solution. It is that they can't see the problem.

G. K. Chesterton. *The Scandal of Father Brown* 'The point of a Pin'.

Growing specialization and diversification have brought a host of monographs and textbooks on increasingly specialized topics. However, the "tree" of knowledge of mathematics and related fields does not grow only by putting forth new branches. It also happens, quite often in fact, that branches which were thought to be completely disparate are suddenly seen to be related.

Further, the kind and level of sophistication of mathematics applied in various sciences has changed drastically in recent years: measure theory is used (non-trivially) in regional and theoretical economics; algebraic geometry interacts with physics; the Minkowsky lemma, coding theory and the structure of water meet one another in packing and covering theory; quantum fields, crystal defects and mathematical programming profit from homotopy theory; Lie algebras are relevant to filtering; and prediction and electrical engineering can use Stein spaces. And in addition to this there are such new emerging subdisciplines as "experimental mathematics", "CFD", "completely integrable systems", "chaos, synergetics and large-scale order", which are almost impossible to fit into the existing classification schemes. They draw upon widely different sections of mathematics. This programme, Mathematics and Its Applications, is devoted to new emerging (sub)disciplines and to such (new) interrelations as exempla gratia:

- a central concept which plays an important role in several different mathematical and/or scientific specialized areas;
- new applications of the results and ideas from one area of scientific endeavour into another;

- influences which the results, problems and concepts of one field of enquiry have and have had on the development of another.

The Mathematics and Its Applications programme tries to make available a careful selection of books which fit the philosophy outlined above. With such books, which are stimulating rather than definitive, intriguing rather than encyclopaedic, we hope to contribute something towards better communication among the practitioners in diversified fields.

Because of the wealth of scholarly research being undertaken in the Soviet Union, Eastern Europe, and Japan, it was decided to devote special attention to the work emanating from these particular regions. Thus it was decided to start three regional series under the umbrella of the main MIA programme.

When one is dealing with real numbers it is clear when one is larger (better) than another. When one is dealing with several numerical criteria things become much harder though of course the notion of an (Pareto) optimal vector still makes sense. Things become still harder when one has to compare and evaluate the merits of different statistical models of a given set of data and their predictive power.

For this type of problem Akaike developed a criterion based on information theoretic ideas which has attracted enormous attention especially from users of mathematics such as time-series analysts, electrical and other engineers, econometricians and (applied) statisticians: This is a first systematic book in English about this criterion and its uses when doing data analysis and modeling, and it contains all that is needed to use the criterion in practice.

The unreasonable effectiveness of mathematics in science ...

Eugene Wigner

Well, if you know of a better 'ole, go to it.

Bruce Bairnsfather

What is now proved was once only imagined.

William Blake

As long as algebra and geometry proceeded along separate paths, their advance was slow and their applications limited.

But when these sciences joined company they drew from each other fresh vitality and thenceforward marched on at a rapid pace towards perfection.

Joseph Louis Lagrange.

Bussum, March 1986 Michiel Hazewinkel

Preface to the English Edition

This is the translation of our book published originally in Japanese in 1983 as a volume of Information Sciences Series (Kyoritsu Publishing Company, Tokyo, Japan, T. Kitagawa, ed.) Fortunately, the Japanese edition has had a large circulation. It is our sincere hope that this English edition will get even more readers.

The publication of this English edition was realized with the support of many of our seniors and colleagues. We first wish to express our appreciation to Prof. S. Amari of Tokyo University for encouraging us to publish this book from D. Reidel Publishing Company. Our hearty thanks are dued to Prof. P. Thomson of Victoria University, New Zealand, for his careful reading of, and valuable comments on, the draft version. Dr. D. F. Findley of U. S. Bureau of the Census and Prof. W. Gersch of University of Hawaii also encouraged us to translate into English and gave us useful comments. We finally wish to express our thanks to Prof. Dr. M. Hazewinkel of the Centre for Mathematics and Computer Science (Amsterdam, The Netherlands) and to Mr. K. Oshida of the publisher for their encouragement and editorial cooperation.

Y. Sakamoto
M. Ishiguro
G. Kitagawa

Tokyo
August 1985

Editor's Preface (for Japanese Edition)

Statistics can provide the breakthrough for the birth of information sciences. Such was the case for the initiation of cybernetics by N. Wiener and information theory by C. Shannon. We also know cases where information sciences are indebted to modern statistical theories, not only for their beginning but also for their further development. Typical examples are pattern recognition and learning theory, which owe much to statistical decision theory and multivariate analysis; the close relation between control theory and the theory of stochastic processes. It is because we thought much of this relation that we added the subseries of statistical theories as A. 5 in our series.* In the subseries, we have already published several volumes, and this new volume A. 5. 4 will include a recent noteworthy result in this area.

This volume ranges from introductory materials to applications, with the distinctive feature that it takes a unique approach, based on the information quantity, to all of statistical analysis. The concept of the information quantity was introduced to, and utilized in, statistics by R. A. Fisher, one of the originators of mathematical statistics. But systematic application of the information quantity to statistical problems was realized by the AIC (Akaike Information Criterion) introduced by Prof. Akaike of the Institute of Statistical Mathemat-

---

* Information Sciences Series, Kyoritsu Publishing Company, Tokyo, Japan,

ics, Tokyo, who found an essential relation between the information quantity and Boltzmann's entropy. It is now well recognized, at home and abroad, that his theory has a wide range of applications.

The three authors of this volume are members of the group doing research on this subject. It is our great pleasure that by the publication of this volume we have an opportunity to increase users, sympathizers and researchers of this theory and approach.

Based on their experience and expertise in actual data analysis, the authors have attempted to treat all statistical problems by a unified statistical procedure using the information criterion. In this sense this volume is a challenge to conventional statistics as well as a proposal of a new approach to statistical analysis. The reader may find some aspects of the approach controversial insofar as they imply a criticism of conventional mathematical statistics, such as the use of statistical tests, individual sampling distribution theory and statistical tables. As the editor, I would like to say that I do not consent to all of their opinions. We, however, should recognize that many problems are now arising from the boundary region of information science and statistics, such as datalogy, etc.. It seems to me that the above frank criticism of conventional techniques by the researchers of the Institute of Statistical Mathematics is quite important and may include clues for fertile future development and meets the demands of the times to surpass the realm of statistics introduced by R. A. Fisher, J. Neyman and A. Wald.

Editor* Tosio Kitagawa

Authors' Preface

An information criterion AIC was introduced in 1971 by Prof. H. Akaike of the Institute of Statistical Mathematics, Tokyo and has attracted much attention from statisticians and engineers. Already a large number of articles have appeared that refer to, or apply, AIC. Readers of this book may well have read some of those papers. Fortunately, we have been closely associated with Prof. H. Akaike and have seen the diverse developments and applications in this field. It was a desire to share our conviction of the usefulness of AIC that made us decide to write this book.

In the course of the analysis of data there is a moment when a given set of data begins to speak its own story. By applying AIC to the data analysis we have repeatedly experienced such thrilling moments. The principal aim of this book is to convey this sensation of discovery. The use of AIC is illustrated by applying it to typical problems, most of which are dealt in the conventional statistics textbooks. Of course these methods may also be useful for a first analysis of data. However, we would like to put our emphasis on the model building process and hope that those examples will provide guidence for the development of new models when readers encounter different problems. Our ultimate aim will be achieved when readers succeed in grasping the stochastic structure of the data by their own models and have the feeling that they have "let the data speak"

We first would like to acknowledge the guidance and the encouragement of Dr. C. Hayashi, the director-general of the

Institute of Statistical Mathematics. Our hearty thanks are extended to Prof. H. Akaike who drew our attention to the theory of AIC and kept encouraging us to finish the manuscript. Here readers should remember that we wrote this book following his methods, but independently of him. Any mistakes, misunderstandings or superficial knowledge are due to us alone. We suggest that readers read his original papers for further details of the theory of AIC. We also would like to express our appreciation to Dr. T. Kitagawa, Professor Emeritus of Kyushu University, for recommending us to write this book. Furthermore, we are grateful to Prof. R. Shimizu, Prof. Y. Ogata and Dr. Y. Hamada of the Institute of Statistical Mathematics for their critical reading of the manuscript. We wish to thank Prof. K. Tanabe for his helpful suggestions and Mr. K. Katsura for his help in programming. We are also deeply indebted to Mr. S. Naniwa of the Bank of Japan, Dr. M. Ooe of the International Latitude Observatory of Mizusawa and Prof. T. Ozaki of the Institute of Statistical Mathematics for their help in collecting the references. We finally wish to express our thanks to Mr. K. Sato of the Kyoritsu Publishing Company in Tokyo who patiently kept on encouraging us to finish the manuscript.

Yosiyuki Sakamoto
Makio Ishiguro
Genshiro Kitagawa

October 1982

Viewpoint and Framework of this Book

An important aim of science is to find the underlying law of a phenomenon in order to make an inference on its future behavior. The purpose of statistics is to build a model (probability distribution) of stochastic phenomenon in order to estimate the future distribution, and finally to predict and control that phenomenon.

For this purpose the following are essential:

(1) the construction of an appropriate probability distribution, i. e. statistical model, in accordance with the researcher' s objective;

(2) the introduction of a criterion to evaluate the goodness of the assumed model.

It must be well recognized that classical statistical test procedures that are too closely and implicitly related to specific models are often useless to make selection between newly developed models.

Akaike showed an information theoretic interpretation of the likelihood and pointed out that an objective evaluation of the goodness of the assumed models becomes possible by extending the concept of likelihood. The criterion is now well-known as AIC (Akaike Information Criterion). The introduction of this objective criterion is quite important since it enables the objective comparison of models that are usually selected subjectively by the analysts and also stimulates the development of more appropriate models.

In this book we review many statistical procedures that have been regarded as estimation problems, statistical tests and descriptive methods of data analysis from a consistent viewpoint, namely the construction of a model and its evaluation by the information quantity. This viewpoint makes our method free from individual theory of sampling distributions and various statistical tables. We recommend readers who are interested in our viewpoint to read the entire book. However those who just need a practical modeling procedure may start with any section in PART II of this book.

[*Construction of this book*]

As seen from the contents, this book consists of three parts. The inter-relationships of the chapters are shown in the following figure.

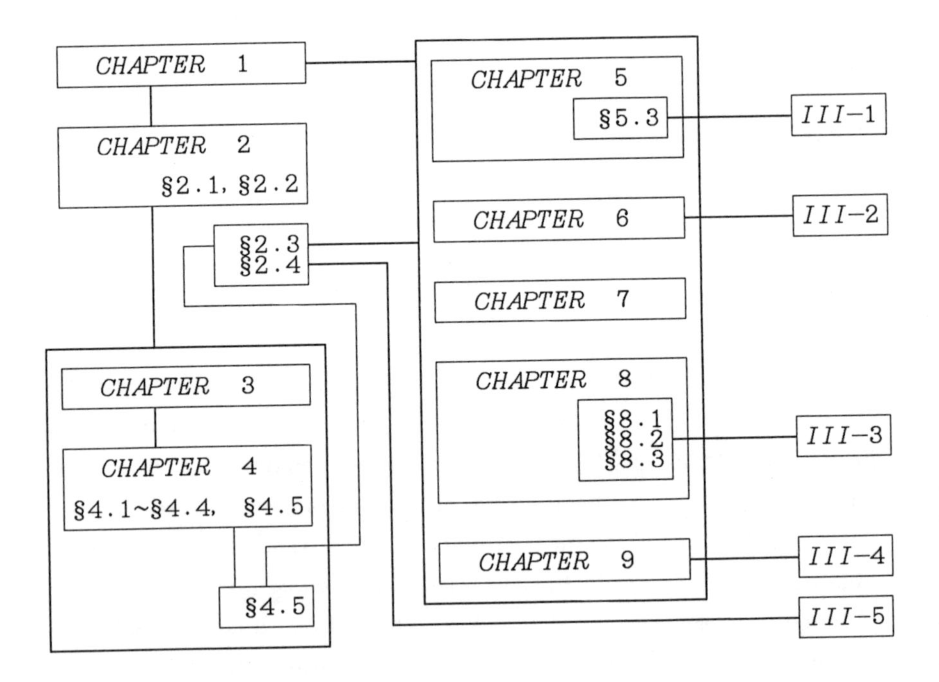

Most of Chapter 1 and 2 are devoted to preparation for the subsequent chapters and are written following the style of a conventional statistics textbook. In contrast, successive chapters are written from our original viewpoint. Especially, in PART II, solutions to typical statistical problems are given from our viewpoint. Therefore, a reader having a general knowledge of statistics may start with Chapter 3. A beginner had better start with Chapter 1. Sections with the mark * in their section numbers may be skipped on the first reading.

The PART I is a product of the combined efforts of three authors and the remaining chapters were allotted as follows:

Sakamoto ------ Chapter 5, Chapter 6, III-1 and III-2

Ishiguro ------ Chapter 7, 8.4 in Chapter 8, Chapter 9, III-4, and III-5

Kitagawa ------ Chapter 8 and III-3

PART I

BASIC THEORY

CHAPTER 1

PROBABILITY AND RANDOM VARIABLES

In every day conversation we use expressions such as "ten to one", "by chance" or "likelihood." These terms refer to the fraction of trials in which a specified event occurs. More specifically we use the expression "the probability that one wins the first prize in a public lottery." In statistics, the concept of probability plays an important role. A few examples of referring to the probability are:

[*Example 1.1*] Suppose there are three lottery tickets one of which has a winning number. What is the probability of drawing the winning number when a ticket is selected at random?

[*Example 1.2*] What is the probability of throwing head when a single coin is tossed?

[*Example 1.3*] When a fair top is spun, what is the probability that it falls down within an angle of $30°$ from a given base line?

[*Example 1.4*] A fair die is rolled. Find the probability of obtaining an ace?

[*Example 1.5*] What is the probability of drawing an ace from a pack of 52 ordinary playing cards?

These examples have a feature common to all. They are concerned with the fraction of trials in which a specified event occurs when the trial or experiment is repeated over and over again under the same conditions. In most statistical problems we are more interested in probability defined in this sense. In this chapter we present the mathematical elements of probability theory.

1.1 Events and Probability

Suppose that an experiment can result in one of s outcomes $\omega_1, \ldots, \omega_s$. The set of all possible outcomes of the experiment is called the *sample space* and is denoted by Ω. In [Example 1.1], there are three ticket. So the possible outcomes can be expressed by ω_1, ω_2 and ω_3 one of which corresponds to the winning number. Therefore the sample space Ω is defined by $\{\omega_1, \omega_2, \omega_3\}$. A subset of the sample space is called an *event* and is denoted by E. We say that the event E occurs if the experiment results in an element of E. In [Example 1.1], the event E_1 of drawing a winning number is $\{\omega_1\}$ and the event E_2 of drawing a blank is $\{\omega_2, \omega_3\}$. As it can be seen from this example, an event E may be defined by

$$E = \{\omega \mid \textit{some conditions on } \omega\}.$$

An event consisting of only one outcome is called an *elementary event*. An event containing no elements is called the *empty event* and is denoted by the null set $\emptyset$. The *complement* E^c of an event E is the set of elementary events in Ω which are not in E.

The *union* of two events E_1 and E_2 is the set of elementary events contained in E_1 or E_2, or both, and is expressed as $E_1 \cup E_2$. The set of elementary events in the sample space Ω contained in both E_1 and E_2 is called the *intersection* of E_1 and

E_2 and is written as $E_1 \cap E_2$. If the events E_1 and E_2 can not simultaneously occur, that is $E_1 \cap E_2 = \phi$, then the events E_1 and E_2 are said to be mutually exclusive.

These concepts can be extended for an infinite sequence of events $E_1, E_2, \ldots$; their union is denoted by $\bigcup_i E_i$ and their intersection by $\bigcap_i E_i$. We say that they are pairwise mutually exclusive if $E_i \cap E_j = \phi$.

If all elementary events in E_1 are also contained in E_2, E_1 is said to be a subevent of E_2, and is expressed as $E_1 \subset E_2$. If $E_1 \supset E_2$ as well as $E_1 \subset E_2$, then we say that E_1 and E_2 are equal and write $E_1 = E_2$.

Suppose we ask for the probability in [Example 1. 1]. The most natural answer to this question is *1/3*. Under the assumption that every elementary event is equi-probable, this answer is obtained by dividing the number of elementary events belonging to the event "drawing a winning number" by the total number of outcomes in the sample space Ω. However this method is not applicable in all cases. Even though we assume equi-probability in [Example 1. 3], we cannot determine the proba-lity as a rate since the sample space consists of an infinite number of elementary events. To deal with such problems, we will define the probability as follows:

A real value $P(E)$ is called the ***probability*** of an event E if $P(E)$ is assigned to every E in Ω and satisfies the following ***axioms of probability***:

(*i*) *If E is an event in Ω, then $0 \leqq P(E) \leqq 1$* (*1. 1*)

(*ii*) $P(\Omega) = 1$ (*1. 2*)

(*iii*) *If $E_1, E_2, \cdots$ is an infinite sequence of mutually exclusive events, then we have*

$$P(E_1 \cup E_2 \cup \cdots) = P(E_1) + P(E_2) + \cdots . \quad (1.\ 3)$$

From these axioms, it follows that $E \cup E^c = \Omega$ and $E \cap E^c = \phi$ and we obtain

$$P(E)+P(E^c) = P(E \cup E^c) = P(\Omega) = 1,$$

that is,

$$P(E^c) = 1 - P(E). \tag{1.4}$$

This relation is called the *law of complementation*. As a special case of this, when $E=\phi$ we have $P(\phi)=0$.

The following example illustrates the concepts of a probability model based on the axioms of probability.

[*Example 1.6*] If we denote by ω_1 and ω_2 the elementary events corresponding to the occurrence of a head or a tail, respectively in [Example 1.2], then the possible events are defined by $\Omega = \{\omega_1, \omega_2\}$, $E_1 = \{\omega_1\}$, $E_2 = \{\omega_2\}$ and ϕ. If we assign probabilities as $P(\phi) = 1$, $P(E_1) = P(E_2) = 1/2$ and $P(\phi) = 0$, it is seen that P satisfies the axioms of probability. This is the model for fair coin tossing. However, we can show that $P(\Omega) = 1$, $P(E_1) = 2/3$, $P(E_2) = 1/3$ and $P(\phi)=0$ also satisfies the axioms of probability. This however is a model for unfair coin tossing.

1.2 *Conditional Probability and Independence*

Conditional probability and independence are basic concepts arise in probability theory. Consider two events E and F with $P(F) > 0$ and define

$$P(E|F) = \frac{P(E \cap F)}{P(F)}. \tag{1.5}$$

$P(E|F)$ is called the conditional probability of an event E given F. The conditional probability determines the probability that the event E occurs given that the event F has occurred. If

$$P(E|F) = P(E), \qquad (1.6)$$

the occurrence of the event F does not affect the probability of the event E. In this case E and F are said to be *independent*. When E and F are independent, from (1.5), it follows that

$$P(E \cap F) = P(E)P(F); \qquad (1.7)$$

that is the probability of their joint occurrence is identical to the product of the individual probabilities of E and F.

Now let us illustrate the concepts of conditional probability and independence. In [Example 1.4], let ω_i be the elementary event that a face "i" occurs. If we denote by E and F the events of an ace and an odd number occurring, respectively, we have

$$E = \{\omega_1\}$$
$$F = \{\omega_1, \omega_3, \omega_5\}.$$

Since $E \cap F = E$, we get $P(E|F) = \frac{P(E \cap F)}{P(F)} = \frac{1/6}{1/2} = \frac{1}{3}$, which shows that the probability of getting an ace is equal to $1/3$ if an odd number is known to have occurred. Similarly, if we put $E_1=\{\omega_2\}$, we have $P(E_1|F) = 0$. Thus conditional probabilities are obtained from a sample space reduced by the knowledge that certain event have occurred.

[*Example 1.7*] If $E = \{ace\}$ and $F = \{heart\}$ in [Example 1.5], then it follows that $P(E)=4/52=1/13$ and $P(F)=13/52=1/4$. Since $E \cap F$ is the event of getting the ace of hearts, $P(E \cap F) = 1/52$. Hence, from (1.5),

$$P(E|F) = \frac{\frac{1}{52}}{\frac{1}{4}} = \frac{1}{13} = P(E),$$

which shows that the events E and F are independent.

1.3 *Random Variable and Cumulative Distribution Function*

If a real valued function $X(\omega)$ is defined on the sample space Ω and if the probability of the events

$$\{\omega | X(\omega) \leqq x\}$$

can be defined for each real number x, then $X(\Omega)$ is called a *random variable*. The *distribution function* of X is defined by

$$F(x) = P(\{\omega | X(\omega) \leqq x\}). \qquad (1.8)$$

The probability in the righthand side of (1.8) will be expressed briefly by $P(X \leqq x)$. When we observe $X(\omega) = x$ in an experiment, x is called a realization.

[*Example 1.8*] In the coin tossing of [Example 1.2], define

$$X(\omega) = \begin{cases} 1 & \omega = \omega_1 \\ 0 & \omega = \omega_2 . \end{cases} \qquad (1.9)$$

It follows that

$$\{\omega | X(\omega) \leqq x\} = \begin{cases} \phi & -\infty < x < 0 \\ \{\omega_2\} & 0 \leqq x < 1 \\ \{\omega_1, \omega_2\} & 1 \leqq x < \infty . \end{cases}$$

and X is a random variable. The corresponding distribution function is given by

$$F(x)=\begin{cases} 0 & -\infty\leqq x<0 \\ \frac{1}{2} & 0\leqq x<1 \\ 1 & 1\leqq x<\infty . \end{cases}$$

If a random variable takes a finite or a countablly infinite number of values, it is called a *discrete random variable*. Its distribution function is a step function. If a *probability mass function*, which represents the probability that a random variable $X(\omega)$ takes a value x_i $(i=1, 2, \ldots)$, is defined by

$$p_{x_i} = P(X=x_i), \tag{1.10}$$

we can write

$$F(x) = \sum_{x_i \leqq x} p_{x_i} \tag{1.11}$$

On the other hand, if a random variable takes values on the continuum, then the corresponding distribution function is continuous. In this case the random variable is called a *continuous random variable*. The distribution function $F(x)$ is given by

$$F(x) = \int_{-\infty}^{\infty} f(y)dy, \tag{1.12}$$

where the function $f(x)$ satisfies

$$\begin{aligned} f(x) &\geqq 0 \qquad -\infty<x<\infty, \\ \int_{-\infty}^{\infty} f(x)dx &= 1. \end{aligned} \tag{1.13}$$

The function $f(x)$ is called the (*probability*) *density function* of X. We usually specify a continuous random variable by its density function.

In general, a distribution function is non-decreasing, is continuous from the right and has the properties that

$$\lim_{x \to -\infty} F(x) = 0$$
$$\lim_{x \to \infty} F(x) = 1. \tag{1.14}$$

1.4 *Expectation*

Let a random variable X take on values $x_1, x_2, \ldots$ with the respective probabilities $p_1, p_2, \ldots$. The expected value of X is given by

$$\mu = E[X] = \sum_{i=1}^{\infty} x_i p_i. \tag{1.15}$$

It is also called the *expectation* or *mean value* of the random variable X. More generally if $g(X)$ is a function of X, then the expectation of $g(X)$ is given by

$$E[g(X)] = \sum_{i=1}^{\infty} g(x_i) p_i. \tag{1.16}$$

If we put $g(X) = (X-\mu)^2$, then we have the *variance* of X,

$$\sigma^2 = E[(X-\mu)^2] = \sum_{i=1}^{\infty} (x_i-\mu)^2 p_i. \tag{1.17}$$

The square root σ of σ^2 is called the *standard deviation* of X.

Similarly, in the case of a continuous random variable X having the density function $f(x)$, the expectation and the variance of X are respectively defined by

$$\mu = E[X] = \int_{-\infty}^{\infty} x f(x) dx \tag{1.18}$$

$$\sigma^2 = E[(X-\mu)^2] = \int_{-\infty}^{\infty} (x-\mu)^2 f(x) dx. \tag{1.19}$$

Generally the expectation of a function $g(X)$ of X takes the form

$$E[g(X)] = \int_{-\infty}^{\infty} g(x)f(x)dx. \tag{1.20}$$

The variance of X has an alternative expression that is more convenient for hand calculation

$$\sigma^2 = E[(X-\mu)^2] = E[X^2]-\mu^2. \tag{1.21}$$

For example, in the case of the random variable defined in (1.9), we have $E[X] = 1 \cdot 1/2 + 0 \cdot 1/2 = 1/2$, $E[X^2] = 1^2 \cdot 1/2 + 0^2 \cdot 1/2 = 1/2$, which shows the expectation of X is $1/2$ and, from (1.21), the variance of X is $1/4$.

1.5 *Multidimensional Probability Distributions*

Suppose that we consider the joint probability distribution of a set of two or more random variables. Then $X(\omega) = (X_1(\omega), \ldots, X_k(\omega))$ is called k dimensional random variable, or more generally a *multidimensional random variable*. For k real numbers $x_1, \ldots, x_k$ the function defined by

$$F(x_1, \ldots, x_k) = P(X_1 \leqq x_1, \ldots, X_k \leqq x_k) \tag{1.22}$$

is called the (joint) distribution function of X. Also, in the case of a k dimensional random variable, there are two important types: discrete and continuous.

In the case of a discrete random variable X, the *multidimensional probability mass function* representing the probability that X takes on a value $(x_1, \cdots, x_k)$, is defined by

$$p_{x_1, \ldots, x_k} = P(X_1 = x_1, \ldots, X_k = x_k) \tag{1.23}$$

which statisfies

(*i*) $\quad p_{x_1, \ldots, x_k} \geqq 0$

$$(ii) \quad \sum_{x_1} \cdots \sum_{x_k} p_{x_1, \ldots, x_k} = 1. \tag{1.24}$$

The distribution function can be written as

$$F(x_1, \ldots, x_k) = \sum_{\substack{y_i \leq x_i \\ i=1,\cdots,k}} p_{y_1, \ldots, y_k}. \tag{1.25}$$

In the case of a continuous random variable X, we define a (*joint*) *density function* $f(x_1, \ldots, x_k)$ which satisfies

$$\begin{aligned} &(i) \quad f(x_1, \ldots, x_k) \geq 0 \\ &(ii) \quad \int_{-\infty}^{\infty} \cdots \int_{-\infty}^{\infty} f(x_1, \ldots, x_k)\, dx_1 \cdots dx_k \end{aligned} \tag{1.26}$$

and the distribution function by

$$F(x_1, \ldots, x_k) = \int_{-\infty}^{x_1} \cdots \int_{-\infty}^{x_k} f(t_1, \ldots, t_k)\, dt_1 \cdots dt_k. \tag{1.27}$$

Given a k dimensional distribution function $F(x_1, \ldots, x_k)$, a $k-1$ dimensional distribution function is given by

$$G(x_1, \ldots, x_{k-1}) \equiv F(x_1, \ldots, x_{k-1}, \infty),$$

which is equal to $P(X_1 \leq x_1, \cdots, X_{k-1} \leq x_{k-1})$ and is called a *marginal distribution* with respect to $(X_1, \ldots, X_{k-1})$. In the case of a discrete random variable, the marginal distribution is defined by

$$p_{x_1, \ldots, x_{k-1}} = \sum_{x_k} p_{x_1, \ldots, x_{k-1}, x_k}. \tag{1.28}$$

In the continuous case, the density function corresponding to the distribution function $G(x_1, \ldots, x_{k-1})$ is given by

$$g(x_1, \ldots, x_{k-1}) = \int_{-\infty}^{\infty} f(x_1, \ldots, x_k)\, dx_k. \tag{1.29}$$

If it always holds that

$$g(x_1, \ldots, x_{k-1}) > 0$$

then the density function of X_k given $(x_1, \ldots, x_{k-1})$ can be defined by

$$f(x_k | x_1, \ldots, x_{k-1}) \equiv \frac{f(x_1, \ldots, x_{k-1}, x_k)}{g(x_1, \ldots, x_{k-1})}$$

and is called a *conditional density function*. From (1.29), for any $(x_1, \ldots, x_k)$, it holds that

$$\int_{-\infty}^{\infty} f(x_k | x_1, \ldots, x_{k-1}) dx_k = 1.$$

Conversely, given a density function $g(x_1, \ldots, x_{k-1})$ of $(X_1, \ldots, X_{k-1})$ and a conditional density function $f(x_k | x_1, \cdots, x_{k-1})$, the joint density function of $(X_1, \ldots, X_k)$ can be defined by

$$f(x_1, x_2, \ldots, x_k) \equiv f(x_k | x_1, \ldots, x_{k-1}) g(x_1, \ldots, x_{k-1}).$$

A similar definition is possible for a discrete random variable.

If the two events $\{\omega | X(\omega) \leqq x\}$ and $\{\omega | Y(\omega) \leqq y\}$ are independent for all x and for y, we say that the random variables X and Y are independent. Thus if it holds that

$$P(X \leqq x | Y \leqq y) = P(X \leqq x) \tag{1.30}$$

or,

$$P(X \leqq x, Y \leqq y) = P(X \leqq x) P(Y \leqq y), \tag{1.31}$$

then the two variables X and Y are independent. If we denote the distribution functions of X and Y by $F_1(x)$, $F_2(y)$ respectively, the (joint) distribution function takes the form

$$F(x, y) = F_1(x) F_2(y). \tag{1.32}$$

If X and Y are discrete random variables, the joint probability function is determined by the product of two marginal probability functions. If both X and Y are continuous variables, the density function of X and Y is determined by the product of two marginal density functions. It is easily seen that if X and Y are independent, it holds that

$$E[g(X)h(Y)] = E[g(X)]E[h(Y)]. \tag{1.33}$$

Suppose the mean values of X and Y are μ_X and μ_Y, respectively. Then the quantity defined by

$$C(X, Y) = E[(X-\mu_X)(Y-\mu_Y)] \tag{1.34}$$

is called the *covariance* of X and Y. If X and Y are independent, we have $C(X, Y)=0$ from (1.33). The converse, however, is not always true. If we denote the variances of X and Y by σ_X^2 and σ_Y^2, respectively, then ρ_{XY} defined by

$$\rho_{XY} = \frac{C(X, Y)}{\sqrt{\rho_X^2 \rho_Y^2}} \tag{1.35}$$

is called the *correlation coefficient* between X and Y.

1.6* *Transformation of Variables*

Let $F(x)$ be the distribution function of a random variable X and $g(x)$ be a monotonically increasing function of x. If we define a new random variable Y by $Y=g(X)$, then the distribution function of Y is obtained by

$$\begin{aligned} F(y) &= P(Y \leq y) = P(X \leq g^{-1}(y)) \\ &= \int_{-\infty}^{g^{-1}(y)} f(x)\,dx. \end{aligned} \tag{1.36}$$

Using the inverse transformation $x=g^{-1}(u)$, we have $\frac{dx}{du} = \frac{dg^{-1}}{du}$.

From this we find

$$F(y) = \int_{-\infty}^{y} \frac{dg^{-1}(u)}{du} f(g^{-1}(u))du,$$

which shows that the density function $h(y)$ of Y is given by

$$h(y) = \frac{dg^{-1}(y)}{dy} f(g^{-1}(y)).$$

Since the same argument is valid also in the case where g is a monotonically decreasing function, it can be shown that if g is a strictly monotonic function, we have the density function of Y

$$h(y) = \left|\frac{dg^{-1}(y)}{dy}\right| f(g^{-1}(y)). \tag{1.37}$$

[*Example 1.9*] Let $Y = g(X) = aX+b$ $(a \neq 0)$ and the density of X be given by $f(x)$. Since

$$g^{-1}(y) = a^{-1}(y-b), \qquad \frac{dg^{-1}}{dy} = a^{-1},$$

we get $h(y) = |a|^{-1}f(a^{-1}(y-b))$.

[*Example 1.10*] Let $Y = g(X) = e^X$. Since

$$g^{-1}(y) = \log y, \qquad \frac{dg^{-1}}{dy} = \frac{1}{y},$$

the density of Y is given by $h(y) = |y|^{-1}f(\log y)$.

More generally, suppose that the density of k dimensional random variable $X=(X_1, \ldots, X_k)$ be $f(x_1, \ldots, x_k)$ and that the random variable $Y=(Y_1, \ldots, Y_k)$ be defined by the one-to-one transformation

$$y_i = g_i(x_1, \ldots, x_k) \qquad i=1, \ldots, k.$$

Then the density of Y is obtained by

$$h(y) = |det\, J|\ f(g_1^{-1}(y), \ldots, g_k^{-1}(y)), \qquad (1.38)$$

where g_i^{-1} is the inverse transformation and J is the Jacobian matrix defined by

$$J(y_1, \ldots, y_n) = \begin{bmatrix} \frac{\partial g_1^{-1}}{\partial y_1} & \cdots & \frac{\partial g_1^{-1}}{\partial y_k} \\ \vdots & & \vdots \\ \frac{\partial g_k^{-1}}{\partial y_1} & \cdots & \frac{\partial g_k^{-1}}{\partial y_k} \end{bmatrix}. \qquad (1.39)$$

[*Example 1.11*] Let A be a nonsingular matrix and the random variable Y is defined by $Y = AX+b$. Then the density function of Y is given by

$$h(y) = |det\, A^{-1}|\ f(A^{-1}(y-b)), \qquad (1.40)$$

where f denotes the density function of X.

CHAPTER 2

PROBABILITY DISTRIBUTIONS AND STATISTICAL MODELS

In this chapter we give some examples of the probability distribution that were defined in Chapter 1. (Discrete distributions are given in section 2.1 and continuous distributions in section 2.2.) In section 2.3 we show some techniques that are used in constituting a statistical model by using these probability distributions. In section 2.4 simulation methods are discussed briefly.

2.1 Discrete Distributions

2.1.1 Binomial Distribution

Suppose that an experiment consists of n independent trials. Each trial results in one of two mutually exclusive outcomes E and E^c with probabilities p and $1-p$. Then the probability that the evnt E occurs k times is given by

$$b(k \mid n, p) = \binom{n}{k} p^k (1-p)^{n-k}, \qquad k=0, 1, \ldots, n \qquad (2.1)$$

The distribution with probability mass function given by (2.1) is called a *binomial distribution*. The following are considered to be examples of binomial random variables.

(i) The number of times x that an odd number appears in ten tosses of a fair die.

(ii) The number of people y that support a certain policy when 100 persons are selected at random from an infinite population in which the rate supporting the population is θ.

For (i) the probability distribution $b(x\,|\,10, \frac{1}{2})$ is given by

$$b(x\,|\,10, \tfrac{1}{2}) = \binom{10}{x}\left(\tfrac{1}{2}\right)^{x}\left(1-\tfrac{1}{2}\right)^{10-x}$$

and for (ii) that is given by

$$b(y\,|\,100, \theta) = \binom{100}{y}\theta^{y}(1-\theta)^{100-y}.$$

It is easy to show that (2. 1) is actually a probability mass function since

$$\sum_{k=0}^{n} b(k\,|\,n, p) = \sum_{k=0}^{n}\binom{n}{k}p^{k}(1-p)^{n-k} = [p+(1-p)]^{n} = 1.$$

From

$$\sum_{k=0}^{n} k\, b(k\,|\,n, p) = np \tag{2. 2}$$

$$\sum_{k=0}^{n} k^{2} b(k\,|\,n, p) = np+n(n-1)p^{2}, \tag{2. 3}$$

we can see that the mean of the binomial distribution is np and from (1. 21) the variance is $np(1-p)$.

Suppose that K_i is a random variable representing the result at the ith trial;

$$K_i = \begin{cases} 1 & \textit{if a certain event occures at the } i^{th} \textit{ trial} \\ 0 & \textit{otherwise} \end{cases}$$

such that probability that $k_i = 1$ equals to p. Then (2. 1) is the probability mass function of the random variable defined as the sum of K_i's, $K = \sum_{i=1}^{n} K_i$. In this case the joint probability that the sequence of n trials results in $(k_1, \ldots, k_n)$ where $k_i = 0$ or 1, is given by

$$b_1(k_1, \ldots, k_n | n, p) = p^k (1-p)^{n-k}, \qquad (2.1')$$

where $k = \sum_{i=1}^{n} k_i$. In this book we will also use (2.1') frequently.

2.1.2 *Poisson Distribution*

The probability distribution with the probability mass function given by (2.4) is called a *Poisson distribution*,

$$p(k|\lambda) = \frac{\lambda^k}{k!} e^{-\lambda} \qquad k = 0, 1, 2, \ldots \qquad (2.4)$$

Here e is the base of natural logarithm and is given by $e = 2.71828$. This probability mass function is obtained as the limit of the binomial probability mass function when $p \to 0$ and $n \to \infty$ while satisfying $np = \lambda$. Indeed

$$\begin{aligned} b(k|n, p) &= \binom{n}{k} p^k (1-p)^{n-k} \\ &= \frac{n(n-1)\cdots(n-k+1)}{k!} \left(\frac{\lambda}{n}\right)^k \left(1-\frac{\lambda}{n}\right)^n \left(1-\frac{\lambda}{n}\right)^{-k} \\ &= \frac{\lambda^k}{k!} \frac{n(n-1)\cdots(n-k+1)}{n^k} \left(1-\frac{\lambda}{n}\right)^n \left(1-\frac{\lambda}{n}\right)^{-k} \\ &= \frac{\lambda^k}{k!} \left[\left(1-\frac{\lambda}{n}\right)^{-\frac{n}{\lambda}}\right]^{-\lambda} \left\{1\left(1-\frac{1}{n}\right)\cdots\left(1-\frac{k-1}{n}\right)\left(1-\frac{\lambda}{n}\right)^{-k}\right\}. \end{aligned}$$

Since $\lim_{m\to\infty}(1+\frac{1}{m})^m = e$, and each term in the braces of the last line of the above tends to 1 as n goes to infinity, we obtain the probability mass function given in (2.4). From this fact, if n is large and p is small, we can approximate the probability of a binomial distribution by a Poisson distribution. In other words, if the probability that a certain event occurs is very small and we have a large number of independent observations, then it is reasonable to consider that the event is distributed as a Poisson distribution. Since

$$E[k] = \sum_{k=0}^{\infty} k\frac{e^{-\lambda}\lambda^{k}}{k!} = \lambda\sum_{k=1}^{\infty}\frac{e^{-\lambda}\lambda^{k-1}}{(k-1)!} = \lambda\sum_{k=0}^{\infty}\frac{e^{-\lambda}\lambda^{k}}{k!} = \lambda \tag{2.5}$$

$$E[k^2] = \sum_{k=0}^{\infty} k^2\frac{e^{-\lambda}\lambda^{k}}{k!} = \lambda^2\sum_{k=2}^{\infty}\frac{e^{\lambda}\lambda^{k-2}}{(k-2)!} + \lambda\sum_{k=1}^{\infty}\frac{e^{-\lambda}\lambda^{k-1}}{(k-1)!} \tag{2.6}$$

$$= \lambda^2+\lambda,$$

both the mean and the variance of a Poisson distribution is given by λ.

2.1.3 *Multinomial Distribution*

The binomial distribution arose when a sequence of n independent and identical experiments are performed each of which results in one of two mutually exclusive events. Suppose now that each trial can result in any one of c mutually exclusive outcomes, with respective probabilities $p_1, \cdots, p_c$, $\sum_{i=1}^{c} p_i=1$. The probability that the first event E_1 occurs k_1 times, the second event E_2 occurs k_2 times, $\ldots$, c^{th} event E_c occurs k_c times is

$$m(k_1, \ldots, k_c \mid p_1, \ldots, p_c) = \frac{n!}{k_1!\cdots k_c!}p_1^{k_1}\cdots p_c^{k_c}, \tag{2.7}$$

with $0 \le k_i \le n$ $(i=1, \ldots, c)$ and $\sum_{i=1}^{c} k_i=n$. The probability distribution defined by this probability mass function is called the *multinomial distribution* and is frequently used for the analysis of discrete type data. As in the case of a binomial distribution, it is easy to see that (2.7) is actually a probability mass function since

$$\sum_{k_1, \cdots, k_c} m(k_1, \ldots, k_c \mid p_1, \ldots, p_c) = \sum_{k_1, \cdots, k_c}\frac{n!}{k_1!\cdots k_c!}p_1^{k_1}\cdots p_c^{k_c}$$

$$= (p_1+\cdots+p_c)^n = 1.$$

If we denote by X_i the random variable that takes the value x when the ith trial results in the event E_x $(x=1, \ldots, c)$, then the probability that we obtain a sequence of observations $(x_1, \ldots, x_n)$, $x_i=1, \ldots, c$, in n independent trials is given by

$$m_1(x_1, \ldots, x_n \mid p_1, \ldots, p_c) = p_1^{k_1} \cdots p_c^{k_c}. \qquad (2.7')$$

Here, k_x is the number of occurrences of the event E_x (or the number of X_i's that take the value x). We will frequently use (2.7') as well as (2.7).

2.2 *Continuous Distributions*

2.2.1 *Uniform Distribution*

A probability distribution is said to be a *uniform distribution* over the interval [0, 1] if its probability density function is given by

$$u(x) = \begin{cases} 1 & 0 \leq x \leq 1 \\ 0 & x<0, \ x>1. \end{cases} \qquad (2.8)$$

We denote a random variable by $U(0, 1)$ if its probability density function is given by (2.8). Intuitively, the random variable $U(0, 1)$ takes on any value between 0 and 1 with equal probablity. More precisely, the probability that $U(0, 1)$ is in any particular subinterval of [0, 1] equals the length of that subinterval. The mean and the variance of $U(0, 1)$ are given by

$$E[U] = \int_0^1 x \, dx = \frac{1}{2},$$

$$\sigma_U^2 = \int_0^1 x^2 dx - \left(\frac{1}{2}\right)^2 = \frac{1}{3} - \frac{1}{4} = \frac{1}{12}.$$

The uniform random variable $U(0, 1)$ plays an important role in simulation as will be seen in section 2.4.

2.2.2 *Normal Distribution*

A function defined by

$$f(x|\mu, \sigma^2) = \frac{1}{\sqrt{2\pi\sigma^2}} exp\left\{-\frac{1}{2\sigma^2}(x-\mu)^2\right\} \tag{2.9}$$

is a probability density function since for any x, μ and $\sigma^2>0$,

$$f(x|\mu, \sigma^2) > 0 \tag{2.10a}$$

$$\int_{-\infty}^{\infty} f(x|\mu, \sigma^2)\,dx = 1. \tag{2.10b}$$

We call the probability distribution defined by (2.9) a *normal* distribution with mean μ and variance σ^2, and denote it by $N(\mu, \sigma^2)$. If a random variable X is distributed as a normal distribution $N(\mu, \sigma^2)$ then we will denote this by

$$X \sim N(\mu, \sigma^2).$$

For simplicity we will sometimes use the notation $f(x)$ for $f(x|\mu, \sigma^2)$.

Partial differentiation with respect to μ of both sides of (2.10b) yields

$$\int_{-\infty}^{\infty} \frac{(x-\mu)}{\sigma^2} f(x)\,dx = 0. \tag{2.11}$$

It follows that the mean is given by

$$E[X] = \int_{-\infty}^{\infty} xf(x)\,dx = \mu\int_{-\infty}^{\infty} f(x)\,dx = \mu. \tag{2.12}$$

Since partial differentiation with respect to μ of (2.11) yields

$$\int_{-\infty}^{\infty} \frac{(x-\mu)^2}{\sigma^4} f(x)\,dx - \frac{1}{\sigma^2}\int_{-\infty}^{\infty} f(x)\,dx = 0,$$

we have

$$E[(X-\mu)^2] = \int_{-\infty}^{\infty} (x-\mu)^2 f(x)\,dx = \sigma^2. \qquad (2.13)$$

This shows that the variance of X is actually given by σ^2.

From (1.38), we know that if X is normally distributed with mean μ and variance σ^2, $X \sim N(\mu, \sigma^2)$, then $Y=(X-\mu)/\sigma$ is normally distributed with mean 0 and variance 1, namely $Y \sim N(0, 1)$. We say that Y has the standard normal distribution and denote by $\phi(y)$. Fig 2.1 shows the shape of $\phi(y)$.

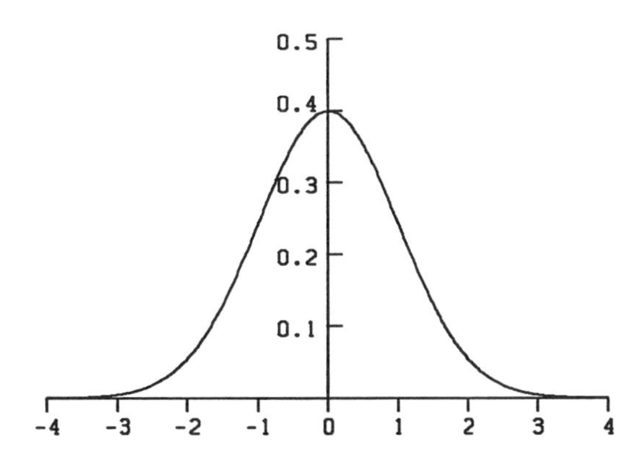

Fig. 2.1 Standard normal distribution

2.2.3 *Multidimensional Normal Distribution*

The function

$$f_k(x) = \left(\frac{1}{2\pi}\right)^{\frac{k}{2}} (\det \Sigma)^{-\frac{1}{2}} \exp\left\{-\frac{1}{2}(x-\mu)^T \Sigma^{-1} (x-\mu)\right\} \qquad (2.14)$$

defined at each k dimensional vector $x=(x_1, x_2, \ldots, x_k)^T$ satisfies $f_k(x)>0$ for any x and

$$\int_{-\infty}^{\infty}\cdots\int_{-\infty}^{\infty} f(x)\,dx_1\cdots dx_k = 1.$$

In (2.14), Σ is a k dimensional positive definite matrix, and $det\,\Sigma$, Σ^T and Σ^{-1} denote the determinant, transpose and inverse of Σ, respectively. A probability distribution is called a k *dimensional normal distribution* and is denoted by $N(\mu, \Sigma)$ if its probability density function is given by (2.14). If a k dimensional random variable $X=(X_1, X_2, \ldots, X_k)^T$ is distributed as a k demensional normal distribution, then it is denoted by

$$X \sim N(\mu, \Sigma).$$

(2.14) is an extension of (2.9).

By the same method that we obtained (2.12) and (2.13), we have

$$E[X] = \mu \qquad (2.15)$$

$$E[(X-\mu)(X-\mu)^T] = \Sigma. \qquad (2.16)$$

μ and Σ are called the mean vector and the covariance matrix of the random variable X.

If Σ is a diagonal matrix with the diagonal elements σ_1^2, σ_2^2, $\ldots$, σ_k^2, and $\mu=(\mu_1, \ldots, \mu_k)^T$, (2.14) reduces to

$$f_k(x) = \prod_{i=1}^{k} f(x_i \mid \mu_i, \sigma_i^2). \qquad (2.17)$$

This means that the random variables X_1, X_2, $\cdots$, X_k are mutually independent and are distributed as $N(\mu_1, \sigma_1^2)$, $N(\mu_2, \sigma_2^2)$, $\ldots$, $N(\mu_k, \sigma_k^2)$, respectively.

Since Σ is a symmetric positive definite matrix, there exists a lower triangular matrix L that factorizes Σ as

$$\Sigma = LL^T. \tag{2.18}$$

Now if we define the random variable $Y = (Y_1, Y_2, \ldots, Y_k)^T$ by

$$Y = L^{-1}(X-\mu), \tag{2.19}$$

then from (2.14) and (1.40) the probability density function, $g_k(y)$, of Y is obtained as

$$\begin{aligned} g_k(y) &= |\det L| f_k(Ly+\mu) \\ &= \left(\frac{1}{2\pi}\right)^{\frac{k}{2}} \exp\left\{-\frac{1}{2}y^T L^T \Sigma^{-1} L y\right\} \\ &= \left(\frac{1}{2\pi}\right)^{\frac{k}{2}} \exp\left\{-\frac{1}{2}y^T y\right\} \\ &= \prod_{i=1}^{k} f(y_k \mid 0, 1). \end{aligned} \tag{2.20}$$

This shows that $Y_1, Y_2, \ldots, Y_k$ are independently distributed as $N(0, 1)$.

2.2.4 *Chi-square Distribution*

Suppose $X_1, X_2, \ldots, X_k$ are independent random variables with $X_i \sim N(0, 1)$. The random variable χ_k^2 defined as the square of the Euclidean norm of $X = (X_1, \ldots, X_k)^T$,

$$\chi_k^2 = |X|^2 = X^T X$$

is non-negative valued. The random variable χ_k^2 is called the *chi-square* random variable with k degrees of freedom. Its probability density function is given by

$$f_k(\chi_k^2) = \frac{1}{2^{\frac{k}{2}}\Gamma(\frac{k}{2})}(\chi_k^2)^{\frac{k}{2}-1} e^{-\chi_k^2/2}, \tag{2.21}$$

where Γ is the *gamma function* defined by

$$\Gamma(k) = \int_0^\infty e^{-x} x^{k-1} dx.$$

The expected values of χ_k^2 and $(\chi_k^2)^2$ are given as

$$\int_0^\infty \chi_k^2 f_n(\chi_k^2) d\chi_k^2 = k \tag{2. 22}$$

$$\int_0^\infty (\chi_k^2)^2 f_n(\chi_k^2) d\chi_k^2 = k(k+2). \tag{2. 23}$$

Thus the mean of the chi-square random variable is equal to its degrees of freedom and, from (1. 21), the variance is twice its degrees of freedom.

If X is a k dimensional normal random variable, $X \sim N(\mu, \Sigma)$, with $\Sigma = LL^T$ and $Y = L^{-1}(X-\mu)$ then $|Y|^2$ or

$$(X-\mu)^T \Sigma^{-1} (X-\mu) \tag{2. 24}$$

is a chi-square random variable with k degrees of freedom.

2. 2. 5 *Limit Theorems*

Let $X_1, X_2, \ldots$ be a sequence of independent and identically distributed random variables, each having finite mean μ and variance σ^2. Under this situation we have the following limit theorems. (For details see, for example, Rao(1965).)

[*The law of Large Numbers*]

$$\frac{1}{n} S_n \to E(X) = \mu \qquad as \ n \to \infty$$

[*The Central Limit Theorem*]

$$\frac{1}{\sqrt{n}\sigma}(S_n - n\mu) = \frac{1}{\sqrt{n}\sigma}\sum_{k=1}^{n}(X_k - \mu) \to N(0, 1) \qquad as \ n \to \infty$$

The law of large numbers states that the sample mean S_n/n

of a sequence of independent random variables having a common distribution will converge to the mean of that distribution. On the other hand, the central limit theorem states that the sum of the deviation from the mean, $\sum_{k=1}^{n}(X_k-\mu)$, divided by $\sqrt{n}\sigma$ converge to a standard normal random variable. Note that the divisor is proportional to $\sqrt{n}$, not n. If it were n then the quantity concerned would converge to 0 by the law of large numbers.

2. 3* *Statistical Models*

In this section we introduce some techniques that are useful in developing models for actual data using probability distributions.

2. 3. 1 *Parametric Models*

The selection of a model starts with considering given the data x as a realization of a random variable X corresponding to a certain probability distribution. The true distribution of X is denoted by $G(x)$. Our objective is to estimate $G(x)$ from the data x. In the following, the probability distribution $G(x)$ is sometimes expressed by G or $G(\cdot)$.

A typical approach to this problem is to consider a family of probability distribution functions of X that are characterized by K parameters

$$F(\cdot\,|\,\theta), \qquad \theta \in \Theta \tag{2. 25}$$

where θ is the parameter space defined by $\Theta = \{\theta=(\theta_1, \ldots, \theta_K) \mid$ constraint on $\theta\}$. We then select the parameters so that F becomes the best approximation to G. (Refer to the maximum likelihood method in Chapter 3.) F is called a parametric model (or simply a model) of X defined on the parameter space Θ.

When X is a continuous random variable, a model is usually defined by using a family of probability density functions

$$f(\cdot|\theta), \qquad \theta \in \Theta \tag{2.26}$$

and is denoted by $f(\cdot|\Theta)$. When X is discrete, the model is usually defined by using a family of probability mass function

$$p(\theta) = (p_1(\theta), p_2(\theta), \ldots), \qquad \theta \in \Theta \tag{2.27}$$

and is denoted by $P(\theta)$.

The development of the model $F(\cdot|\theta)$ is called *parameterization*.

[*Example 2. 1*]

$$f(x|\theta) = \frac{1}{\sqrt{2\pi\theta_2}} e^{-\frac{(x-\theta_1)^2}{2\theta_2}} \qquad \theta \in \Theta = \{(\theta_1, \theta_2) | \theta_2 > 0\} \tag{2.28}$$

is an example of a continuous model and

$$p(\theta) = (\theta_1, \ldots, \theta_k),$$
$$\theta \in \Theta = \left\{ (\theta_1, \ldots, \theta_k) | \theta_i \geq 0, \ i = 1, \ldots, K, \ \sum_{i=1}^{K} \theta_i = 1 \right\} \tag{2.29}$$

is an example of a discrete model.

Due to constraints imposed on the parameters, the values of all the parameters may be uniquely determined from the values of only k of the parameters. In this case, k is said to be the number of free parameters contained in the model.

The numbers of free parameters contained in the models (2.28) and (2.29) are 2 and $K-1$, respectively.

2. 3. 2 *Composition of Models*

A. Mixture Model

Let $F(\cdot|\Theta_F)$ and $G(\cdot|\Theta_G)$ be models for a random variable X. If we define a new probability distribution function by

$$H(\cdot|\alpha, \theta, \theta') = \alpha F(\cdot|\theta) + (1-\alpha)G(\cdot|\theta'). \qquad (2.30)$$

H is a model for the random variable defined on the parameter space

$$\Theta_H = \{(\alpha, \theta, \theta')|0 \leqq \alpha \leqq 1, \theta \in \Theta_F, \theta' \in \Theta_G\}$$

and its number of free parameters is given by

$$1 + (\textit{number of free parameters of } F(\cdot|\Theta_F)) + (\textit{number of free parameters of } G(\cdot|\Theta_G)).$$

This model is called the mixture of F and G.

B. Independent Model

Suppose $F(\cdot|\Theta_F)$ is a model of X and $G(\cdot|\Theta_G)$ is a model of Y. Then

$$H(x, y|\theta, \theta') = F(x|\theta)G(y|\theta'), \quad \theta \in \Theta_F, \theta' \in \Theta_G \qquad (2.31)$$

is a model for the joint distribution of (X, Y) and its number of free parameters is given by

$$(\textit{number of free parameters of } F(\cdot|\Theta_F)) + (\textit{number of free parameters of } G(\cdot|\Theta_G)).$$

This model is called *independent model*, since this model gives the joint distribution of X and Y when they are independent.

2.3.3 Restricted Model

Suppose a model $F(\cdot|\Theta)$ is given. By restricting the parameter space of $F(\cdot|\Theta)$ to be

$$\Theta' = \{\theta \in \Theta \mid additional\ constraints\ on\ \theta\}, \qquad (2.32)$$

we obtain the model $F(\cdot|\Theta')$. We shall consider that $F(\cdot|\Theta)$ and $F(\cdot|\Theta')$ to be different models. The number of free parameters contained in $F(\cdot|\Theta')$ is less than or equal to that of $F(\cdot|\Theta)$ depending on the type of additional restrictions on the parameter space.

[*Example* 2.2] Let $f(\cdot|\Theta)$ be the model given by (2.28). If we put

$$\Theta' = \{(\theta_1, \theta_2) \in \Theta \mid \theta_2 = a\theta_1^2\}$$

for some positive a, we obtain a restricted model $f(\cdot|\Theta')$:

$$f(x|\theta_1, a\theta_1^2) = \frac{1}{\sqrt{2\pi a\theta_1^2}} e^{-\frac{(x-\theta_1)^2}{2a\theta_1^2}}.$$

The number of free parameters of this model is only one.

2.3.4 *Reparameterization*

Suppose there is a model $F(\cdot|\Lambda)$ of X defined on a parameter space Λ and $\lambda = r(\theta)$ is a function defined on

$$\Theta = \{\theta = (\theta_1, \ldots, \theta_L) \mid constraint\ on\ \theta\} \qquad (2.33)$$

which takes values in Λ. Then

$$F(\cdot|r(\theta)), \qquad \theta \in \Theta \qquad (2.34)$$

is a model for X defined on Θ. Parametoric models are often considered by this method. This is called *reparameterization*.

[*Example* 2.3] The poisson distribution (2.4) can be

considered as a reparameterized model $p(r(\theta))$ that is obtained from a model with infinite degrees of freedom

$$p(\lambda) = (\lambda_0, \lambda_1, \lambda_2, \ldots),$$
$$\lambda \in \Lambda = \left\{(\lambda_0, \lambda_1, \ldots) \mid \lambda_i \geqq 0, \ i=0, 1, \ldots, \sum_{i=0}^{\infty} \lambda_i = 1\right\}$$

and reparameterization

$$r(\theta) = \left(e^{-\theta}, \theta e^{-\theta}, \frac{\theta^2}{2}e^{-\theta}, \ldots, \frac{\theta^i}{i!}e^{-\theta}, \ldots\right) \qquad \theta \in \Theta = \{\theta \mid \theta \geqq 0\}.$$

The number of free parameter of this model is one.

By reparameterization of a model we can reduce the number of free parameters in the model. The restriction of model (Section 2.3.3) is an example of this technique.

[*Example 2.4*] In the model (2.29), we furthermore restrict the parameter space to

$$\Theta' = \left\{\Theta = (\theta_1, \ldots, \theta_K) \mid \theta_i \geqq 0, \ i=1, \ldots, K, \ \sum_{i=1}^{K} \theta_i = 1, \ \theta_1 = \theta_2\right\}$$

and define a model $p(\Theta')$. This model is equivalent to a model $p(r(\Lambda))$ that is obtained by the reparameterization

$$\Theta = r(\Lambda) = (\lambda_1, \lambda_1, \lambda_2, \ldots, \lambda_{K-1})$$
$$\lambda \in \Lambda = \left\{(\lambda_1, \ldots, \lambda_{K-1}) \mid \lambda_i \geqq 0, \ i=1, \ldots, K-1, \ 2\lambda_1 + \sum_{i=2}^{K-1} \lambda_i = 1\right\}.$$

2.3.5 *Conditional Distribution Model*

Suppose we want to know the effect of a random variable X on the random variable Y, in particular, the the conditional distribution of Y given a realization x of X.

For example, suppose Y is a random variable corresponding to the weather of a particular day and X the meteorological data up to the previous day. If we know the conditional distribution of Y given the realization x of X then it will be useful for weather forecasting.

In this case, by regarding the parameter λ of the model $G(\cdot|\Lambda)$ of Y as a function of x, we obtain a model

$$G(\cdot \mid \lambda = r(x;\theta)),$$
$$\theta \in \Theta = \{\theta = (\theta_1, \ldots, \theta_L) \mid constraint\ on\ \Theta\} \qquad (2.35)$$

for which the distribution of Y changes with the value of x. Here Θ is a parameter that specifies the function r. We call (2.35) a *conditional distribution model* and denote it by $G(r(X;\Theta))$. The parameter spacc of the model $G(\cdot|r(X;\Theta))$ is Θ. X is called the *explanatory variable* and Y is called the *objective variable*. The construction of $G(\cdot|(X;\Theta))$ is also an example of reparameterization.

[*Example* 2.5] Fig. 2.2 shows the plot of independent realizations (x_i, y_i), $i=1, \ldots, 100$ of a random variable (X, Y). To get information about the dependence of Y on X, let us construct a conditional distribution model with explanatory variable X and the objective variable Y.

From Fig. 2.2 we know that the observations are scattered around the dashed curve. Therefore we will consider a normal distribution model of Y in which the mean value μ depends on x, such as

$$g(y \mid \mu(x), \sigma^2) = \frac{1}{\sqrt{2\pi\sigma^2}} e^{\frac{(y-\mu(x))^2}{2\sigma^2}}, \qquad (2.36)$$

Since x varies between 0 and 1, we approximate $\mu(x)$ by a Fourier series

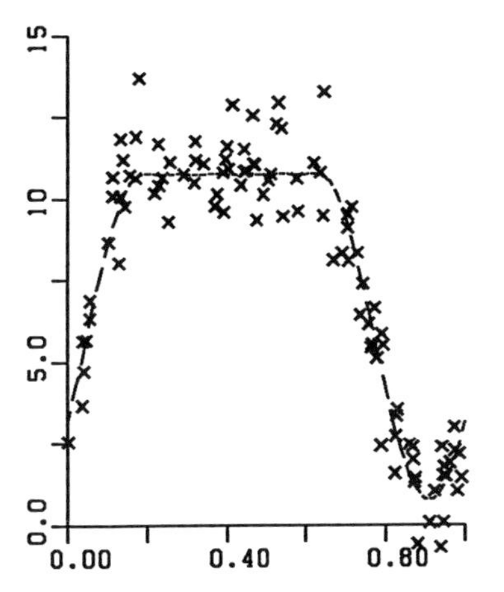

Fig. 2. 2 A realization of (X_i, Y_i), $i=1, \ldots, 100$

$$\mu(x) = S(x; a) = a_0 + \sum_{m=1}^{M} (a_{2m-1} \sin 2m\pi x + a_{2m} \cos 2m\pi x)$$

$$a = (a_0, a_1, \ldots, a_{2M}). \qquad (2.37)$$

By putting, $K=2M+2$ and

$$\Theta = \{\theta = (\sigma^2, a) \mid \sigma^2 > 0\}$$
$$(\mu(x), \sigma^2) = r(x; \theta) = (S(x; a), \sigma^2), \qquad (2.38)$$

we obtain a conditional distribution model $g(\cdot \mid r(X; \Theta))$ that is defined on Θ.

If we have a conditional distribution model $g(\cdot \mid r(X; \Theta))$ of an objective variable Y with an explanatory variable X as well as the model $h(\cdot \mid \Theta')$ of the explanatory variable X, then the model of the joint distribution of (X, Y) is given by

$$f(x, y \mid \theta, \theta') = h(x \mid \theta') \, g(y \mid r(x; \theta)),$$
$$\theta \in \Theta, \ \theta' \in \Theta'. \qquad (2.39)$$

[*Example 2. 6*] If we assume that the model of the explanatory variable X in [Example 2. 5] is the uniform distribution over

[0, 1], then by using the conditional distribution model given in (2.36) and (2.37), we obtain the following joint distribution model for (X, Y)

$$f(x, y \mid \theta) = \begin{cases} \frac{1}{\sqrt{2\pi\sigma^2}} e^{-\frac{(y-S(x;a))^2}{2\sigma^2}} & 0 \leqq x \leqq 1 \\ 0 & x<0, \ x>1 \end{cases} \tag{2.40}$$

$$\theta \in \Theta = \{(\sigma^2, a) \mid \sigma^2 > 0\},$$

2.4* Random Numbers and Simulation

2.4.1 Random Numbers and a Table of Random Numbers

Let X_i $(i=1, 2, \ldots, n)$ be independent and identically distributed random variables that follow a distribution F. A realization of $X=(X_1, \ldots, X_n)$ is denoted by $x=(x_1, \ldots, x_n)$. If F is the uniform distribution over [0, 1], x is called a sequence of *uniform random numbers*. If F is the standard normal distribution then they are called *normal random numbers*. If X_i takes a value from $\{0, 1, \ldots, 9\}$ and F is specified by the probability mass function

$$p = (p_1, p_2, \ldots, p_{10}) = (0.1, 0.1, \ldots, 0.1),$$

then a table of realizations that arranges x is called a table of random numbers. Tables of random numbers and normal random numbers are given in section III-5.

2.4.2 Generation of Uniform Random Numbers

If we divide a table of random numbers into subsets of m numbers and read each subset as a number with m digits after the decimal point, we obtain a pseudo uniform random number. In other words, let $x=(x_1, x_2, \ldots, x_n)$ be a table of random numbers and

$0 \leqq j \leqq n/m$. Then $y_0, y_1, y_2, \ldots$ with

$$y_j = \sum_{i=1}^{m} \frac{x_{mj+i}}{10^i} \tag{2.41}$$

is a sequence of pseudo uniform random numbers.

A simple algorithm to generate a pseudo uniform random number by a computer is

$$II = MOD(\ II \times CONST,\ RMD\ )$$
$$y_i = \frac{II}{RMD},$$

where *RMD* is the largest prime number expressible in the computer (e.g., 2147483647.0D0 for a 32 bit machine.) The selection of *CONST* is not so essential (Use 31415925.0D0, for example.) Use a large odd integer (say, 1983011711) for the initial *II* and repeat the above two steps as many times as you need.

2.4.3 *Simulation Based on a Probability Mass Function*

To generate a realization, k_i $(i = 1, 2, \ldots, n)$, of independent discrete random variables that take values on $\{1, 2, \ldots\}$ with probabilities $\{p_1, p_2, \ldots\}$, we first obtain the cumuluative probability distribution

$$S_j = \sum_{i=1}^{j} p_i.$$

Then we obtain the k_i from a sequence of uniform numbers x by forming

$$k_i = min\{\ j \mid x_i \leqq S_j\}. \tag{2.42}$$

Here, as j increased we determine k_i as the first j that satisfies $x_i \leqq S_j$.

2.4.4 *Simulation Based on a Distribution Function*

For simplicity, we assume that $G(y)$ is a monotone increasing continuous distribution function and $y = G^{-1}(x)$ is the inverse transformation. In this case an independent sequence of realizations y_i $(i = 1, 2, \ldots, n)$ of a random variable Y that follows the distribution function G is generated from a sequence of uniform random numbers x_i $(i = 1, 2, \ldots, n)$ by

$$y_i = G^{-1}(x_i). \tag{2.43}$$

This method can be verified since

$$P\{y \leqq a\} = P\{G^{-1}(x) \leqq a\} = P\{x \leqq G(a)\} = G(a).$$

PROBLEM

2.1 Check that the probability distributions introduced in sections 2.1 and 2.2 are the parametric models defined in this section and count the number of free parameters contained in each model.

CHAPTER 3

ESTIMATION

Any phenomenon with uncertainty can be regarded as the realization of a random variable that follows a certain probability distribution. In this chapter, following Akaike (1973, 1974), we shall develop a method of evaluating the goodness of fit of a model that approximates the probability distribution (hereafter, we shall simply call a *model*) and of estimating the model from the given data.

A model is usually expressed in the form of a probability distribution. Therefore, the goodness of the model can be evaluated by the similarity of the probability distribution specified by the model to the true probability distribution that generated the data. Furthermore, fitting a model to the data can be regarded as estimating the true probability distribution from the data. The viewpoint that both the models and the true structure are probability distributions is our basic standpoint for the evaluation and the estimation of a model.

3.1 *Information Quantity and Entropy*

We first consider discrete distribution. Let the true distribution be given by $p=\{p_1, p_2, \ldots, p_m\}$ where p_i is the probability that the event ω_i occurs and satisfies $p_i > 0$ and $p_1+\cdots+p_m=1$. For the moment, we will assume that we know this true distribution. Our problem here is as follows; when there are models probability distributions that approximate this true distribution, how do we evaluate the goodness of the approximation of these models to the true distribution?

[*Example 3.1*] Two baseball commentators, Mr. A and Mr. B, predicted that the probability of a certain team winning will be 0.7 and 0.5, respectively. (These are the binomial models $q_A = \{0.7, 0.3\}$ and $q_B = \{0.5, 0.5\}$, respectively.) If the true probability of winning is 0.4, it is obvious that which commentator was right. Then, if the true probability of winning turns out to be 0.6, which prediction was closer to the true one?

To answer this problem, we need an objective criterion that measures the "distance" between the true distribution and the model. Suppose $p = \{p_1, p_2, \ldots, p_m\}$ is the true distribution and $q = \{q_1, q_2, \ldots, q_m\}$ a discrete distribution model, $\log p/q$ is a random variable that takes the value $\log p_i/q_i$ when the event ω_i occurs. The expectation of $\log p/q$,

$$I(p;q) = E \log \frac{p}{q} = \sum_{i=1}^{m} p_i \log \frac{p_i}{q_i} \qquad (3.1)$$

is called the *Kullback-Leibler quantity of information* (Hereafter called the *K-L information quantity*) of the true distribution p with respect to the model q. Throughout this book *log* denotes the natural logarithm. The K-L information quantity has the following properties.

Assume p and q are probability distributions satisfying $p_i>0$, $q_i>0$ $(i=1, \cdots, m)$, $\sum_{i=1}^{m} p_i=1$ and $\sum_{i=1}^{m} q_i=1$. Then, $I(p;q)$ satisfies the following;

$$(i) \quad I(p;q) \geqq 0, \qquad (3.2)$$

$$(ii) \quad I(p;q) = 0 \Leftrightarrow p_i = q_i \quad (i=1, \ldots, m).$$

[*Proof*] Let $f(x) = \log x - x + 1$ for $x > 0$. $f(x)$ attains its maximum value 0 only at $x=1$. Thus $\log x \leqq x-1$ and the equality holds only at $x=1$. Therefore, by putting $x = q_i/p_i$, we have

$$\log \frac{q_i}{p_i} \leqq \frac{q_i}{p_i} - 1 \quad (i=1, \ldots, m),$$

it follows that

$$\sum_{i=1}^{m} p_i \log \frac{q_i}{p_i} \leqq \sum_{i=1}^{m} p_i \left(\frac{q_i}{p_i} - 1 \right) = \sum_{i=1}^{m} q_i - \sum_{i=1}^{m} p_i = 1 - 1 = 0.$$

Therefore we get

$$\sum_{i=1}^{m} p_i \log \frac{p_i}{q_i} = - \sum_{i=1}^{m} p_i \log \frac{q_i}{p_i} \geqq 0.$$

It is obvious that the equality holds only at $q_i = p_i$. Q. E. D.

We adopt this K-L information quantity as the criterion for the similarity of two distributions. Namely, the smaller (and hence the closer to 0) the value of $I(p;q)$, the closer we consider the model q is to the true distribution.

Let us compare two models given in [Example 3. 1].

$$\begin{aligned} I(p;q_A) &= 0.6 \log \frac{0.6}{0.7} + 0.4 \log \frac{0.4}{0.3} \\ &= -0.0925 + 0.1151 = 0.0226, \\ I(p;q_B) &= 0.6 \log \frac{0.6}{0.5} + 0.4 \log \frac{0.4}{0.5} \\ &= 0.1094 - 0.0893 = 0.0201. \end{aligned}$$

Therefore, according to the K-L information quantity, the prediction by Mr. B was closer to the truth.

Fig. 3. 1 illustrates the K-L information quantity of the model $q=(q, 1-q)$ when the true model is assumed to be $p = (0.5, 0.5)$, $(0.6, 0.4)$, $(0.7, 0.3)$ and $(0.8, 0.2)$, respectively. Note that except for the case $p=(0.5, 0.5)$, the K-L information quantity is asymmetric.

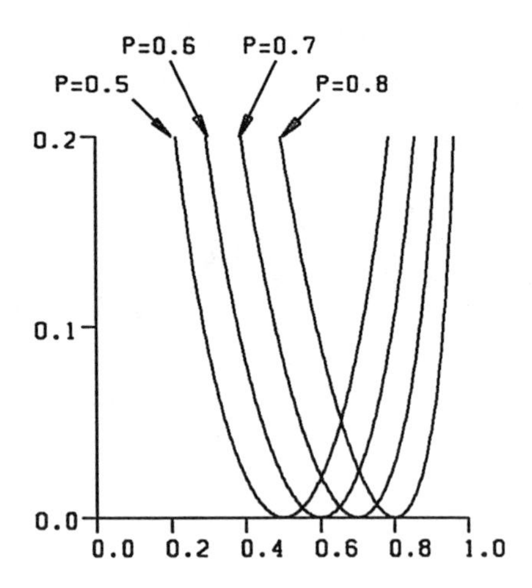

Fig. 3. 1 K-L information quantities $I(p;q)$

The negative of the K-L information quantity

$$B(p;q) = \sum_{i=1}^{m} p_i \log\frac{q_i}{p_i}, \tag{3. 3}$$

is called the *entropy*. The following property of entropy is the one reason that the K-L information quantity was adopted as the criterion for the goodness of fit of the model. This is a generalization of the stochastic view of entropy due to Boltzmann (1877).

[*Interpretation*] Entropy is approximately equal to 1/n times the logarithm of the probability that the relative frequency distribution of n observations obtained from the assumed model equals the true distribution.

[*Proof*] For simplicity, consider the situation where $p_i=n_i/n$ for some integers $n_1, \ldots, n_m$, $n_1+\cdots+n_m=n$ and $q=\{q_1, \ldots, q_m\}$ is our model. The probability that we obtain the frequency distribution $\{n_1, \ldots, n_m\}$ from n observations that follow this distribution is

$$W = \frac{n!}{n_1! \cdots n_m!} q_1^{n_1} \cdots q_m^{n_m}.$$

By taking logarithm we have

$$\log W = \log n! - \sum_{i=1}^{m} \log n_i! + \sum_{i=1}^{m} n_i \log q_i.$$

Here, assuming that each n_i is sufficiently large and applying Stirling's formula, we obtain

$$\log n_i! \fallingdotseq \log \sqrt{2\pi} + \left(n_i + \frac{1}{2}\right) \log n_i - n_i,$$

which yields

$$\log W \fallingdotseq \left(n + \frac{1}{2}\right) \log n + \log \sqrt{2\pi} - n$$
$$- \left\{ \sum_{i=1}^{m} \left(n_i + \frac{1}{2}\right) \log n_i + m \log \sqrt{2\pi} - \sum_{i=1}^{m} n_i \right\} + \sum_{i=1}^{m} n_i \log q_i.$$

Hence

$$\begin{aligned}
\log W &= n \log n - \Sigma n_i \log n_i + \Sigma n_i \log q_i + O(\log n) \\
&= -\Sigma n_i \log \frac{n_i}{n} + \Sigma n_i \log q_i + O(\log n) \\
&= -n\left(\Sigma \frac{n_i}{n} \log \frac{n_i}{n} - \Sigma \frac{n_i}{n} \log q_i\right) + O(\log n) \\
&= -n\left(\Sigma p_i \log p_i - \Sigma p_i \log q_i\right) + O(\log n) \\
&= -n \Sigma p_i \log \frac{p_i}{q_i} + O(\log n) \\
&= nB(p; q) + O(\log n)
\end{aligned}$$

Thus it follows that

$$\frac{1}{n} \log W \fallingdotseq B(p;q). \qquad Q.E.D.$$

Next consider the case where our model is continuous and has a probability density function. Let $g(x)$ be the true density function and $f(x)$ be the density function that specifies the model. In this case the K-L information quantity of the true distribution with respect to the model is defined by

$$\begin{aligned} I(g;f) &= E_X \log \frac{g(X)}{f(X)} \\ &= \int_{-\infty}^{\infty} \log \frac{g(x)}{f(x)}\, g(x)\, dx. \end{aligned}$$

$I(g;f)$ has the following properties;

(i) $I(g;f) \geqq 0$

(ii) $I(g;f) = 0 \Leftrightarrow g(x) = f(x)$ (a. e.).

[*Example* 3. 2] Suppose the true distribution is the standard normal distribution $N(0, 1)$. Which model, $N(0.5, 1)$ or $N(0, 1.5)$, is closer to the true distribution?

We first generalize the problem and compute the K-L information quantity when the true distribution is $N(\mu, \tau^2)$ and our model is $N(\xi, \sigma^2)$. Namely

$$\begin{aligned} g(x) &= \frac{1}{\sqrt{2\pi\sigma^2}} e^{-\frac{(x-\mu)^2}{2\sigma^2}} \\ f(x) &= \frac{1}{\sqrt{2\pi\tau^2}} e^{-\frac{(x-\xi)^2}{2\tau^2}}. \end{aligned}$$

From (3. 4), the K-L information quantity is obtained by

$$I(g;f)=\int_{-\infty}^{\infty} \log \frac{g(x)}{f(x)}\, g(x)\,dx$$
$$=\int_{-\infty}^{\infty} \log g(x)g(x)\,dx-\int_{-\infty}^{\infty} \log f(x)g(x)\,dx$$

Here, using $E_X(X-\xi)^2=\sigma^2+(\mu-\xi)^2$, we have that

$$\int_{-\infty}^{\infty} \log f(x)\, g(x)\,dx = \int_{-\infty}^{\infty}\left\{-\frac{1}{2}\log 2\pi\tau^2-\frac{(x-\xi)^2}{2\tau^2}\right\}\frac{1}{\sqrt{2\pi\sigma^2}}e^{-\frac{(x-\mu)^2}{2\sigma^2}}dx$$
$$= -\frac{1}{2}\log 2\pi\tau^2-\frac{\sigma^2+(\mu-\xi)^2}{2\tau^2}.$$

By putting $\tau^2=\sigma^2$, $\mu=\xi$, we have

$$\int_{-\infty}^{\infty} \log g(x)g(x)\,dx = -\frac{1}{2}\log 2\pi\sigma^2 - \frac{1}{2}.$$

Thus by (3.6), the K-L information quantity is given by

$$I(g;f) = \frac{1}{2}\left\{\log\frac{\tau^2}{\sigma^2}+\frac{\sigma^2+(\mu-\xi)^2}{\tau^2}-1\right\}.$$

In particular, by putting $\mu = 0$ and $\sigma^2 = 1$, the K-L information quantity for [Example 3.2] is given by

$$I(g;f) = \frac{1}{2}\left\{\log\tau^2+\frac{1+\xi^2}{\tau^2}-1\right\}.$$

Fig. 3.2 shows the contours of the K-L information quantity using τ^2 and ξ as axes. Let us compare the two models in [Example 3.2] by using this K-L information quantity.

$$I(g;f_1) = \frac{1}{2}\left(\log 1+\frac{1+0.25}{1}-1\right) = 0.125,$$
$$I(g;f_2) = \frac{1}{2}\left(\log 1.5+\frac{1+0}{1.5}-1\right) = 0.036.$$

This indicates that the normal distribution model $N(0, 1.5)$ is closer to the true distribution, $N(0, 1)$, than $N(0.5, 1)$. This decision can be visually justified from the figure shown in Fig. 3. 3.

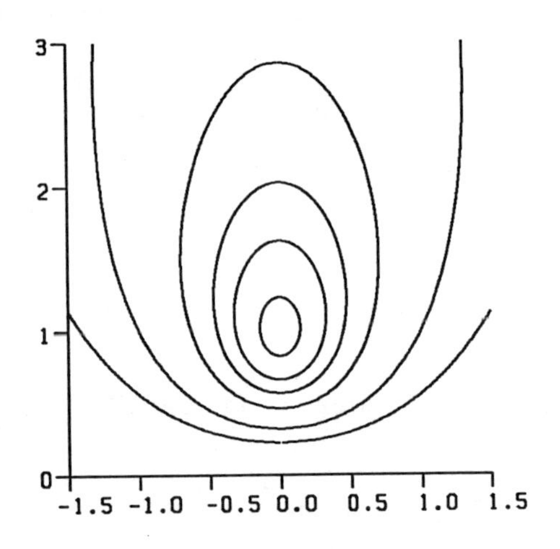

Fig. 3. 2 K-L information quantity $I(p;q)$ for normal distribution model.

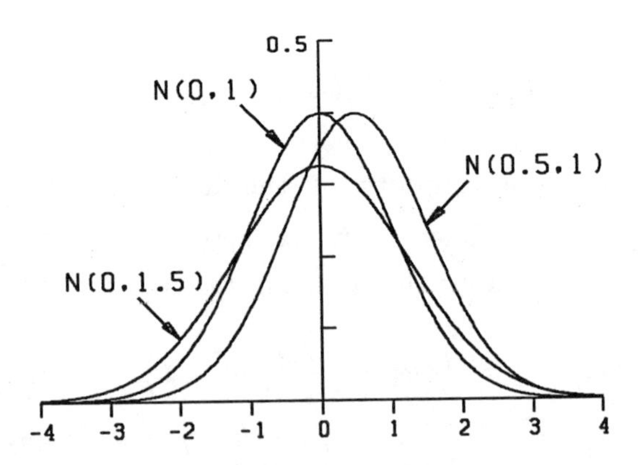

Fig. 3. 3 Densities of N(0, 1), N(0, 1. 5) and N(0. 5, 1)

3.2 Estimate of K-L Information Quantity; Log Likelihood

As shown in section 3.1, we can compare the goodness of fit of various models by using the K-L information quantity provided we know the true distribution. But, in actuality, the true distribution is usually unknown and we only have data obtained from the true distribution. In this situation, where we do not know the true distribution, how can we compare the goodness of fit of the models that approximate the true distribution? For example, let us consider the following problem that corresponds to Example 3.1.

[*Example 3.1'*] Two baseball commentators, Mr. A and Mr. B, predicted prior to the season that the probability of a certain team winning will be 70% and 50%, respectively. After 100 games the actual record was 65% (65 won and 35 lost). Whose prediction was more correct?

First of all, note that [Example 3.1'] is a much more realistic example than [Example 3.1]. Generally we assume that we have n independent observations $\{x_1, x_2, \ldots, x_n\}$ that are obtained from the true distribution $p=\{p_1, p_2, \ldots, p_m\}$. Each observation results in one of the events $\omega_1, \ldots, \omega_m$. If we define n_i as the number of occurrences of the event ω_i, then we have $n_1+n_2+\cdots+n_m = n$. Based on these data we shall try to estimate the K-L information quantity with respect to the model, namely

$$I(p;q) = \sum_{i=1}^{m} p_i \log \frac{p_i}{q_i} = \sum_{i=1}^{m} p_i \log p_i - \sum_{i=1}^{m} q_i \log q_i. \qquad (3.7)$$

Here the first term on the right hand side is a constant that depends on the true distribution p only. Therefore, the larger the second term of (3.7), the smaller the K-L information quantity, $I(p;q)$, becomes. This means that for the comparison of models by the K-L information quantity, we just need to estimate the value of $\sum_{i=1}^{m} p_i \log q_i$. Since p_i is the true probability, $\sum_{i=1}^{m} p_i \log q_i$ is the expectation $E \log q$ of the random variable $\log q$ that takes the value $\log q_i$ when the event ω_i occurs. (This is called the expected log likelihood.) By the law of large numbers

$$\frac{1}{n}\sum_{l=1}^{m} \log q_{x_l}, \tag{3.8}$$

where $q_{x_l} = q_i$ if the event ω_i occurs at the lth observation x_l, converges to the expected log likelihood $E \log q$ as $n \to \infty$. Since the number of times that the random variable $\log q$ takes the value $\log q_i$ is given by n_i, (3.8) is identical to 1/n times

$$\sum_{i=1}^{m} n_i \log q_i. \tag{3.9}$$

Thus, if the data is given, we can essentially estimate the K-L information quantity with respect to the model q by (3.9). This is called the *log likelihood* of the model q and is denoted by $l(q)$. We consider that the larger the log likelihood is, the better the model is.

Consider [Example 3.1']. The actual record was that they won $n_1=65$ games and lost $n_2=35$ games. Thus the log likelihood is given by

$$l(q) = 65 \log q_1 + 35 \log q_2.$$

Mr. A predicted that $q_1=0.7$ and $q_2=0.3$. The log likelihood of this model is

$$l(q_A) = 65 \log 0.7 + 35 \log 0.3 = -65.32.$$

On the other hand, the model of Mr. B is $\{0.5, 0.5\}$. The log likelihood of this model is

$$l(q_B) = 65 \log 0.5 + 35 \log 0.5 = -69.31.$$

Since $l(q_A) > l(q_B)$, we consider that the prediction by Mr. A was better.

Since this judgement is based on the log likelihood that is (n times) an estimate of the expected log likelihood (and, in essence, the K-L information quantity), it is inevitable that it has some risk of error. Let us take a look at Fig. 3.4. The dashed curve expresses the expected log likelihood of the true model $p=\{0.6, 0.4\}$ given in [Example 3.1], and this is our true criterion. On the other hand the solid curve shows $1/n$ times the log likelihood. We can see that $1/n$ times the log likelihood is approximately given by shifting the expected log likelihood. According to the law of large numbers, $1/n$ times the log likelihood converges to the expected log likelihood as $n \to \infty$.

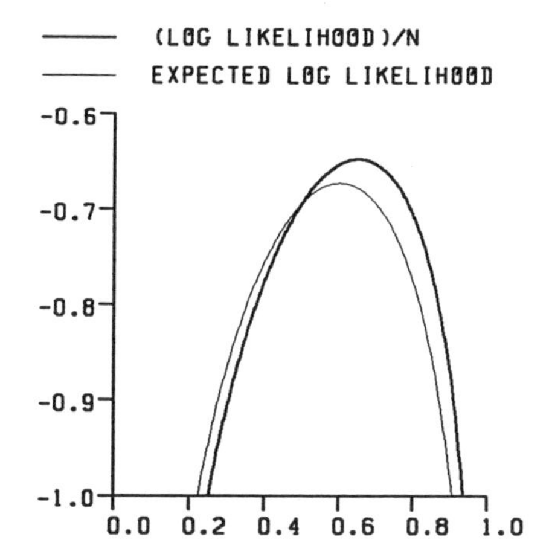

Fig. 3.4 expected log likelihood and (log likelihood)/n

Similarly, in the case of continuous distributions we can

estimate the K-L information quantity from the data. Suppose $g(x)$ and $f(x)$ are the probability density functions of the true distribution and a model, respectively. From the definition of the K-L information quantity

$$I(g;f) = \int_{-\infty}^{\infty} \log \frac{g(x)}{f(x)} g(x)dx \qquad (3.10)$$

$$= \int_{-\infty}^{\infty} \log g(x)\, g(x)dx - \int_{-\infty}^{\infty} \log f(x)\, g(x)dx.$$

As in the case of discrete distributions, since the first term of the right hand side is a constant, we only have to estimate the expected log likelihood

$$\int_{-\infty}^{\infty} \log f(x)\, g(x)dx. \qquad (3.11)$$

We call

$$\sum_{i=1}^{n} \log f(x_i) \qquad (3.12)$$

the log likelihood of the model. If we have n independent observations $\{x_1, x_2, \ldots, x_n\}$, the expected log likelihood can be approximated by $\frac{1}{n}$ times the log likelihood

$$\frac{1}{n} \sum_{i=1}^{n} \log f(x_i). \qquad (3.13)$$

By taking account of the negative sign, it is considered that the larger the log likelihood the closer the model is to the true distribution. Consider the following example.

[*Example* 3.3] Suppose we have 10 observations -1.10, -0.40, -0.20, -0.02, 0.02, 0.71, 1.35, 1.46, 1.74, 3.89, and two models;

$$MODEL(1): \quad f_1(x) = \frac{1}{\sqrt{2\pi}} e^{-\frac{x^2}{2}}$$

$$MODEL(2): \quad f_2(x) = \frac{1}{\pi(x^2+1)}$$

Which one is closer to the true distribution? Here, $f_2(x)$ is called the probability density function of the Cauchy distribution.

The log likelihood of $MODEL(1)$ is

$$\sum_{i=1}^{10} \log f_1(x) = -5 \log 2\pi - \frac{1}{2}\sum_{i=1}^{10} x_i^2$$
$$= -9.189 - 0.5 \times 23.629 = -21.00.$$

On the other hand, that of $MODEL(2)$ is

$$\sum_{i=1}^{10} \log f_2(x) = -10 \log \pi - \sum_{i=1}^{10} \log(x_i^2+1)$$
$$= -11.447 - 7.743 = -19.19.$$

Therefore we consider that the Cauchy distribution model is closer to the true distribution than the standard normal distribution model. Note that, as is this example, it becomes possible to compare the goodness of the approximation of different types of models by considering the log likelihood to be an estimate of the expected log likelihood.

Assume here that a joint density function $f(x_1, \ldots, x_n | \theta)$ of the random variable $(X_1, X_2, \ldots, X_n)$ is given. θ is a parameter that specifies the density function. When the observations $(x_1, \ldots, x_n)$ are given, f is a function of θ called the likelihood and is denoted by $L(\theta)$, namely

$$L(\theta) = f(x_1, \ldots, x_n | \theta).$$

In particular, if the random variables are independent, then the joint probability density function of $(X_1, \ldots, X_n)$ is given as the product of each density function of X_i $(i=1, \ldots, n)$. Thus we have

$$L(\theta) = f(x_1 | \theta) f(x_2 | \theta) \cdots f(x_n | \theta).$$

Taking the logarithm on both sides yields

$$l(\theta) = \sum_{i=1}^{n} \log f(x_i | \theta).$$

So far, we have derived the log likelihood directly as an estimate of the expected log likelihood. However, if a model is given in the form of a probability distribution, it is easier to define the likelihood as the joint distribution of the observations and then obtain the log likelihood as the logarithm of the likelihood. Even when $(X_1, \ldots, X_n)$ are not independent we can define the log likelihood as the logarithm of the likelihood; i. e.

$$l(\theta) = \log f(x_1, \ldots, x_n | \theta).$$

3. 3 *The Method of the Maximum Likelihood*

In section 3. 2, we showed that the comparison of the K-L information quantity is essentially equivalent to the comparison of the log likelihoods estimated from the data. If we have several fixed models, we can expect to get the best model that is closest to the true distribution by selecting the model with the largest value of log likelihood. Similarly, if our model contains some adjustable parameters, we can get a good model by selecting the values of parameters that maximize the log likelihood. This method of estimating parameters contained in a model is called the *method of the maximum likelihood*. The estimator derived by this method is called the *maximum likelihoood estimator* (*MLE*). The model specified by the MLE is the *maximum likelihood model*. The value of the log likelihood of the maximum likelihood model is called the *maximum log likelihood*. In the following subsections, we will determine the maximum likelihood estimators for some typical probability distributions.

3. 3. 1 Binomial Distribution

Given observations $\{k_1, \ldots, k_n\}$, we can obtain the number of occurrences n_1 and n_2 $(n_1+n_2=n)$ of the type of events. From (3. 9), the log likelihood of binomial distribution $\{p, 1-p\}$ is

$$l(p) = n_1 \log p + n_2 \log (1-p). \tag{3. 14}$$

Here we have ignored a constant that does not depend on p. The necessary condition that the parameter maximizes the log likelihood is

$$\frac{dl}{dp} = \frac{n_1}{p} - \frac{n_2}{1-p} = 0$$

and thus

$$n_1(1-p) = n_2 p.$$

Therefore

$$\hat{p} = \frac{n_1}{n} \tag{3. 15}$$

is the maximum likelihood estimator of p. Substitution of this into (3. 14) yields the maximum log likelihood

$$l(\hat{p}) = n_1 \log n_1 + n_2 \log n_2 - n \log n. \tag{3. 16}$$

In the case of [Example 3. 1'], since $n_1=65$ and $n_2=35$, the maximum likelihood estimate of the winning probability is obtained as

$$\hat{p} = \frac{65}{65+35} = 0.65.$$

3.3.2 *Multinomial Distribution*

Given observations $\{k_1, \ldots, k_n\}$, count the number of occurrences n_i $(n_1+\cdots+n_c=n)$ of each event ω_i. The log likelihood of the multinomial distribution $q=\{q_1, \ldots, q_c\}$ is

$$l(q) = \sum_{i=1}^{c} n_i \log q_i. \tag{3.17}$$

Here, since q is a probability distribution, $\sum_{i=1}^{c} q_i=1$. By substituting $q_c = 1-\sum_{i=1}^{c-1} q_i$ into (3.17), we have

$$l(q) = \sum_{i=1}^{c-1} n_i \log q_i + n_c \log\left\{1-\sum_{i=1}^{c-1} q_i\right\}.$$

The necessary condition for q to maximize this log likelihood is

$$\frac{\partial l}{\partial q_i} = \frac{n_i}{q_i} - \frac{n_c}{1-\sum_{j=1}^{c-1} q_j} = 0 \qquad (i=1, \ldots, c-1).$$

From this we have

$$\frac{n_i}{q_i} = \frac{n_c}{1-\sum_{j=1}^{c-1} q_j} = \frac{n_c}{q_c} \qquad (i=1, \ldots, c-1).$$

and it follows that

$$\frac{n_i}{q_i} = K \qquad (i=1, \ldots, c)$$

for some constant K. Since

$$n = \sum_{i=1}^{c} n_i = K \sum_{i=1}^{c} q_i = K,$$

the maximum likelihood estimate is given by

$$\hat{q}_i = \frac{n_i}{n}. \qquad (3.18)$$

Thus the maximum likelihood model is $q = \{n_1/n, \ldots, n_c/n\}$ and the maximum log likelihood is given by

$$l(\hat{q}) = \sum_{i=1}^{c} n_i \log n_i - n \log n. \qquad (3.19)$$

3.3.3 *Poisson Distribution*

Given n independent observations $\{k_1, k_2, \ldots, k_n\}$, the log likelihood of the Poisson distribution (2.4) is given by

$$l(\lambda) = \sum_{i=1}^{n} (-\lambda + k_i \log \lambda - \log k_i!). \qquad (3.20)$$

From

$$\frac{\partial l}{\partial \lambda} = \sum_{i=1}^{n} \left(-1 + \frac{k_i}{\lambda}\right) = 0,$$

the maximum likelihood estimator is given by

$$\hat{\lambda} = \frac{1}{n} \sum_{i=1}^{n} k_i. \qquad (3.21)$$

This is the sample mean of $\{k_1, \ldots, k_n\}$. By substituting $\hat{\lambda}$ into (3.20), the maximum log likelihood is given by

$$l(\hat{\lambda}) = -\sum_{i=1}^{n} k_i + \left\{\sum_{i=1}^{n} k_i\right\} \log \left\{\frac{1}{n} \sum_{i=1}^{n} k_i\right\} - \sum_{i=1}^{n} \log k_i!. \qquad (3.22)$$

3. 3. 4 Normal Distribution

When both the mean and the variance are unknown, the log likelihood of the normal distribution (2. 8) is given by

$$l(\mu, \sigma^2) = -\frac{n}{2}\log 2\pi\sigma^2 - \frac{1}{2\sigma^2}\sum_{i=1}^{n}(x_i-\mu)^2. \tag{3. 23}$$

Thus

$$\frac{\partial l}{\partial \mu} = \frac{1}{\sigma^2}\sum_{i=1}^{n}(x_i-\mu) = 0$$

$$\frac{\partial l}{\partial \sigma^2} = -\frac{n}{2\sigma^2} + \frac{1}{2\sigma^4}\sum_{i=1}^{n}(x_i-\mu)^2 = 0$$

are necessary conditions for μ and σ^2 to be the maximum likelihood estimators. By solving these equations

$$\sum_{i=1}^{n} x_i = n\mu \tag{3. 24}$$

$$\sum_{i=1}^{n}(x_i-\mu)^2 = n\sigma^2, \tag{3. 25}$$

we obtain

$$\hat{\mu} = \frac{1}{n}\sum_{i=1}^{n} x_i \tag{3. 26}$$

$$\hat{\sigma}^2 = \frac{1}{n}\sum_{i=1}^{n}(x_i-\hat{\mu})^2. \tag{3. 27}$$

Note that $\hat{\mu}$ is identical to the sample mean. Substitution of these into (3. 23) yields the maximum log likelihood of the model

$$l(\hat{\mu}, \hat{\sigma}^2) = -\frac{n}{2}\log 2\pi\hat{\sigma}^2 - \frac{n}{2} = -\frac{n}{2}\log\left\{\frac{2\pi}{n}\sum_{i=1}^{n}(x_i-\hat{\mu})^2\right\} - \frac{n}{2}. \tag{3. 28}$$

3.3.5 *Properties of the Maximum Likelihood Estimators*

Suppose $X_1, \ldots, X_n$ are independent and identically distributed random variables that follow a probability density function $f(x|\theta^*)$. Under certain regularity conditions for $f(x|\theta)$, the maximum likelihood estimator $\hat{\theta}$ converges to a normal random variable as $n \to \infty$ (asymptotic normality). Namely, for large n it is approximately true that

$$\hat{\theta} \sim N\left(\theta^*, \frac{1}{n}I^{-1}\right),$$

where I is the *Fisher information matrix*, the (i, j) element of which is given by

$$E_X\left[\frac{\partial}{\partial\theta_i}log\ f(X|\theta)\frac{\partial}{\partial\theta_j}log\ f(X|\theta)\right]_{\theta=\theta^*}.$$

In particular, as n goes to infinity, $\hat{\theta}$ converges to the true value θ^* (the consistency), the bias of $\hat{\theta}$ as an estimate of θ^* vanishes (the asymptotic unbiasedness) and $\hat{\theta}$ has the smallest variance among the calss of unbiased estimators (asymptotic efficiency).

CHAPTER 4

AKAIKE INFORMATION CRITERION

In this chapter, we introduce AIC, the Akaike Information Criterion as a basis of comparison and selection among several models. Those who need only an overview and the definition of AIC may proceed to Part II after reading Sections 4.1 and 4.6 where some practical advice on the use of AIC is given.

Sections 4.2 through 4.5 are devoted to the derivation of AIC. Definitions and assumptions necessary for the development of this chapter are given in Section 4.2 and followed by a numerical example to illustrate the necessity and role of AIC. In Section 4.3, AIC is derived under the assumption that the true distribution can be described by the given model when its parameters are suitably adjusted. Section 4.4 concerns the more realistic case where this assumption does not hold. AIC for conditional distribution models is discussed in Section 4.5.

AIC was introduced by Akaike (1973). His epoch-making paper showed the important role of the Kullback-Leibler information quantity in statistics and also derived AIC as its estimator. The derivation of AIC given in this chapter follows his original paper, though some parts are modified to improve the clarity of the reasoning.

4.1 General Idea

It was shown in the previous chapter that the goodness of values of parameters of a specific model can be measured by the expected log likelihood, namely, the larger the expected log likelihood the better the values of parameters. The log likelihood can be regarded as an estimator of the expected log likelihood.

We introduce the *mean expected log likelihood* as a measure for the goodness of fit of a model. This quantity is defined as the mean, with respect to the data x, of the expected log likelihood of the maximum likelihood model. The larger the mean expected log likelihood the better the fit of the model. At first sight, it would seem that the mean expected log likelihood can be estimated by the maximum log likelihood. The maximum log likelihood, however, is shown to be a biased estimator of the mean expected log likelihood. The maximum log likelihood has a general tendency to over estimate the true value of the mean expected log likelihood. This tendency is more prominent for models with a larger number of free parameters. This means that if we choose the model with the largest maximum log likelihood, a model with an unnecessarily large number of free parameters is likely to be chosen.

By a close examination of the relationship between the bias and the number of free parameters of a model, we will find that

$$(\textit{maximum log likelihood of a model}) \\ -(\textit{number of free parameters of the model})$$

is an asymptotically unbiased estimator of the mean expected log likelihood. Taking historical reasons into account, minus twice of this value

$$AIC = -2\times(\textit{maximum log likelihood of the model}) \\ +2\times(\textit{number of free parameters of the model}) \quad (4.1)$$

is proposed as the criterion for model selection. A model which minimizes the *AIC* (*minimum AIC estimate, MAICE*) is considered to be the most appropriate model. Definition (4.1) implies that when there are several models whose values of maximum likelihood are about the same level, we should choose the one with the smallest number of free parameters. In this sense AIC realizes, so called, the principle of parsimony.

4.2 *Mean Expected Log likelihood*

4.2.1 *Definitions and Assumptions*

Let $x=(x_1, x_2, \cdots, x_n)$ be a realization of a random variable $X=(X_1, X_2, \cdots, X_n)$. The X_i' s are assumed to be independent and identically distributed with the probability density function $g(\cdot)=f(\cdot\,|\,\theta^*)$, where f is a model with K parameters, i.e.,

$$MODEL(K):\ f(\cdot\,|\,\theta),\quad \theta=(\theta_1, \theta_2, \ldots, \theta_K). \qquad (4.2)$$

We assume that there are no constraints on the parameters, i.e., the number of free parameters in the model is K. We call $\theta^*=(\theta_1^*, \theta_2^*, \cdots, \theta_K^*)$ the vector of true parameters. In this section, the true parameters and the true mean expected log likelihood are distinguished by *. All vectors are row vectors if not specified otherwise.

The model defined by restricting the parameter space as

$$\Theta_k=\{\,\theta\in\Theta_K\,|\,\theta_{k+1}=\theta_{k+2}=\ldots=\theta_K=0\,\} \qquad (4.3)$$

is denoted by $MODEL(k)$. For l such that $k\leq l$, the $MODEL(k)$ can be obtained by restricting $MODEL(l)$. For these type of restricted models,

$$(\textit{number of free parameters of } MODEL(k)) = k. \qquad (4.4)$$

Since $\Theta_1\subset\Theta_2\subset\ldots\subset\Theta_K$, there exists a minimum among the k' s that satisfy $\theta^* \in \Theta_k$. This minimum is called the *true number of free parameters* and is denoted by k^*.

Given the data x, the maximum likelihood estimate $\hat{\theta}_k$ of the parameters of *MODEL*(k) is obtained by maximizing the log likelihood

$$l(\theta) = \sum_{i=1}^{n} \log f(x_i \mid \theta) \tag{4.5}$$

over $\theta \in \Theta_k$. The maximum log likelihood is given by

$$l(\hat{\theta}_k) = \max_{\theta \in \Theta_k} l(\theta). \tag{4.6}$$

Here f is a probability mass function or a probability density function depending on whether X_i is a discrete or continuous randam variable. In the following argument, we will assume that X_i is a continuous random variable. For a discrete random variable the density function should be replaced by a probability mass function and the integration by the summation.

The expected log likelihood of the distribution $f(\cdot \mid \theta)$ is defined by

$$E_Z\{\log f(Z \mid \theta)\} = \int f(z \mid \theta^*) \log f(z \mid \theta) dz, \tag{4.7}$$

where Z is a random variable with the same distribution as X_i and is independent of X. Corresponding to the log likelihood $l(\theta)$, we define $l^*(\theta)$ as n times the expected log likelihood, namely

$$l^*(\theta) = nE_Z\{\log f(Z \mid \theta)\}. \tag{4.8}$$

The larger the value of $l^*(\theta)$ the better the approximation of the distribution $f(\cdot \mid \theta)$ to the true one $f(\cdot \mid \theta^*)$. For simplicity, hereafter we will omit the modifier 'n times' and will call $l^*(\theta)$ the expected log likelihood.

The goodness of fit of the maximum likelihood model can be evaluated by $l^*(\hat{\theta}_k)$. However, since this quantity is dependent on the realization x of the random variable X, we will evaluate the model by its mean value called the *mean expected log likelihood*,

$$l_n^*(k) \equiv E_X\{ l^*(\hat{\theta}_k) \} = \int l^*(\hat{\theta}_k) \prod_{i=1}^{n} g(x_i) dx. \tag{4.9}$$

The mean expected log likelihood no longer depends on a particular realization, and depends only on the true model, the assumed model and the sample size. The model with larger mean expected log likelihood is considered to be the better one.

4.2.2 Numerical Example

The data in [Example 2.5] shown in Fig. 2.2 are 100 independent realizations from the model in [Example 2.6]:

$$MODEL(K): \quad f(x, y|\theta) = \begin{cases} \dfrac{1}{\sqrt{2\pi\sigma^2}} e^{-(y-S(x:a))^2/2\sigma^2} & 0 \leq x \leq 1 \\ 0 & x<0, \ x>1 \end{cases} \tag{4.10}$$

$$S(x;a) = a_0 + \sum_{m=1}^{M} (a_{2m-1} sin2m\pi x + a_{2m} cos2m\pi x) \tag{4.11}$$

$$\theta \in \Theta = \{ (\theta_1, \ldots, \theta_K) = (\sigma^2, a_0, \ldots, a_{2M}) \mid \sigma^2 > 0 \} \tag{4.12}$$

$$K = 2M + 2$$

with $K=22$ and $\theta=\theta^*=(1, a^*)$. The value of the true parameter θ^* is shown in the rightmost column in Table 4.1. This table also shows the maximum likelihood estimate $\hat{\theta}_k=(\hat{\sigma}_k^2, \hat{a}_k)$ and the maximum log likelihood for the restricted models

$$MODEL(k): \ f(\cdot|\Theta_k), \quad \Theta_k = \{ \theta \in \Theta \mid \theta_{k+1} = \theta_{k+2} = \ldots = \theta_K = 0 \} \\ k = 2, 4, \ldots, 22 \tag{4.13}$$

Table 4.1 Maximum Likelihood Estimates, Maximum Log Likelihood and Mean Expected Log Likelihood

θ	$\hat{\theta}_2$	$\hat{\theta}_4$	$\hat{\theta}_6$	$\hat{\theta}_8$	$\hat{\theta}_{10}$	$\hat{\theta}_{12}$	$\hat{\theta}_{14}$	$\hat{\theta}_{16}$	$\hat{\theta}_{18}$	$\hat{\theta}_{20}$	$\hat{\theta}_{22}$	1.000
θ_1	14.220	4.145	1.121	0.984	0.974	0.970	0.968	0.966	0.956	0.955	0.949	1.000
θ_2	8.012	7.983	8.043	8.010	8.010	8.012	8.013	8.013	8.014	8.013	8.011	8.000
θ_3		2.470	2.491	2.488	2.484	2.484	2.479	2.479	2.470	2.468	2.465	2.415
θ_4		-3.684	-3.869	-3.854	-3.850	-3.851	-3.853	-3.852	-3.847	-3.847	-3.847	-3.806
θ_5			2.252	2.238	2.242	2.248	2.248	2.247	2.247	2.253	2.249	2.119
θ_6			-0.984	-0.987	-0.987	-0.995	-0.994	-0.997	-1.005	-1.006	-1.009	-0.997
θ_7				0.523	0.516	0.515	0.518	0.518	0.512	0.515	0.515	0.545
θ_8				-0.022	-0.013	-0.012	-0.010	-0.007	0.002	0.004	0.007	0.069
θ_9					-0.129	-0.128	-0.130	-0.126	-0.133	-0.136	-0.139	-0.094
θ_{10}					-0.074	-0.074	-0.074	-0.073	-0.072	-0.071	-0.073	-0.078
θ_{11}						0.070	0.076	0.079	0.079	0.081	0.089	-0.021
θ_{12}						0.041	0.041	0.038	0.024	0.024	0.021	-0.065
θ_{13}							-0.058	-0.058	-0.061	-0.061	-0.060	-0.011
θ_{14}							0.032	0.036	0.042	0.047	0.044	0.042
θ_{15}								-0.065	-0.067	-0.069	-0.075	-0.011
θ_{16}								0.003	-0.010	-0.010	-0.011	0.010
θ_{17}									0.145	0.146	0.142	0.020
θ_{18}									-0.024	-0.024	-0.016	-0.004
θ_{19}										-0.034	-0.035	0.002
θ_{20}										0.041	0.039	0.001
θ_{21}											0.050	-0.006
θ_{22}											-0.089	-0.009
$l(\hat{\theta}_k)$	-1373.14	-1064.97	-738.13	-705.46	-702.93	-701.91	-701.41	-700.90	-698.35	-697.84	-696.30	
$l^*_{500}(k)$ †	-1371.11	-1051.87	-750.07	-716.43	-715.73	-716.28	-717.20	-718.29	-719.36	-720.57	-721.80	

† Since it is difficult to evaluate the mean expected log likelihood analytically, it is estimated by the sample mean of $l^*(\hat{\theta}_k)$ in 1000 repetition of the experiment.

fitted to *500* observations obtained by augmenting *400* additional observations to the hundred shown in Fig. 2. 2. The values of the mean expected log likelihoods are shown on the bottom of the same table. According to the figures for the mean expected log likelihood, *MODEL(10)* attains the best fit. This might seem to be inconsistent with the fact that the true distribution can never be expressed without employing *MODEL(22)*. This result, however, shows that there are situations where we have to use a model with a smaller number of free parameters, even when the true number of free parameters is known.

Let $S(\cdot\,;\hat{a}_k)$ be the estimator of the regression curve $S(\cdot\,;a^*)$ when we adopt *MODEL(k)*. Fig. 4. 1 shows the estimates obtained by assuming $k=10$ and 22. Comparing these with the true regression curve $S(\cdot\,;a^*)$ shown by the dashed curve, we see that $S(\cdot\,;\hat{a}_{10})$ is actually better than $S(\cdot\,;\hat{a}_{22})$.

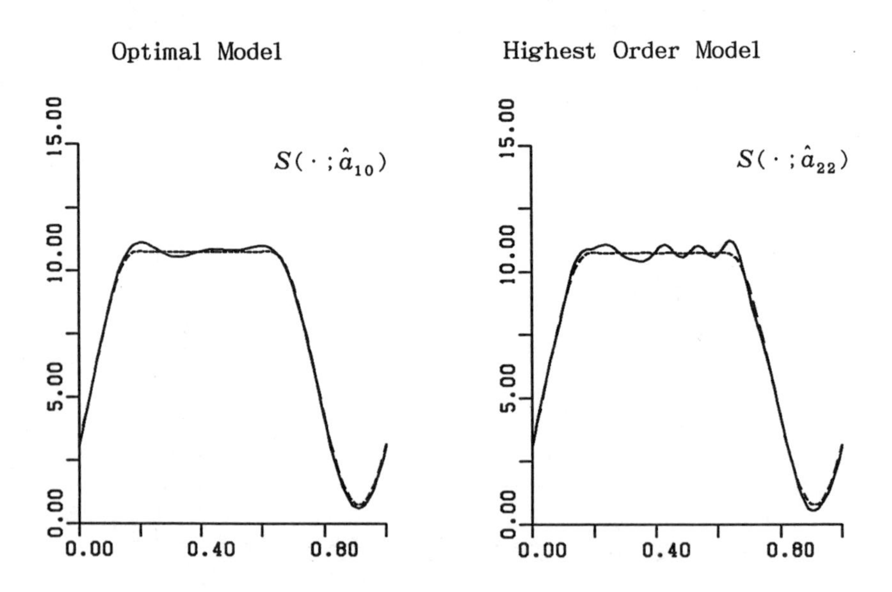

Figure 4. 1 Regression curve estimation

Table 4. 2 The mean of the maximum log likelihood and the difference from $l^*_{500}(k)$ evaluated by a Monte Carlo experiment

	2	4	6	8	10	12
Mean of $l(\hat{\theta}_k)$	-1368. 85	-1047. 52	-744. 53	-708. 55	-705. 61	-703. 94
Mean of $l(\hat{\theta}_k)-l_{500}*(k)$	2. 26	4. 35	5. 54	7. 88	10. 12	12. 34

	14	16	18	20	22
Mean of $l(\hat{\theta}_k)$	-702. 68	-701. 65	-700. 61	-699. 55	-698. 49
Mean of $l(\hat{\theta}_k)-l_{500}*(k)$	14. 52	16. 64	18. 75	21. 02	23. 31

Table 4. 2 summarizes the mean values of the maximum log likelihood, $l(\hat{\theta}_k)$, and that of the difference between $l(\hat{\theta}_k)$ and the mean expected log likelihood $l^*_{500}(k)$. Here, since it is difficult to obtain the mean expected log likelihood analytically, it was approximated by a Monte Carlo experiment consisting of 1000 repetitions.

From this table, it is evident that the mean value of the maximum log likelihood increases monotonically with the increase of the number of free parameters k. It also can be seen that the bias of the maximum log likelihood as the estimator of the mean expected log likelihood is approximately equal to the number of free parameters contained in the model. Since $\Theta_1 \subset \Theta_2 \subset \ldots \subset \Theta_K$, it follows that the maximum log likelihoods of the models, $MODEL(k)$ $k=1, \ldots, K$, satisfy

$$l(\hat{\theta}_1) \leqq l(\hat{\theta}_2) \leqq \ldots \leqq l(\hat{\theta}_K). \qquad (4.\ 14)$$

The remarkable point with the table is that it suggests that we might be able to estimate the mean expected log likelihood of a model by

$$\begin{aligned}&(\textit{maximum log likelihood})\\&-(\textit{number of free parameters}).\end{aligned} \tag{4.15}$$

We will investigate this possibility in the following sections.

4.3 *AIC*

In this section we will derive the *AIC* of the *MODEL*(K) that contains the true model.

4.3.1 *Estimation Error of Parameters*

Let $[\partial h/\partial\theta]$ denote the K dimensional column vector whose jth element is given by $\partial h/\partial\theta_j$ and $[\partial^2 h/\partial\theta^2]$ the $K\times K$ matrix with its (j, l) element defined by $\partial^2 h/\partial\theta_j\partial\theta_l$. We define a matrix J_* by

$$J_*=-E_Z\left[\frac{\partial^2}{\partial\theta^2}\log f(Z\mid\theta)\right]_{\theta^*} \tag{4.16}$$

where $[\ ]_{\theta^*}$ means that we evaluate the expression within the bracket at $\theta=\theta^*$. Under the present assumption that the true model is contained in our model, J_* is identical to the Fisher information matrix.

As shown in section 3.3.5, under certain regularity conditions, the maximum likelihood estimator $\hat{\theta}_K$ has the asymptotic normality

$$\sqrt{n}(\hat{\theta}_K-\theta^*) \rightarrow N(0, J_*^{-1}) \quad (n\rightarrow\infty). \tag{4.17}$$

The Taylor expansion of the expected log likelihood

$$l^*(\theta) = nE_Z\{\log f(Z\mid\theta)\} \tag{4.18}$$

around the true parameter θ^* yields the approximation

$$l^*(\theta)=l^*(\theta^*)+n(\theta-\theta^*)E_Z\left[\frac{\partial log f(Z|\theta)}{\partial\theta}\right]_{\theta^*}$$

$$+\frac{1}{2}n(\theta-\theta^*)E_Z\left[\frac{\partial^2 log f(Z|\theta)}{\partial\theta^2}\right]_{\theta^*}(\theta-\theta^*)^T. \qquad (4.19)$$

From the properties of the K-L information quantity (3.5) the maximum value of $l^*(\theta)$ occurs at $\theta=\theta^*$, and the second term of the right hand side vanishes. Thus we have

$$l^*(\theta)=l^*(\theta^*)-\frac{1}{2}\sqrt{n}(\theta-\theta^*)J_*\sqrt{n}(\theta-\theta^*)^T. \qquad (4.20)$$

From the asymptotic normality of the maximum likelihood estimator and the relation between normal random variables and chi-square random variables mentioned in section 2.2.4, $n(\theta-\theta^*)J_*(\theta-\theta^*)^T$ is asymptotically distributed as a chi-square distribution with K degrees of freedom. Consequently, since the mean of the chi-square random variable is K, it is approximately true, for large n, that

$$l_n^*(K)\equiv E_X\{l^*(\hat{\theta}_K)\}=l^*(\theta^*)-\frac{K}{2} \qquad (4.21)$$

(see Fig. 4.2). This relation reveals that, due to the estimation error of the parameters, the mean expected log likelihood of a model is less than the expected log likelihood of the true model by $K/2$.

[*Example 4.1*] Let x_i, $i=1,\cdots,n$ be n independent realizations from $N(0,1)$. The log likelihood and the expected log likelihood of a normal distribution model $N(\mu,1)$ are given respectively by

$$l(\mu)=\sum_{i=1}^{n} log\ f(x_i|\mu)=\sum_{i=1}^{n} log\left\{\frac{1}{\sqrt{2\pi}}e^{-(x_i-\mu)^2/2}\right\}$$

$$=-\frac{n}{2}log\ 2\pi-\frac{1}{2}\sum_{i=1}^{n}(x_i-\mu)^2 \qquad (4.22)$$

and

$$l^*(\mu)=n\int_{-\infty}^{\infty} log\left\{\frac{1}{\sqrt{2\pi}}e^{-(z-\mu)^2/2}\right\}\frac{1}{\sqrt{2\pi}}e^{-z^2/2}dz$$

$$=-\frac{n}{2}log2\pi-\frac{n}{2}\int_{-\infty}^{\infty}(z^2-2\mu z+\mu^2)\frac{1}{\sqrt{2\pi}}e^{-z^2/2}dz$$

$$=-\frac{n}{2}log2\pi-\frac{n}{2}(1+\mu^2)$$

$$=-\frac{n}{2}log2\pi-\frac{n}{2}-\frac{n}{2}\mu^2. \qquad (4.23)$$

In this case, from

$$J_*=-E_Z\left[\frac{d^2log\, f(Z\mid\mu)}{d\mu^2}\right]_{\mu=0}$$

$$=-E_Z\left[\frac{d^2}{d\mu^2}\left(-\frac{1}{2}log\; 2\pi-\frac{(Z-\mu)^2}{2}\right)\right]_{\mu=0}$$

$$=-E_Z\left[\frac{d}{d\mu}(Z-\mu)\right]_{\mu=0}=1, \qquad (4.24)$$

we have exactly

$$l^*(\mu) = l^*(0)-\frac{1}{2}\sqrt{n}(\mu-0)J_*\sqrt{n}(\mu-0). \qquad (4.25)$$

Since the maximum likelihood estimator of μ is given by

$$\hat{\mu} = \frac{1}{n}\sum_{i=1}^{n}x_i,$$

it follows that

$$E_X\{\sqrt{n}\hat{\mu}\} = \sqrt{n}\int_{-\infty}^{\infty}\cdots\int_{-\infty}^{\infty}\hat{\mu}\prod_{i=1}^{n}f(x_i\mid 0)\,dx_1\cdots dx_n = 0 \qquad (4.26)$$

and using the independence of the observations,

$$E_X\{n\hat{\mu}^2\} = E_X\left\{\frac{1}{n}\sum_{i=1}^{n}\sum_{j=1}^{n}x_ix_j\right\}$$

$$= E_X\left\{\frac{1}{n}\sum_{i=1}^{n}x_i^2\right\} = 1. \qquad (4.27)$$

Here, $\hat{\mu}$ is a normal random variable, and we have the exact relationship

$$\sqrt{n}\hat{\mu} \sim N(0, J_*^{-1}),$$

which corresponds to the asymptotic normality (4.17). By taking expectation of both sides of the Taylor expansion (4.25) after substituting $\hat{\mu}$ for μ and using (4.27), we have

$$\begin{aligned} E_X\{ l^*(\hat{\mu}) \} &= l^*(0) - \frac{1}{2}E_X\{ n\hat{\mu}^2 \} \\ &= l^*(0) - \frac{1}{2}. \end{aligned} \quad (4.28)$$

For this example, since the model has only one parameter we see that (4.21) holds exactly.

4.3.2 *AIC(K)*

Let us consider the relation between the expected log likelihood $l^*(\theta^*)$ of the true model and the maximum log likelihood $l(\hat{\theta}_K)$ of *MODEL(K)*. The Taylor expansion of the log likelihood

$$l(\theta) = \sum_{i=1}^{n} \log f(x_i \mid \theta) \quad (4.29)$$

around the maximum likelihood estimate $\hat{\theta}_K$ yields the approximation

$$l(\theta) = l(\hat{\theta}_K) + (\theta - \hat{\theta}_K)\left[\frac{\partial l}{\partial \theta}\right]_{\hat{\theta}_K} + \frac{1}{2}(\theta - \hat{\theta}_K)\left[\frac{\partial^2 l}{\partial \theta^2}\right]_{\hat{\theta}_K}(\theta - \hat{\theta}_K)^T. \quad (4.30)$$

Here the second term in the right hand side vanishes since the log likelihood (4.30) takes its maximum value at $\hat{\theta}_K$.

By the law of large number (section 2.2.5)

$$\frac{1}{n}\left[\frac{\partial^2 l}{\partial\theta^2}\right]_{\theta^*} = \frac{1}{n}\sum_{i=1}^{n}\left[\frac{\partial^2}{\partial\theta^2}\log f(x_i\,|\,\theta)\right]_{\theta^*}$$

$$\to E_Z\left[\frac{\partial^2}{\partial\theta^2}\log f(Z\,|\,\theta)\right]_{\theta^*} = -J_* \qquad (4.31)$$

as $n\to\infty$. Moreover, since $\hat{\theta}_K\to\theta^*$ as $n\to\infty$, we have

$$\frac{1}{n}\left[\frac{\partial^2 l}{\partial\theta^2}\right]_{\hat{\theta}_K} \to -J_* \quad (n\to\infty). \qquad (4.32)$$

Therefore, it is approximately true for large n that

$$l(\theta)=l(\hat{\theta}_K)-\frac{1}{2}\sqrt{n}(\theta-\hat{\theta}_K)J_*\sqrt{n}(\theta-\hat{\theta}_K)^T. \qquad (4.33)$$

Substituting θ^* for θ and taking expectations of both sides we have, in the same way the expectation (4.21) of $l(\hat{\theta}_K)$ was derived,

$$E_X\{\,l(\theta^*)\,\} = E_X\{\,l(\hat{\theta}_K)\,\}-\frac{K}{2} \qquad (4.34)$$

(see Fig. 4.2). On the other hand, from the independence of the

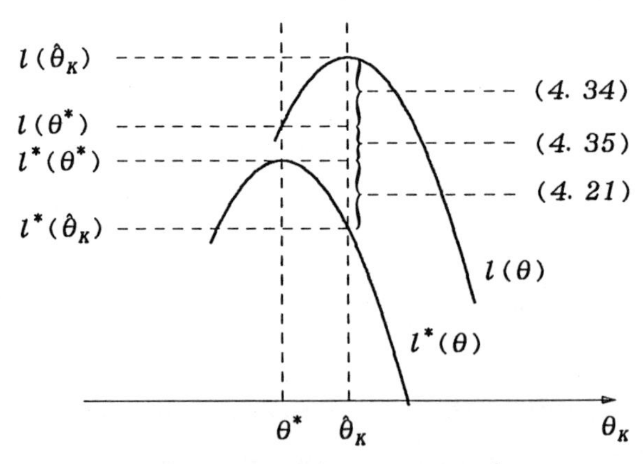

Figure 4.2 Relation between $l^*(\theta)$ and the log likelihood $l(\theta)$

X'_is, we can show the unbiasedness of the likelihood function:

$$l^*(\theta^*) \equiv nE_Z\{\log f(Z|\theta^*)\}$$
$$= E_X\left\{\sum_{i=1}^{n} \log f(X_i|\theta^*)\right\} = E_X\{l(\theta^*)\}. \qquad (4.35)$$

Thus from the two equations above, we have

$$l^*(\theta^*) = E_X\{l(\hat{\theta}_K)\} - \frac{K}{2} \qquad (4.36)$$

From this relation and $l_n^*(K) = l^*(\theta^*) - K/2$, we finally obtain

$$l_n^*(K) = E_X\{l(\hat{\theta}_K)\} - K. \qquad (4.37)$$

This relation reveals that the maximum log likelihood is a biased estimator of the mean expected log likelihood and its bias is equal to the number of free parameters of the model. Thus, correcting this bias, the estimator

$$l(\hat{\theta}_K) - K \qquad (4.38)$$

constitutes an unbiased estimator of the mean expected log likelihood $l_n^*(K)$. This confirms our conjecture made observing Table 4.1. We define *AIC* as minus twice the above quantity (4.38), namely

$$\begin{aligned} AIC(K) &= -2l(\hat{\theta}_K) + 2K \qquad (4.39) \\ &= -2(\textit{maximum log likelihood of the model}) \\ &\quad +2(\textit{number of free parameters of the model}). \end{aligned}$$

The reason why we multiply (4.38) by -2 is that in the

literature the equation (4. 34), for example, is expressed in the form

$$-2E_X\{ l(\theta^*) - l(\hat{\theta}_K) \} = K$$

(for example, see Rao (1965) p349).

Since

$$E_X\{ -\frac{1}{2}AIC(K) - l^*(\hat{\theta}_K) \} = 0,$$

AIC can also be interpreted as an unbiased estimator of the -2 times expected log likelihood $l^*(\hat{\theta}_K)$ of the maximum likelihood model.

[*Example 4. 2*] For the model given in [Example 4. 1], we have

$$\begin{aligned} l(\mu) &= -\frac{n}{2}\log 2\pi - \frac{1}{2}\sum_{i=1}^{n}(x_i-\mu)^2 \\ &= -\frac{n}{2}\log 2\pi - \frac{1}{2}\sum_{i=1}^{n}(x_i-\hat{\mu}+\hat{\mu}-\mu)^2 \\ &= -\frac{n}{2}\log 2\pi - \frac{1}{2}\sum_{i=1}^{n}(x_i-\hat{\mu})^2 - \frac{n}{2}(\hat{\mu}-\mu)^2 \\ &= l(\hat{\mu}) - \frac{n}{2}(\hat{\mu}-\mu)^2. \end{aligned} \tag{4. 40}$$

Namely, in this case the Taylor expansion (4. 33) is exact. Upon putting $\mu=0$ and taking expectation, we have

$$E_X\{ l(0) \} = E_X\{ l(\hat{\mu}) \} - \frac{1}{2}.$$

Thus using the unbiasedness (4. 35) of the likelihood function, it follows that

$$l^*(0) = E_X\{ l(\hat{\mu}) \} - \frac{1}{2}$$

and using the relation (4. 28) between $l^*(\hat{\mu})$ and $l^*(0)$ we have

$$E_X\{ l^*(\hat{\mu}) \} = E_X\{ l(\hat{\mu}) \} - 1. \tag{4. 41}$$

4.4* *AIC of the Restricted Model*

Consider models

$$MODEL(k): \ f(\cdot \mid \theta),\ \theta \in \Theta_k, \qquad 1 \leq k \leq K$$

which are obtained by restricting the $MODEL(K)$. We note here that although we assume $\theta^* \in \Theta_K$, we do not assume that $\theta^* \in \Theta_k$. In other words, the true model is not necessarily expressible by $MODEL(k)$.

In the following we will use the notation

$$\|\theta\|^2 = \theta J_* \theta^T,$$

where J_* is the negative expected Hessian matrix defined by (4.16). Moreover, for simplicity, we shall assume the *orthogonality* (of the parametrization) :

$$J_* \textit{ is the } K \times K \textit{ identity matrix.} \qquad (\dagger)$$

The following arguments can be shown to hold when J_* is a $K \times K$ positive definite matrix, but this relaxed assumption makes the proof longer. Equations (4.43) and (4.46) with † hold under the orthogonality (†). The rest of the equations do not require this assumption.

4.4.1 *Bias of a Model and the Estimation Error of the Parameters*

We define θ_k^* as the value of $\theta \in \Theta_k$ that maximizes the expected log likelihood yielding

$$l^*(\theta_k^*) = \max_{\theta \in \Theta_k} l^*(\theta). \qquad (4.42)$$

θ_k^* is the best parameter for the *MODEL*(k). Using the Taylor expansion (4.20) of l^* and the orthogonality (†), we can show that θ_k^* is given by

$$\theta_k^* = (\theta_1^*, \theta_2^*, \cdots, \theta_k^*, 0, \cdots, 0). \qquad (4.43)^\dagger$$

Thus, for any $\theta \in \Theta_k$, we have the relation

$$\|\theta - \theta^*\|^2 = \|\theta - \theta_k^*\|^2 + \|\theta_k^* - \theta^*\|^2. \qquad (4.44)$$

Substituting this relation into the Taylor expansion (4.20), we have the approximations

$$l^*(\theta) = l^*(\theta_k^*) - \frac{n}{2}\|\theta - \theta_k^*\|^2$$
$$l^*(\theta_k^*) = l^*(\theta^*) - \frac{n}{2}\|\theta_k^* - \theta^*\|^2. \qquad (4.45)$$

The expected log likelihood of the best parameter, $l^*(\theta_k^*)$ will be called the approximation bound of *MODEL*(k), since it gives the upper limit of the expected log likelihood, $l^*(\theta)$, under the restriction that $\theta \in \Theta_k$. Thus the approximation bound may be interpreted as a term that expresses the bias of *MODEL*(k). No matter how much data we may have, the expected log likelihood of the maximum likelihood model, $l^*(\hat{\theta}_k)$, can never be larger than $l^*(\theta_k^*)$.

From the definition of the maximum likelihood estimator and the orthogonality (†) of the parametrization, we have approximately that

$$\hat{\theta}_k = (\hat{\theta}_1, \hat{\theta}_2, \cdots, \hat{\theta}_k, 0, \cdots, 0), \qquad (4.46)^\dagger$$

where $(\hat{\theta}_1, \hat{\theta}_2, \ldots, \hat{\theta}_K) = \hat{\theta}_K$.

From the asymptotic normality, the expressions $(4.43)^\dagger$ and $(4.46)^\dagger$ for $\hat{\theta}_k$ and θ_k^*, and the orthogonality (†), it can be shown that $n\|\hat{\theta}_k-\theta_k^*\|^2$ is approximately distributed as a chi-square random variable with k degrees of freedom and the error assessment

$$nE_X\{\|\hat{\theta}_k-\theta_k^*\|^2\} = k \tag{4.47}$$

is obtained.

Thus, evaluating the expected log likelihood (4.45) at $\theta=\hat{\theta}_k$, and taking expectations, we have the approximate mean expected log likelihood

$$l_n^*(k)\equiv E_X\{l^*(\hat{\theta}_k)\}=l^*(\theta_k^*)-\frac{k}{2}. \tag{4.48}$$

This relation reveals that the mean expected log likelihood of the restricted model is less than the approximation bound by half the number of free parameters of the model. It is obvious that this is due to the estimation error of the parameters.

4.4.2 $AIC(k)$

We will next consider the relation between the approximation bound $l^*(\theta_k^*)$ and the maximum log likelihood of the restricted model.

From the definition (4.6) of the maximum log likelihood and the Taylor expansion (4.33) of the likelihood it is approximately true that

$$\|\theta-\hat{\theta}_K\|^2=\|\theta-\hat{\theta}_k\|^2+\|\hat{\theta}_k-\hat{\theta}_K\|^2 \tag{4.49}$$

for any $\theta\in\Theta_k$. Then we can rewrite the likelihood (4.33) into the form

$$l(\theta)=l(\hat{\theta}_K)-\frac{n}{2}\|\hat{\theta}_k-\hat{\theta}_K\|^2-\frac{n}{2}\|\theta-\hat{\theta}_k\|^2 \tag{4.50}$$

and, setting $\theta=\hat{\theta}_k$,

$$l(\hat{\theta}_K)-\frac{n}{2}\|\hat{\theta}_k-\hat{\theta}_K\|^2=l(\hat{\theta}_k). \tag{4.51}$$

Above two equations and the substitution $\theta=\theta_k^*$ yields

$$l(\theta_k^*)=l(\hat{\theta}_k)-\frac{n}{2}\|\theta_k^*-\hat{\theta}_k\|^2.$$

Taking expectations and using the error assessment (4.47) we have

$$E_X\{l(\theta_k^*)\}=E_X\{l(\hat{\theta}_k)\}-\frac{k}{2}, \tag{4.52}$$

which yields, using again the unbiasedness of the likelihood function (4.35) with θ^* replaced by θ_k^*,

$$l^*(\theta_k^*)=E_X\{l(\hat{\theta}_k)\}-\frac{k}{2}. \tag{4.53}$$

From this and the relation (4.48) between the mean expected log likelihood $l_n^*(k)$ and the approximation bound $l^*(\theta_k^*)$,

$$AIC(k)=-2l(\hat{\theta}_k)+2k \quad (k=1, 2, \ldots, K) \tag{4.54}$$

is an unbiased estimator of minus twice the mean expected log likelihood, $-2l_n^*(k)$.

Using the definition (4.8) of the expected log likelihood, the expression (4.48) of the mean expected log likelihood can be rewritten into the form

$$l_n^*(k)=E_X\{l^*(\hat{\theta}_k)\}=nE_Z\{\log f(Z|\theta_k^*)\}-\frac{k}{2}.$$

This shows that as the sample size n becomes smaller, the contribution of the number of free parameters in the mean expected log likelihood becomes more significant. This means that even though the data may be obtained from the same distribution, MAICE procedure would tend to choose a model with a smaller number of parameters when the sample size is smaller.

4.4.3 *Numerical Example*

Table 4.3 shows -1/2 times the *AIC*'s obtained from the maximum log likelihood's given in Table 4.1 and the mean expected log likelihood's $l^*_{500}(k)$. They both take their maximum at $k=10$, and this indicates that MAICE actually gives the best model in this case.

Table 4.3 -1/2 AIC and Mean Expected Log Likelihood

	2	4	6	8	10	12
$\frac{-1}{2}AIC(k)$	-1375.04	-1068.80	-743.91	-713.62	-712.83	-713.89
$l_{500^*}(k)$	-1371.11	-1051.87	-750.07	-716.43	-715.73	-716.28

	14	16	18	20	22
$\frac{-1}{2}AIC(k)$	-715.30	-716.81	-716.04	-717.62	-718.33
$l_{500^*}(k)$	-717.20	-718.29	-719.39	-720.57	-721.80

Table 4.4 shows the sample mean and the variance (mean square of the deviations from the mean expected log likelihood) of 1/2 *AIC* obtained by 1000 repetations of the same experiment.

Table 4.4 Mean and Variance of -1/2 *AIC*

	2	4	6	8	10	12
Mean	-1370.85	-1051.52	-750.53	-716.55	-715.61	-715.94
Variance	151.64	169.24	255.05	267.82	272.18	275.94

	14	16	18	20	22
Mean	-716.68	-717.65	-718.61	-719.55	-720.49
Variance	276.50	277.85	278.49	279.30	279.30

Comparing with Table 4.3, we see that the mean of $-1/2\ AIC$ is quite close to the mean expected log likelihood. It should also be noticed that the variance of $-1/2\ AIC$ is not so dependent on the number of free parameters and takes almost the same value. For $k \geqq 8$, it takes values between 270 and 280.

4.4.4 *Error of AIC*

We have seen that $-1/2\ AIC(k)$ is asymptotically an unbiased estimator of the mean expected log likelihood $l_n^*(k)$. Here we check the estimation error of $AIC(k)$. Using the relation (4.49) between $\hat{\theta}_K$, $\hat{\theta}_k$ and the parameter space Θ_k, we can express the maximum likelihood (4.51) of $MODEL(k)$ as follows:

$$
\begin{aligned}
l(\hat{\theta}_k) &= l(\hat{\theta}_K) - \frac{n}{2}\|\hat{\theta}_k - \hat{\theta}_K\|^2 \\
&= l(\hat{\theta}_K) - \frac{n}{2}\|\theta_k^* - \hat{\theta}_K\|^2 + \frac{n}{2}\|\theta_k^* - \hat{\theta}_k\|^2 \\
&= l(\hat{\theta}_K) - \frac{n}{2}\|\theta_k^* - \theta^* + \theta^* - \hat{\theta}_K\|^2 + \frac{n}{2}\|\theta_k^* - \hat{\theta}_k\|^2 \\
&= l(\hat{\theta}_K) - \frac{n}{2}\|\theta^* - \hat{\theta}_K\|^2 - n(\theta_k^* - \theta^*)J_*(\theta^* - \hat{\theta}_K)^T \\
&\quad - \frac{n}{2}\|\theta_k^* - \theta^*\|^2 + \frac{n}{2}\|\theta_k^* - \hat{\theta}_k\|^2 .
\end{aligned}
$$

Then, comparing the first two terms of the most right hand side with the Taylor expansion (4.33) of the likelihood, we have

$$
\begin{aligned}
l(\hat{\theta}_k) &= l(\theta^*) - n(\theta_k^* - \theta^*)J_*(\theta^* - \hat{\theta}_k)^T - \frac{n}{2}\|\theta_k^* - \theta^*\|^2 + \frac{n}{2}\|\theta_k^* - \hat{\theta}_k\|^2 \\
&= l^*(\theta^*) - \frac{n}{2}\|\theta_k^* - \theta^*\|^2 + l(\theta^*) - l^*(\theta^*) \\
&\quad - n(\theta_k^* - \theta^*)J_*(\theta^* - \hat{\theta}_K)^T + \frac{n}{2}\|\theta_k^* - \hat{\theta}_k\|^2
\end{aligned}
$$

and so, using the expression (4.45) of the expected log likelihood, we have

$$l(\hat{\theta}_k) = l^*(\theta_k^*) + \{l(\theta^*) - l^*(\theta^*)\}$$
$$- n(\theta_k - \theta^*) J_*(\theta^* - \hat{\theta}_K)^T + \frac{n}{2} \| \theta_k^* - \hat{\theta}_k \|^2 .$$

Therefore using the relation (4.48) between the mean expected log likelihood $l_n^*(k)$ and the approximation measure $l^*(\theta_k^*)$ of the model, $-1/2\ AIC(k)$ can be expressed as

$$-\frac{1}{2} AIC(k) = l(\hat{\theta}_k) - k$$
$$= l_n^*(k) + \{l(\theta^*) - l^*(\theta^*)\} - n(\theta_k^* - \theta^*) J_*(\theta^* - \hat{\theta}_K)^T$$
$$+ \frac{1}{2}\{n \| \theta_k^* - \hat{\theta}_k \|^2 - k\}. \qquad (4.55)$$

Here, from the unbiasedness (4.35) of the likelihood function, the asymptotic normality (4.17) and the error assessment (4.47), the error terms

$$l(\theta^*) - l^*(\theta^*),$$
$$n(\theta_k^* - \theta^*) J_*(\theta^* - \hat{\theta}_K)^T$$

and

$$n \| \theta_k^* - \hat{\theta}_k \|^2 - k$$

are (asymptotically) distributed as random variables with mean zero.

Since $l(\theta^*) - l^*(\theta^*)$, the difference between the log likelihood and the expected log likelihood of the true model, does not depend on k, we will call this the common error. If we call the sum of the last two terms the individual error, we get the decomposition

$$-1/2\ AIC(k) = (\textit{mean expected log likelihood}) + (\textit{common error}) + (\textit{individual error}).$$

Note that no matter how large the common error may be it will not interfere with the selection of the model.

Among the individual errors, $1/2\ \{n\|\theta_k^*-\hat{\theta}_k\|^2 - k\}$ has variance k due to the property of the chi-square distribution. Thus this error increases as the number of free parameters is increased. On the other hand, assuming the orthogonality (†) we have

$$n(\theta_k^*-\theta^*)J_*(\theta^*-\hat{\theta}_K)^T=n\sum_{j=k+1}^{K}\theta_j^*(\hat{\theta}_j-\theta_j^*).$$

Thus from the asymptotic normality of $\hat{\theta}_K$, we have asymptotically

$$E_X\{n^2\,|\,(\theta_k^*-\theta^*)J_*(\theta^*-\hat{\theta}_K)^T\,|^2\}=n\sum_{j=k+1}^{K}(\theta_j^*)^2,$$

and this means that the error $n(\theta_k^*-\theta^*)J_*(\theta^*-\theta_K)^T$ decreases as k increases. Therefore, the individual error that is the sum of these two errors may either increase or decrease with the increase of the number of free parameters. Table 4.4 shows that, for this specific case, these two terms balance and the sum of the two terms takes almost the same value independent of the number of free parameters.

As mentioned before, the common error does not affect model selection. Table 4.3 is a typical example. For $k \geqq 8$, the values of $-1/2\ AIC(k)$ are consistently larger than $l_{500}^*(k)$ by 2 to 3.

In this example, to evaluate the individual error we compute the individual variances, the mean squares of

$$\textit{Individual error} = -\frac{1}{2}AIC(k) - l^*_{500}(k) - \textit{common error},$$

and listed in Table 4.5. We see the variance becomes very small for $k \geqq 6$. These values are smaller than the ones anticipated when the terms in (4.55) are mutually independent. This is due to the fact that $AIC(j)$ and $AIC(k)$ are dependent.

Table 4.5 Individual Variance of AIC

	2	4	6	8	10	12
Individual Variance	401.62	319.41	80.07	9.00	3.94	1.56

	14	16	18	20	22
Individual Variance	0.92	0.69	0.89	1.61	2.58

Table 4.6 Frequency Distribution of k_A

	2	4	6	8	10	12
Frequencies	0	0	0	326	342	173

	14	16	18	20	22	Total
Frequencies	91	27	17	14	10	1000

Table 4.7 Frequency Distribution of k_A for n=2000

	2	4	6	8	10	12
Frequencies	0	0	0	6	210	359

	14	16	18	20	22	Total
Frequencies	264	68	58	21	14	1000

Table 4.8 Mean Expected Log Likelihood of Each Model for n=2000

	2	4	6	8	10	12
$l_{2000^*}(k)$	-5482.25	-4201.66	-2991.03	-2852.96	-2846.58	-2845.27

	14	16	18	20	22
$l_{2000^*}(k)$	-2845.33	-2846.27	-2847.14	-2848.20	-2849.21

Table 4.9 Analysis of $n\|\hat{\theta}_k-\theta^*\|^2$

	2	4	6	8	10	12
$n\|\hat{\theta}_k-\theta^*\|^2$	197289.11	14254.06	357.62	30.46	17.41	14.73
$n\|\hat{\theta}_k-\theta_{k^*}\|^2$	171163.07	8445.88	33.61	8.23	10.10	12.09
$n\|\theta_{k^*}-\theta^*\|^2$	26126.05	5808.19	324.02	22.23	7.31	2.64

	14	16	18	20	22
$n\|\hat{\theta}_k-\theta^*\|^2$	14.80	16.65	18.36	20.44	22.43
$n\|\hat{\theta}_k-\theta_{k^*}\|^2$	14.04	16.11	18.24	20.32	22.43
$n\|\theta_{k^*}-\theta^*\|^2$	0.76	0.54	0.12	0.12	0.00

We would expect k_A, the value of k that minimizes $AIC(k)$ to be a fairly good estimate of the k that maximizes the mean expected log likelihood. Table 4.6 shows the frequency distribution of k_A in 1000 experiments. We see that k_A is concentrated around k =10 as expected.

Tables 4.7 and 4.8 respectively show the frequency distribution of k_A and the mean expected log likelihood of each model when 2000 observations are used for estimation. We see from Table 4.7 that the distribution of k_A's shifts towards larger values concentrated around k=12. From Table 4.8 we can see that $MODEL(12)$ is the best model.

Table 4.9 shows the mean values of $n\|\hat{\theta}_k-\theta^*\|^2$, $n\|\hat{\theta}_k-\theta^*_k\|^2$ and $n\|\theta^*_k-\theta^*\|^2$ obtained from the 1000 experiments summarized in

Tables 4.7 and 4.8. The value of $n\|\hat{\theta}_k-\theta^*\|^2$ takes its minimum at $k=12$ corresponding to Table 4.8. It can be seen that for large k the approximation

$$n\|\hat{\theta}_k-\theta_k^*\|^2=k$$

holds for the estimation error of the parameters. The approximation becomes worse for smaller k since $\hat{\theta}_k$ is far from the true parameter θ^*. We see that $n\|\theta_k^*-\theta^*\|^2$ decreases, and therefore the measure of approximation

$$l^*(\theta_k^*) = l^*(\theta^*) - \frac{n}{2}\|\theta_k^*-\theta^*\|^2$$

increases, as the number of free parameters k increases. We can also see that the mean value of $n\|\hat{\theta}_k-\theta^*\|^2$ takes its minimum at $k=12$ giving a compromise between the estimation error and the approximation bound.

4.5* *AIC of a Conditional Distribution Model*

4.5.1 *Conditional Log Likelihood*

Suppose that X and Y are the explanatory and objective variables, respectively, and that $h(\cdot|\Theta')$ is a model for X and $g(\cdot|X,\Theta)$ is a conditional distribution model for Y given X. Given the data (x_i, y_i), $i=1,\ldots,n$, the log likelihood of the joint distribution model is obtained from

$$\begin{aligned} l(\theta,\theta') &= \sum_{i=1}^{n} \log\{h(x_i|\theta')g(y_i|x_i,\theta)\} \\ &= \sum_{i=1}^{n} \log h(x_i|\theta') + \sum_{i=1}^{n} \log g(y_i|x_i,\theta). \qquad (4.56) \end{aligned}$$

If we are only interested in the estimation of the conditional distribution of Y given X, we need only find the value of the parameter θ that maximizes the second term of the right hand side of the above equation, namely

$$l'(\theta) = \sum_{i=1}^{n} \log g(y_i \mid x_i, \theta), \tag{4.57}$$

which is called the *conditional log likelihood*. The maximum likelihood estimator $\hat{\theta}$ satisfies the relation

$$l'(\hat{\theta}) = \max_{\theta \in \Theta} l'(\theta). \tag{4.58}$$

When the true joint distribution of (X, Y) is given by $h(x \mid \theta'^*) g(y \mid x, \theta^*)$ then

$$\begin{aligned} \frac{1}{n} l'(\theta) &\rightarrow E_{(X,Y)} \log g(Y \mid X, \theta) \\ &= \iint h(x \mid \theta'^*) g(y \mid x, \theta^*) \log g(y \mid x, \theta) dy dx \\ &= E_X\{E_{Y|X}\{\log g(Y \mid X, \theta)\}\} \end{aligned} \tag{4.59}$$

holds for $n \rightarrow \infty$. The right hand side is the expectation with respect to the distribution of X of the expected log likelihood of the conditional distribution model of Y given the realization of X. This relation shows that the conditional log likelihood $l'(\theta)$ is an estimate of the conditional expected log likelihood.

4.5.2 AIC of the Conditional Distribution Model

Let $\theta = (\theta_1, \theta_2, \cdots, \theta_K)$ then we have

$$\frac{\partial^2 l(\theta, \theta')}{\partial \theta_j \partial \theta_i} = \frac{\partial^2 l'(\theta)}{\partial \theta_j \partial \theta_i}.$$

Setting

$$J_* = -E_{X,Y}\left[\frac{\partial^2 \log g(Y|X,\theta)}{\partial\theta^2}\right]_{\theta^*} \tag{4.60}$$

we can show that the maximum conditional log likelihood estimator $\hat{\theta}_K$ has the asymptotic normality. This means that if we define l^* by

$$l^*(\theta) = nE_X\{E_{Y|X}\{\log g(Y|X,\theta)\}\},$$

the same arguments as were used in sections 4.3 and 4.4 can be applied to the conditional distribution model. Namely, when a conditional distribution model $g(\cdot|X,\Theta_k)$ is fitted to the data, its AIC is obtained by

$$AIC = -2\log l'(\hat{\theta}_k) + 2k \tag{4.61}$$

where

$$l'(\hat{\theta}_k) = \max_{\theta\in\Theta_k} l'(\theta).$$

This AIC is of course an estimate of $-2n$ times the conditional expected log likelihood, namely

$$-2n\, E_X\{E_{Y|X}\{\log g(Y|X,\hat{\theta}_k)\}\}. \tag{4.62}$$

4.6 *Some Remarks on the Use of AIC*

1° The number of free parameters estimated from data should be less than $2\sqrt{n}$ ($n/2$ at most) where n is the sample size. The reason is that if the number of the free parameters is too large, the relations such as the asymptotic normality (4.17) do not necessarily hold. For a family of nested (or hierarchical) models such as the one given in section 4.2.2, the typical behavior of AIC is shown in Fig. 4.3.

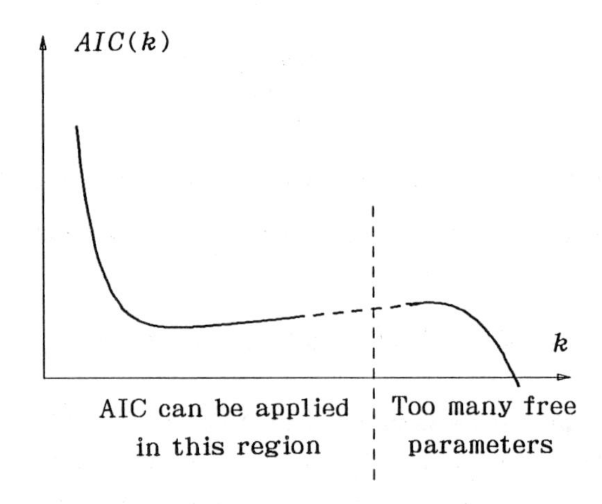

Figure 4. 3 The region where AIC works

2° It is the difference of AIC values that matters and not the actual values themselves. This is because of the fact that $AIC(k)$ is an estimate of the mean expected log likelihood of a model and is not an estimate of the K-L information quantity, $I(g(\cdot);f(\cdot\,|\,\hat{\theta}_k))$. From the relation between AIC and the entropy, if the difference of *AIC*'s for *MODEL*(j) and *MODEL*(k) is larger than 1~2, then the difference is considered to be significant. If

$$|AIC(j)-AIC(k)|\ll 1,$$

then the goodness of the fits of these models are almost the same. For example, if

$$AIC(j) = 1981816.15$$
$$AIC(k) = 1981819.84$$

then *MODEL*(j) is considered to be better than *MODEL*(k). On the other hand, if

$$AIC(j) = 0.00001$$
$$AIC(k) = 0.10000$$

then the goodness of fits of both models are much the same. It is possible that, even if *AIC*'s of two models are nearly equal, the distributions expressed by the models are quite different. In that case it is reasonable to consider that neither of the models is good.

3° When we fit a model such as the *MODEL*(k) fitted in Example 3.5 of section 4.2.2, it is possible that *AIC* may gradually decrease with increasing order and may not have a clear minimum. This phenomenon usually indicates that the parametrization chosen is bad. For example, this happens if we fit the trigonometric series model (4.11) for the true regression curve

$$S(x; a, b) = ax + b.$$

Even when we can get MAICE, if the number of free parameters is too large, it is quite probable that we can get a better fitting model.

4° AIC is not a criterion for the estimation of the true order but the one for the best fit model. This is not a defect of AIC. When we try to estimate the true distribution from a finite number of data, the concept of true order is meaningless.

PART II

MODELS

CHAPTER 5

DISCRETE PROBABILITY DISTRIBUTION MODELS

5.1 *Binomial Distribution Model*

5.1.1 *Checking a Proportion*

[*Example 5.1*] A certain TV program is supposed to have a rating of 10%. According to a poll based on random sampling*, the actual number of viewers of the program was only 28 out of 400 respondents. Would you judge from these data that the program rating went down to 7%?

As stated in Chapter 2, when n trials are performed independently and at each trial only one of two mutually exclusive events occurs, the probability that the event of interest occurs n_1 times out of n trials is given by the binomial distribution

$$b(n_1 | n, p) = \binom{n}{n_1} p^{n_1} (1-p)^{n-n_1}, \tag{5.1}$$

where p is the probability of occurrence of that event. We denote the sample size by n, the program rating in the population by p and the number of viewers by n_1, respectively.

---

* The number of combinations of N individuals taken n at a time is given by ${}_NC_n$. In random sampling every combination of n individuals in the population has the same chance of being selected.

If the population size N is large, then the binomial distribution is applicable to [Example 5. 1]. Based on the data, we wish to distinguish between the following two models: One model is obtained by assuming that the program rating is 10% as previously and is represented by

$$MODEL(0): \; p = 0.1,$$

the other is obtained by assuming that the program rating takes any other possible value and is represented by

$$MODEL(1): \; p = \theta.$$

Regarding (5. 1) as the function of the parameter p, the corresponding log likelihood $l_1(p)$ is, from (5. 1), given by

$$l_1(p)=K_1+n_1 \log p+(n-n_1) \log (1-p), \tag{5.2}$$

where $K_1=\log {}_nC_{n_1}$.

The maximum likelihood estimator $\hat{\theta}$ of θ under the assumption of $MODEL(1)$ is given by $\hat{\theta}=n_1/n$ as is shown in Chapter 3. Since the parameter to be estimated is only θ, the corresponding AIC is given by

$$AIC^*(1)=(-2)\left\{K_1+n_1 \log \frac{n_1}{n}+(n-n_1) \log \left(1-\frac{n_1}{n}\right)\right\}+2\times 1. \tag{5.3}$$

Since in $MODEL(0)$ the value of the parameter p is already assigned and is not necessary to be estimated, using (5. 2), the corresponding AIC is given by

$$AIC^*(0)=(-2)\{K_1+n_1 \log 0.1+(n-n_1) \log (1-0.1)\}+2\times 0. \tag{5.4}$$

The common constant $(-2)K_1$ can be ignored for the comparison between the statistics (5. 3) and (5. 4) and the AIC's can be modified to

$$AIC(1)=(-2)\left\{n_1 \log \frac{n_1}{n}+(n-n_1) \log \left(1-\frac{n_1}{n}\right)\right\}+2 \qquad (5.5)$$

$$AIC(0)=(-2)\{n_1 \log 0.1+(n-n_1) \log (1-0.1)\}. \qquad (5.6)$$

In this and the following chapter we use the notation AIC^* if the statistic AIC contains the constant term, otherwise we use the notation AIC.

Now, if we apply these statistics (5.5) and (5.6) to the data shown in [Example 5.1], we have

$$AIC(1)=(-2)\left\{28 \log \frac{28}{400}+(400-28) \log \left(1-\frac{28}{400}\right)\right\}+2$$
$$=204.91$$
$$AIC(0)=(-2)\{28 \log 0.1+(400-28) \log 0.9\}$$
$$=207.33.$$

From these results we can see that $AIC(1)$ is significantly less than $AIC(0)$, which means that $MODEL(1)$ is better than $MODEL(0)$. It suggests that we should regard the program rating as 7% rather than 10%.

Problems such as this example have been handled with a conventional test procedure of a proportion. However, all these problems can be handled by comparing the value of (5.5) with that of (5.6) where the value 0.1 in (5.6) is replaced by that of the given hypothesis.

5.1.2 *Checking the Difference of Two Proportions*

[*Example 5.2*] In a second poll taken one month after the poll in [Example 5.1], 40 out of an audience of 1000 watched the TV program. Can we conclude that the rating actually went down in the past month?

Table 5.1 Viewers of a TV program

	Watched	Not Watched	Sample Size
First Poll	28	372	400
Second Poll	40	960	1000

Note that this example consists of a comparison of two independent random samples, whereas only one random sample was given in the preceding example. The survey results are summarized in Table 5. 1.

To handle a more general case, we introduce the following notation;

$n(i_1)$: the sample size at the i_1-th poll ($i_1=1, 2$)

$n(i_1, i_2)$: the number of respondents belonging to category i_2 at the i_1-th poll ($i_1=1, 2,\ i_2=1, 2$)

$p(i_2 | i_1)$: the probability that a respondent belongs to category i_2 at the i_1-th poll.

The notation $p(i_2 | i_1)$ is used to represent the conditional probability of i_2 given i_1. Since each respondent must fall into one category, it is clear that

$$\sum_{i_2=1}^{2} n(i_1, i_2)=n(i_1), \qquad i_1=1, 2 \tag{5.7}$$

$$\sum_{i_2=1}^{2} p(i_2 | i_1)=1, \qquad i_1=1, 2 \ . \tag{5.8}$$

Using this notation the probability $P(\{n(i_1, i_2)\} | \{p(i_2 | i_1)\})$ of obtaining the observations $\{n(i_1, i_2)\}$, $i_1=1, 2$, $i_2=1, 2$ as shown in Table 5. 1 is given by the product of the two binomial distributions and is expressed as

$$\begin{aligned} &P(\{n(i_1, i_2)\} | \{p(i_2 | i_1)\}) \\ &=\frac{n(1)!}{n(1, 1)!n(1, 2)!}p(1 | 1)^{n(1, 1)}\{1-p(1 | 1)\}^{n(1, 2)} \\ &\times\frac{n(2)!}{n(2, 1)!n(2, 2)!}p(1 | 2)^{n(2, 1)}\{1-p(1 | 2)\}^{n(2, 2)} \\ &=\prod_{i_1=1}^{2}\left\{\frac{n(i_1)!}{\prod_{i_2=1}^{2} n(i_1, i_2)!}\prod_{i_2=1}^{2} p(i_2 | i_1)^{n(i_1, i_2)}\right\} , \end{aligned} \tag{5.9}$$

where the symbol Π represents a product. By regarding $p(i_2|i_1)$ in (5.9) as the parameter, the log likelihood $l(\{p(i_2|i_1)\})$ is given by

$$l(\{p(i_2|i_1)\})=K_2+\sum_{i_1=1}^{2}\sum_{i_2=1}^{2}n(i_1, i_2)\log p(i_2|i_1) \qquad (5.10)$$

with

$$K_2=\log\frac{\prod_{i_1=1}^{2}n(i_1)!}{\prod_{i_1=1}^{2}\prod_{i_2=1}^{2}n(i_1, i_2)!}. \qquad (5.11)$$

Under the assumption that the two program ratings are equal, the model is given by

$$MODEL(0):\ p(1|1)=p(1|2)=\theta_0\ ;$$

if they are different, it is given by

$$MODEL(1):\ p(1|1)=\theta_1$$

$$p(1|2)=\theta_2\ .$$

We assume $p(1|1)=p(1|2)=\theta_0$ under $MODEL(0)$ and have from (5.8) that

$$p(2|i_1)=1-\theta_0,\quad i_1=1, 2.$$

Substituting these relations into (5.10), the log likelihood $l(\theta_0)$ is

$$l(\theta_0)=K_2+\left\{\sum_{i_1=1}^{2}n(i_1, 1)\right\}\log\theta_0$$

$$+\left[\sum_{i_1=1}^{2}\{n(i_1)-n(i_1, 1)\}\right]\log(1-\theta_0). \qquad (5.12)$$

Since this takes the same form as the log likelihood (5.2) for the binomial distribution in the preceding section, the maximum likelihood estimator $\hat{\theta}_0$ of θ_0 is given by

$$\hat{\theta}_0=\frac{n(1,1)+n(2,1)}{n(1)+n(2)} .$$

$MODEL(0)$ has only one parameter θ_0 to be estimated and the corresponding AIC is

$$AIC^*(0)=(-2)\left[K_2+\sum_{i_1=1}^{2} n(i_1,1)\log\frac{\sum_{i_1=1}^{2} n(i_1,1)}{\sum_{i_1=1}^{2} n(i_1)}\right.$$
$$\left.+\{\sum_{i_1=1}^{2} n(i_1)-\sum_{i_1=1}^{2} n(i_1,1)\}\log\left\{1-\frac{\sum_{i_1=1}^{2} n(i_1,1)}{\sum_{i_1=1}^{2} n(i_1)}\right\}\right]$$
$$+2\times 1 . \qquad (5.13)$$

Under $MODEL(1)$, from (5.8),

$$p(2|i_1) = 1-\theta_{i_1}, \qquad i_1 = 1, 2. \qquad (5.14)$$

If we substitute this relation in (5.10) and differentiate partially with respect to θ_1 and θ_2 and set these derivatives equal to zero, we find that the maximum likelihood estimators $\hat{\theta}_1=n(1,1)/n(1)$ and $\hat{\theta}_2=n(2,1)/n(2)$. This model contains only these two parameters and the corresponding AIC is given by

$$AIC^*(1)=(-2)\left[K_2+\sum_{i_1=1}^{2}\left\{n(i_1,1)\log\frac{n(i_1,1)}{n(i_1)}\right.\right.$$
$$\left.\left.+(n(i_1)-n(i_1,1))\log\left(1-\frac{n(i_1,1)}{n(i_1)}\right)\right\}\right]+2\times 2 . \qquad (5.15)$$

In the same manner as in the preceding section, the common constant $(-2)K_2$ is ignored and the necessary AIC's are reduced to

$$AIC(0)=(-2)\left[\left\{\sum_{i_1=1}^{2} n(i_1, 1)\right\} \log \frac{\sum_{i_1=1}^{2} n(i_1, 1)}{\sum_{i_1=1}^{2} n(i_1)} + \left\{\sum_{i_1=1}^{2} n(i_1, 2)\right\} \log \frac{\sum_{i_1=1}^{2} n(i_1, 2)}{\sum_{i_1=1}^{2} n(i_1)}\right] + 2\times 1 \qquad (5.16)$$

$$AIC(1)=(-2)\sum_{i_1=1}^{2}\sum_{i_2=1}^{2} n(i_1, i_2) \log \frac{n(i_1, i_2)}{n(i_1)} + 2\times 2. \qquad (5.17)$$

Substituting the values given in [Example 5.2] in these formulas, we obtain

$$AIC(0)$$
$$=(-2)\left[(28+40) \log \frac{28+40}{400+1000} + (372+960) \log \frac{372+960}{400+1000}\right] + 2$$
$$=546.00$$
$$AIC(1)=(-2)\left[28 \log \frac{28}{400} + \cdots + 960 \log \frac{960}{1000}\right] + 4$$
$$=542.80 \ .$$

Since $AIC(1)$ is obviously smaller than $AIC(0)$, we say that $MODEL(1)$ is the better one. This suggests that we should estimate the program ratings by $MODEL(1)$ which assumes that there is a difference between those two proportions.

This type of problem can easily be handled by comparing the values of (5.16) and (5.17). Notice that the comparison of (5.16) and (5.17) is identical to that of the two statistics which will be introduced in section 6.1 to check the independence in a 2 × 2 contingency table.

The statistic $\hat{\theta}_1-\hat{\theta}_2$ is a random variable whose realization depends on two independent random samples $n(1)$ and $n(2)$ and according to the Central Limit Theorem it is approximately distributed as a normal distribution for large values of n. This type of problem has been considered as a particular case of the test of the difference of two means and usually solved by the use of this approximation. Such procedures have widely been used to avoid the troublesome computation of probabilities based on the binomial distribution. Since in the present method we can ignore the constant term, the computation is simplified and the normal approximation is not needed.

5.2 *Multinomial Distribution Model*

5.2.1 *Check of Homogeneity*

[*Example* 5.3] A die was thrown 90 times and the results shown in Table 5.2 were obtained. Is it reasonable to consider that the die is fair?

Table 5.2 Distribution of faces of a die

Faces of a Die	Frequencies
1	17
2	13
3	19
4	11
5	18
6	12
Total	90

Suppose that each trial can result in any one of c exclusive classes (or categories) $A_1, \cdots, A_c$, with respective probabilities $p(1), \cdots, p(c)$ $(p(i)>0, \sum_{i=1}^{c} p(i)=1)$. Based on n independent trials, the probability $P(n(1), \cdots, n(c) \mid p(1), \cdots, p(c))$ of getting A_1's $n(1)$ times, $\cdots$, A_c's $n(c)$ times is obtained from the multinomial distribution

$$P(n(1), \cdots, n(c) \mid p(1), \cdots, p(c))$$
$$= \frac{n!}{n(1)! \cdots n(c)!} p(1)^{n(1)} \cdots p(c)^{n(c)}. \tag{5.18}$$

Hence the log likelihood $l(p(1), \cdots, p(c))$ is given by

$$l(p(1), \cdots, p(c)) = K_3 + \sum_{i=1}^{c} n(i) \log p(i) , \tag{5.19}$$

where

$$K_3 = \log \frac{n!}{\prod_{i=1}^{c} n(i)!} .$$

Now if a given die is fair, each face will have the probability $1/6$ of appearing in a single roll. Therefore the question is to check the model assuming that a given set of data is drawn from the multinomial distribution with respective probabilities $1/6$, against the alternative which assumes that the data are drawn from the multinomial distribution in which each face has its own probability. Such a problem can generally be formulated as the comparison of the following two models:

$$MODEL(0): \quad p(i) = \frac{1}{c}$$
$$MODEL(1): \quad p(i) = \theta(i), \qquad i = 1, \cdots, c .$$

From (5.19) the maximum log likelihood $l(1/c)$ for $MODEL(0)$ is given by

$$l(\theta_0) = K_3 + \sum_{i=1}^{c} n(i) \log \frac{1}{c} . \tag{5.20}$$

Since all the parameters in that model are already specified and need not be estimated, the corresponding AIC is

$$AIC^*(0)=(-2)\left\{K_3+\sum_{i=1}^{c} n(i)\ log\ \frac{1}{c}\right\}+2\times 0 \quad . \qquad (5.21)$$

Under $MODEL(1)$, the maximum likelihood estimator $\hat{\theta}(i)$ of a parameter $\theta(i)$ is given by $\hat{\theta}(i)=n(i)/n$ as stated in Chapter 3. Then, from the constraint $\sum_{i=1}^{c} p(i)=\sum_{i=1}^{c} \theta(i)=1$, we have only $c-1$ free parameters in this model. Thus, using (5. 19), the AIC for this model is given by

$$AIC^*(1)=(-2)\left\{K_3+\sum_{i=1}^{c} n(i)\ log\ \frac{n(i)}{n}\right\}+2(c-1) \quad . \qquad (5.22)$$

If we ignore the common constant term in those statistics as before, we can compare the two models by the following AIC's :

$$AIC(0)=(-2)\sum_{i=1}^{c} n(i)\ log\ \frac{1}{c} \qquad (5.23)$$

$$AIC(1)=(-2)\sum_{i=1}^{c} n(i)\ log\ \frac{n(i)}{n}+2(c-1) \quad . \qquad (5.24)$$

In [Example 5. 3] we have $c=6$. Substituting the data into (5. 23) and (5. 24), we have

$$AIC(0)=(-2)\times(17+\cdots+12)\times\ log\ \frac{1}{6}=322.52$$

$$AIC(1)=(-2)\times\left(17\ log\ \frac{17}{90}+\cdots+12\ log\ \frac{12}{90}\right)+2\times 5=328.61 \ .$$

Since $AIC(0) < AIC(1)$, we can regard $MODEL(0)$ as the better one, which implies the die is fair.

From the data in [Example 5. 3] it would appear that the frequencies seem to depend only on whether the face showing is even or odd. To check this observation, we shall consider the model

$$MODEL(2):\quad p(1)=p(3)=p(5)=\theta_1$$
$$p(2)=p(4)=p(6)=\theta_2 \ .$$

From the constraint $\sum_{i=1}^{6} p(i)=1$, θ_1 and θ_2 have the constraint $\sum_{i=1}^{6} p(i)=3(\theta_1+\theta_2)=1$, that is,

$$\theta_1+\theta_2=\frac{1}{3} . \tag{5. 25}$$

Now the log likelihood for *MODEL(2)*, using (5. 19), is given by

$$l(\theta_1, \theta_2)=K_3+\{n(1)+n(3)+n(5)\} \log \theta_1 + \{n(2)+n(4)+n(6)\} \log \theta_2 . \tag{5. 26}$$

By substituting (5. 25) in (5. 26) and differentiating with respect to θ_1 and setting the derivative equal to zero, we find the maximum likelihood estimators

$$\hat{\theta}_1=\frac{n(1)+n(3)+n(5)}{3n}$$

$$\hat{\theta}_2=\frac{n(2)+n(4)+n(6)}{3n} .$$

Due to the constraint (5. 25) this model has only one free parameter and the corresponding *AIC* is finaly given by

$$AIC(2)=(-2)\left[\{n(1)+n(3)+n(5)\} \log \frac{n(1)+n(3)+n(5)}{3n} + \{n(2)+n(4)+n(6)\} \log \frac{n(2)+n(4)+n(6)}{3n}\right]+2 . \tag{5. 27}$$

This statistic yields the value $AIC(2)=320.89$ for the data given in [Example 5. 3]. This value is the smallest of all the three *AIC* values. Thus if we are forced to choose between the two, equiprobable or not, we should choose the equiprobable model. However, if we may include the last hypothesis, we should adopt it.

5. 2. 2 Check of Identity of Several Multinomial Distributions

[*Example 5. 4*] To measure to what extent a voter is concerned about a general election, three independent random samples were drawn in 1972 from three cities, Tokyo, Machida and Kurashiki. The results are summarized in Table 5. 3. Can we conclude that the distributions of voter-concern for the three cities are identical?

Table 5. 3 Voter-concern in the three cities

	Tokyo	Machida	Kurashiki
Deeply Concerned	74	126	69
Fairly Concerned	187	312	254
Not Concerned	179	284	372
Others	28	29	36
Total	468	751	731

(Research Report, No. 31, 1973, The Institute of Statistical Mathematics)

This problem is a generalization of [Example 5. 2]. In this example the number of response categories in each poll is four rather than two and the number of polls is increased from two to three. Therefore, the likelihood function is the product of three multinomial distributions. We can solve the problem by similar methods. We consider this example for the particular reasons that this type of problem is more practical and we wish to clarify the difference between this problem and the problem of comparison of several contingency tables to be discussed in section 6. 2.

To discuss such problems generally, we use the following notations.

$n(i_1)$: the sample size at i_1-th city ($i_1=1, \cdots, c_1$)

$n(i_1, i_2)$: the number of respondents belonging in a category i_2 at i_1-th city ($i_1=1, \cdots, c_1$, $i_2=1, \cdots,$

c_2)

$p(i_2|i_1)$: the probability that a respondent belongs in a category i_2 at i_1-th city.

It is clear that

$$\sum_{i_2=1}^{c_2} n(i_1, i_2)=n(i_1), \quad i_1=1, \cdots, c_1 \tag{5.28}$$

$$\sum_{i_2=1}^{c_2} p(i_2|i_1)=1, \quad i_1=1, \cdots, c_1 . \tag{5.29}$$

Then the probability $P(\{n(i_1, i_2)\}|\{p(i_2|i_1)\})$ of obtaining such results $\{n(i_1, i_2)\}$, $(i_1=1, \cdots, c_1, \ i_2=1, \cdots, c_2)$ as in Table 5.3 is given by the product of three multinomial distributions expressed as

$$P(\{n(i_1, i_2)\}|\{p(i_2|i_1)\}) = \prod_{i_1}\left\{\frac{n(i_1)!}{\prod_{i_2} n(i_1, i_2)!}\prod_{i_2} p(i_2|i_1)^{n(i_1, i_2)}\right\} \tag{5.30}$$

where $\prod_{i_1}=\prod_{i_1=1}^{c_1}$, and $\prod_{i_2}=\prod_{i_2=1}^{c_2}$. If we set

$$K_4= \log\{\prod_{i_1} n(i_1)!/\prod_{i_1}\prod_{i_2} n(i_1, i_2)!\}, \tag{5.31}$$

then, from (5.30), the log likelihood $l(\{p(i_2|i_1)\})$ with respect to the parameters $\{p(i_2|i_1)\}$ may be written as

$$l(\{p(i_2|i_1)\})=K_4+\sum_{i_1}\sum_{i_2} n(i_1, i_2) \log p(i_2|i_1) . \tag{5.32}$$

Assuming that the distributions regarding voter-concern about a general election are the same in the three cities, we can construct a model

$$MODEL(0): \ p(i_2|i_1)=\theta(i_2) \quad i_1=1, \cdots, c_1, \quad i_2=1, \cdots, c_2 .$$

If the distributions are different, the model becomes

$$MODEL(1): \ p(i_2|i_1)=\theta(i_2|i_1) \quad i_1=1, \cdots, c_1, \quad i_2=1, \cdots, c_2 .$$

Since it is assumed under *MODEL(0)* that $p(i_2|i_1)=p(i_2)$, the log likelihood $l(\{\theta(i_2)\})$, from (5.32), is given by

$$l(\{\theta(i_2)\})=K_4+\sum_{i_2}\{\sum_{i_1}n(i_1, i_2)\}\log\theta(i_2). \tag{5.33}$$

This takes the same form as the log likelihood (5.19) in the preceding section and the maximum likelihood estimator is obtained as $\hat{\theta}(i_2)=\sum_{i_1}n(i_1, i_2)/\sum_{i_1}n(i_1)$. Then from (5.29) the number of free parameters is c_2-1 and the corresponding *AIC* is given by

$$AIC^*(0)=(-2)\left[K_4+\sum_{i_2=1}^{c_2}\left\{\sum_{i_1=1}^{c_1}n(i_1, i_2)\right\}\log\frac{\sum_{i_1=1}^{c_1}n(i_1, i_2)}{\sum_{i_1=1}^{c_1}n(i_1)}\right]+2(c_2-1). \tag{5.34}$$

In contrast, the log likelihood under *MODEL(1)* is given by

$$l(\{\theta(i_2|i_1)\})=K_4+\sum_{i_1}\sum_{i_2}n(i_1, i_2)\log\theta(i_2|i_1). \tag{5.35}$$

If we put $\partial l(\{\theta(i_2|i_1)\})/\partial\theta(i_2|i_1)=0$ taking into consideration the constraint (5.29), we have $\hat{\theta}(i_2|i_1)=n(i_1, i_2)/n(i_1)$ as the maximum likelihood estimator of $\theta(i_2|i_1)$. Although this model contains c_1c_2 parameters, it has only $c_1(c_2-1)$ free parameters due to the constraint (5.29). Hence we have

$$AIC^*(1)=(-2)\left\{K_4+\sum_{i_1=1}^{c_1}\sum_{i_2=1}^{c_2}n(i_1, i_2)\log\frac{n(i_1, i_2)}{n(i_1)}\right\}+2c_1(c_2-1). \tag{5.36}$$

Ignoring $(-2)K_4$, the *AIC*'s reduce to

$$AIC(0)=(-2)\sum_{i_2=1}^{c_2}\left\{\sum_{i_1=1}^{c_1} n(i_1, i_2)\right\} \log \frac{\sum_{i_1=1}^{c_1} n(i_1, i_2)}{\sum_{i_1=1}^{c_1} n(i_1)}+2(c_2-1) \quad (5.37)$$

$$AIC(1)=(-2)\sum_{i_1=1}^{c_1}\sum_{i_2=1}^{c_2} n(i_1, i_2) \log \frac{n(i_1, i_2)}{n(i_1)}+2c_1(c_2-1), \quad (5.38)$$

If we substitute the values given in [Example 5.4] into (5.37) and (5.38), we have

$$AIC(0)=(-2)\left\{(74+126+69) \log \frac{74+126+69}{468+751+731}+\cdots\right.$$

$$\left.+(28+29+36) \log \frac{28+29+36}{468+751+731}\right\}+2\times(4-1)$$

$$=4487.11$$

$$AIC(1)=(-2)\left\{74 \log \frac{74}{468}+\cdots+36 \log \frac{36}{731}\right\}+2\times3\times(4-1)$$

$$=4456.98 .$$

From this calculation we should conclude that three distributions are different.

5.2.3 *Goodness of Fit of Hypothesized Distribution*

[*Example 5.5*] Suppose that *100* nails produced by a certain machine are packed into each box. According to a random sample of *100* boxes, actual number of defective nails in each box were as shown in Table 5.4. Can you say that the number of defective ones in each box is distributed as a Poisson distribution?

Table 5.4 Distribution of the number of defective nails in each box

Number of Defective Nails	Number of Boxes
0	12
1	28
2	28
3	18
4	9
5	4
6	1
7 and over	0
Total	100

Given a set of data, by properly choosing a set of classes and then placing each measurement in its appropriate class, we can construct a frequency table as shown in Table 5.4. Such a frequency table can be regarded as an example of a multinomial distribution. Thus we need to check whether the data follows a multinomial distribution without any restriction, or follows one with parameters further restricted by other parameters. Such a problem can be solved similarly to that of section 5.2.1. The difference is that each probability is reparameterized with any other parameters.

Generally, when the data are classified into c possible classes with respective probability $p(i)$, $i=1, \cdots, c$, the probability $P(\{n(i)\}|\{p(i)\})$ of getting the frequency distribution $\{n(i)\}$, $i=1, \cdots, c$ (as in Table 5.4) is given by the multinomial distribution

$$P(\{n(i)\}|\{p(i)\})=\frac{n!}{\prod_{i=1}^{c} n(i)!}\prod_{i=1}^{c} p(i)^{n(i)}. \tag{5.39}$$

Hence the log likelihood $l(\{p(i)\})$ with respect to the parameters $\{p(i)\}$ is defined, as in (5.19), by

$$l(\{p(i)\})=K_3+\sum_{i=1}^{c} n(i) \log p(i) . \tag{5.40}$$

One hypothesis is that each measurement has the multinomial distribution with c items and is simply expressed by

$$MODEL(1):\ p(i)=\theta(i),\quad i=1, \cdots, c .$$

Of course, this model has the constraint

$$\sum_{i=1}^{c} \theta(i)=1 . \tag{5.41}$$

The other hypothesis assuming that each measurement is

obtained from a Poisson distribution means that the parameters $p(i)$ are considered to be reparameterized as follows:

$$MODEL(0): \quad p(i)=\frac{\lambda^{i-1}}{(i-1)!}e^{-\lambda} \quad (i=1, \cdots, c-1)$$

$$p(c)=1-\sum_{i=1}^{c-1}\frac{\lambda^{i-1}}{(i-1)!}e^{-\lambda} \ . \qquad (5.42)$$

From the discussion in section 5.2.1, the *AIC* for *MODEL* (*1*) is given by

$$AIC^*(1)=(-2)\left\{K_3+\sum_{i=1}^{c}n(i)\ log\ \frac{n(i)}{n}\right\}+2(c-1) \ . \qquad (5.43)$$

To find the maximum likelihood estimator $\hat{\lambda}$ of the parameter λ of *MODEL*(*0*), we have to substitute (5.42) for (5.40) and find the value of λ for which it is a maximum. However this entails a numerical maximization of the likelihood which is beyond the scope of this book. If we adopt the maximum likelihood estimator $\hat{\lambda}$ based on each measurement instead of an estimator based on a frequency table such as Table 5.4, $\hat{\lambda}$ is given by the sample mean $\bar{x}$ as stated in Chaper 3. Since the only parameter to be

estimated is λ, the corresponding *AIC* is expressed as

$$AIC^*(0)=(-2)\left\{K_3+\sum_{i=1}^{c-1}n(i)\ log\ \frac{\bar{x}^{i-1}}{(i-1)!}e^{-\bar{x}}\right.$$

$$\left.+n(c)\ log\left(1-\sum_{i=1}^{c-1}\frac{\bar{x}^{i-1}}{(i-1)!}e^{-\bar{x}}\right)\right\}+2\times 1. \qquad (5.44)$$

By neglecting the common constant $(-2)K_3$ as usual, the necessary *AIC*'s follow

$$AIC(1)=(-2)\sum_{i=1}^{c}n(i)\ log\ \frac{n(i)}{n}+2(c-1) \qquad (5.45)$$

$$AIC(0)=(-2)\left\{\sum_{i=1}^{c-1} n(i) \log \frac{\bar{x}^{i-1}}{(i-1)!}e^{-\bar{x}} + n(c) \log \left(1-\sum_{i=1}^{c-1}\frac{\bar{x}^{i-1}}{(i-1)!}e^{-\bar{x}}\right)\right\}+2. \quad (5.46)$$

Putting $c=7$, we shall calculate the *AIC* values for [Example 5.5]. Since $\bar{x}=2.0$, from (5.45) and (5.46) we have

$$AIC(1)=(-2)\left\{12 \log \frac{12}{100}+\cdots+ \log \frac{1}{100}\right\}+2\times 6$$
$$=345.50$$
$$AIC(0)=(-2)\left\{12 \log \frac{2.0^0}{0!}e^{-2.0}+\cdots+ \log (1-0.9834)\right\}+2$$
$$=336.09\ .$$

From these computations we see that $AIC(0)$ is less than $AIC(1)$ and we conclude that the data follows a Poisson distribution.

5.3 *Histogram Model*

5.3.1 *Comparison of Two Histograms*

[*Example 5.6*] Based on a random sample of *100* measurements, two frequency tables shown in Tables 5.5 and 5.6 were constructed. We wish to judge which of the two is the better.

Table 5.5

Class Boundaries	Frequency	Relative Frequency
- 0.45	1	0.01
0.45 - 0.85	2	0.02
0.85 - 1.25	7	0.07
1.25 - 1.65	10	0.10
1.65 - 2.05	16	0.16
2.05 - 2.45	12	0.12
2.45 - 2.85	15	0.15
2.85 - 3.25	16	0.16
3.25 - 3.65	7	0.07
3.65 - 4.05	9	0.09
4.05 - 4.45	4	0.04
4.45 -	1	0.01
Total	100	1.00

Table 5.6

Class Boundaries	Frequency	Relative Frequency
- 0.85	3	0.03
0.85 - 1.65	17	0.17
1.65 - 2.45	28	0.28
2.45 - 3.25	31	0.31
3.25 - 4.05	16	0.16
4.05 -	5	0.05
Total	100	1.00

As stated in the preceding section, tables such as Table 5.5 and Table 5.6 are called frequency tables, which are constructed to determine the nature of the population distribution. To construct a frequency table, we have to choose a set of class boundaries then find the frequencies in each class. The frequencies divided by the sample size are called the relative frequencies and the graph of the frequencies is called the frequency histogram. In a frequency histogram the vertical axis is often graduated in frequencies or in relative frequencies. We adopt here the axis graduated in units of relative frequencies per unit interval width. That is, we first choose the $c+1$ cutting points $a=a_0<a_1<\cdots<a_c=b$ $(a_i=a+i\Delta a)$ in the interval $(a, b]$ and then divide the frequency $n(i)$ of a class $I_i=(a_{i-1}, a_i]$ by the product of the sample size n and the interval length Δa, which gives the height. If we finally draw rectangles, each of which has the base Δa and the corresponding height, we have graphical representations, such as Figure 5.1 and Figure 5.2. We will refer to such graphical representations as frequency histograms. Then the area of a rectangle on I_i is $n(i)/\{n\cdot\Delta a\}\cdot\Delta a=n(i)/n$ and the area over the interval $(a, b]$ is totaled to $\sum_{i=1}^{c} n(i)/n=1$.

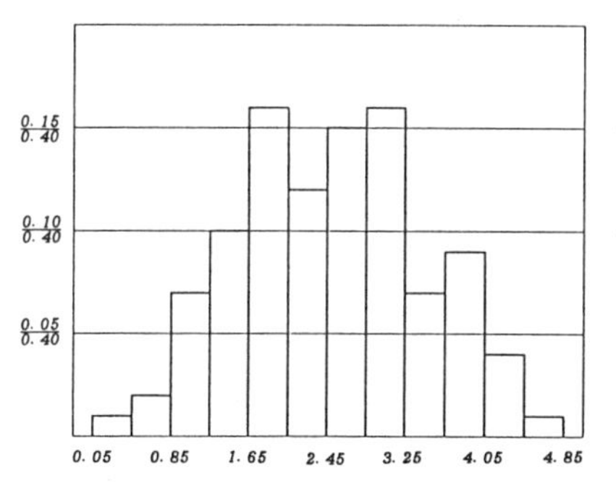

Figure 5.1
Histogram for Table 5.5

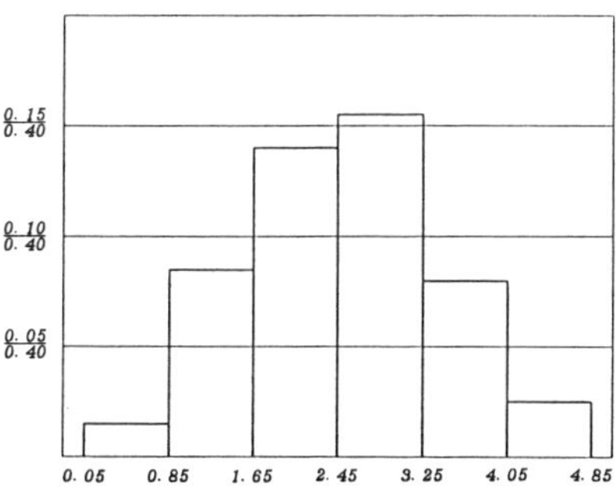

Figure 5.2
Histogram for Table 5.6

Now if some criterion to determine class boundaries is already given, such as in the case of " the distribution of the number of households by size of household ", we can easily draw the necessary frequency histogram. However, if we have no information on how we might carry out the categorization, we have to determine the necessary classes on the basis of a suitably chosen criterion. In such a case, the main purpose of the frequency histogram is to get a rough idea of the shape of the population distribution from which the data are drawn. From this point of view, we may regard as better, a frequency histogram whose size and number of classes facilitate more accurate estimation of the underlying population distribution. A frequency histogram with too few classes compared with the sample size will lose the information supplied by data. Thus we need to determine the frequency histogram with an optimal number of classes. These considerations suggest that given a frequency histogram with many classes, if two or more classes in which we can assume homogeneous probabilities are pooled into one class, the search for an optimal frequency histogram will be realized.

If we regard a given frequency table as a multinomial distribution and denote by c the maximum number of classes in the table and further denote by $p(i)$ and $n(i)$ $(i=1, \cdots, c)$ the probability and cell frequency in each class respectively, then the probability $P(\{n(i)\} \mid \{p(i)\})$ of obtaining the frequencies $\{n(i)\}$ as shown in Table 5.5 is given by

$$P(\{n(i)\} \mid \{p(i)\}) = \frac{n!}{\prod_{i=1}^{c} n(i)!} \prod_{i=1}^{c} p(i)^{n(i)} \quad . \tag{5.47}$$

Hence the log likelihood $l(\{p(i)\})$ is represented by

$$l(\{p(i)\}) = K_3 + \sum_{i=1}^{c} n(i) \log p(i) \quad , \tag{5.48}$$

where $c=12$ for our example.

From the above stated idea, the comparison between Table 5.5 and Table 5.6 can be considered as that of the following two models.

$$MODEL(1): \quad p(i)=\theta_1(i), \quad i=1, \cdots, c$$

$$MODEL(2): \quad p(2j-1)=p(2j)=\theta_2(j), \quad j=1, \cdots, \frac{c}{2},$$

where, for simplicity, c is assumed to be an even number. From the constraint $\sum_{i=1}^{c} p(i)=1$, the two models have the respective constraint

$$\sum_{i=1}^{c} \theta_1(i)=1 \tag{5.49}$$

$$2\sum_{j=1}^{c/2} \theta_2(j)=1 . \tag{5.50}$$

In $MODEL(2)$ it is considered that Figure 5.2 and Figure 5.3 are identical since they express the same probability distribution inspite of the difference of the number of classes.

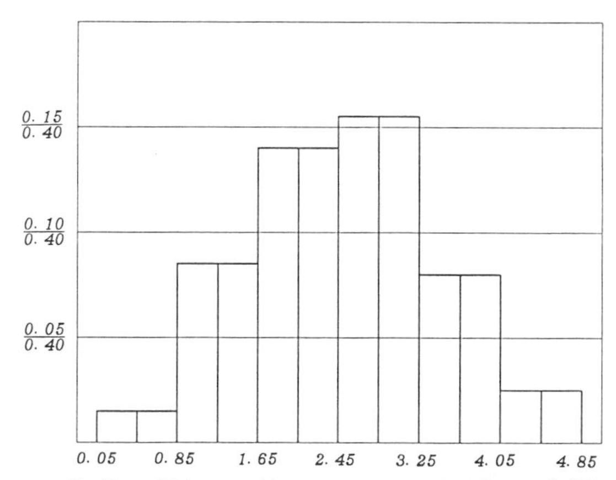

Figure 5.3 Alternative representation of Figure 5.2

The similar calculation to that in the preceding section gives the AIC for $MODEL(1)$

$$AIC^*(1)=(-2)\left[K_3+\sum_{i=1}^{c}n(i)\log\frac{n(i)}{n}\right]+2(c-1). \qquad (5.51)$$

From the conditions under $MODEL(2)$ and (5.48), we find the log likelihood

$$l(\{\theta_2(j)\})=K_3+\sum_{j=1}^{c/2}\{n(2j-1)+n(2j)\}\log\theta_2(j)\ . \qquad (5.52)$$

Taking into account the constraint (5.50), as stated in the case of $MODEL(2)$ in section 5.2.1, the corresponding maximum likelihood estimator is given by

$$\hat{\theta}_2(j)=\frac{n(2j-1)+n(2j)}{2n}\ .$$

Then, since the number of parameters estimated is $c/2-1$ the AIC is given by

$$AIC^*(2)=(-2)\left[K_3+\sum_{j=1}^{c/2}\{n(2j-1)+n(2j)\}\log\frac{n(2j-1)+n(2j)}{2n}\right]$$
$$+2\left(\frac{c}{2}-1\right). \qquad (5.53)$$

By neglecting the common constant in (5.51) and (5.53) as before, we can compare the models by

$$AIC(1)=(-2)\sum_{i=1}^{c}n(i)\log\frac{n(i)}{n}+2(c-1) \qquad (5.54)$$

$$AIC(2)=(-2)\sum_{j=1}^{c/2}\{n(2j-1)+n(2j)\}\log\frac{n(2j-1)+n(2j)}{2n}$$
$$+2\left(\frac{c}{2}-1\right). \qquad (5.55)$$

If we introduce the new notations $n'(j)$ and c' which denote the frequencies and the number of classes in Table 5.6 (not in Table 5.5) the statistic (5.55) is reduced to

$$AIC'(2)=(-2)\sum_{j=1}^{c'} n'(j)\log\frac{n'(j)}{2n}+2(c'-1)\ , \qquad (5.56)$$

which shows that the value of AIC can be calculated only from Table 5.6.

To solve [Example 5.6], we apply (5.54) to Table 5.5 and apply (5.56) to Table 5.6 and then we get

$$AIC(1)=(-2)\left[\log\frac{1}{100}+2\log\frac{2}{100}+\cdots+\log\frac{1}{100}\right]+2\times 11$$
$$=470.76$$
$$AIC'(2)=(-2)\left[3\log\frac{3}{2\times 100}+\cdots+5\log\frac{5}{2\times 100}\right]+2\times 5$$
$$=462.41\ .$$

It is obvious that $AIC'(2) < AIC(1)$, which implies that Table 5.6 is better than Table 5.5 for the data in [Example 5.6]. The truth is that the data were originally taken from the normal distribution with mean *2.4* and variance *1*. It is seen in Figure 5.2 that the irregularities present in Figure 5.1 here disappeared, which will facilitate the estimation of the population distribution.

5.3.2 *Search for An Optimal Histogram*

[*Example 5.7*] The following data are a random sample of *100* measurements. Find an optimal frequency histogram.

28.67	40.29	10.61	33.85	36.19	20.63	9.64	15.26	15.53	73.62
63.29	32.77	32.28	11.90	54.16	4.73	24.67	17.66	25.84	22.89
15.68	5.48	36.41	20.33	44.58	57.23	65.89	57.91	2.39	9.15
10.27	3.04	12.35	32.78	44.23	31.14	6.03	27.90	28.73	42.09
3.99	9.74	6.85	0.16	9.26	7.72	34.42	32.77	6.80	10.45
29.80	5.89	13.56	50.55	0.51	0.19	7.19	5.94	11.24	32.32
15.27	29.64	10.03	2.01	13.89	20.83	27.49	14.46	8.22	27.81
33.65	38.57	8.66	1.40	23.97	15.11	63.32	7.76	1.58	48.66
44.46	0.02	38.12	18.51	101.75	34.16	27.99	5.22	1.82	8.22
4.89	97.50	2.10	26.19	10.11	8.39	25.83	1.05	25.63	18.35

We shall consider the procedure of constructing an optimal histogram by generalizing the ideas in the preceding section.

Let $x_1, \cdots, x_n$ be a sample of n measurements from a certain population and let $x_{(1)}$ be the smallest value in the sample, $x_{(n)}$ the largest. We divide the interval $[x_{(1)}-0.5d, x_{(n)}+0.5d]$ into c classes, where d denotes the precision of each measurement and the i-th class contains $n(i)$ observations. One way to determine the category size c is to take $[2n^{1/2}-1]$, where the symbol [ ] denotes integer part. This defines the initial categorization. The considerations of the preceding section suggest that we check the goodness of fit of various models to the initial frequency table to reduce the number of free parameters. In this manner we will be able to obtain an optimal frequency table, or an optimal frequency histogram.

The probability of getting the frequencies under the initial categorization is given by (5.47) and its log likelihood is given by (5.48). Needless to say

$$\sum_{i=1}^{c} p(i)=1 . \tag{5.57}$$

As in the usual procedure, consider the frequency histogram which has equal length intervals except for the two at both ends. If we denote by c_1 and c_2 the number of classes to be pooled in the low and high side in the initial frequency table and denote by r the number of classes to be pooled in the central part, then we can identify a model by using the notation $MODEL(c_1, r, c_2)$. The corresponding model is expressed by

$MODEL\ (c_1, r, c_2)$:

$p(1)=p(2)= \cdots =p(c_1)=\theta(1)$

$$p(c_1+(j-2)r+1)=\cdots=p(c_1+(j-1)r)=\theta(j),$$

$$j=2,\ \cdots,\ \frac{c-c_1-c_2}{r}+1$$

$$p(c-c_2+1)=\cdots=p(c)=\theta\left(\frac{c-c_1-c_2}{r}+2\right).$$

From (5. 57) this model has the constraint

$$c_1\theta(1)+r\sum_j\theta(j)+c_2\theta\left(\frac{c-c_1-c_2}{r}+2\right)=1 \ . \qquad (5.58)$$

The maximum log likelihood estimators of $\theta(j)$, $j=1, \cdots, (c-c_1-c_2)/r+2$ can be found by the same procedure as in section 5. 2. 1 and in the preceding section. The resulting *AIC* is

$$\begin{aligned}
&AIC(c_1, r, c_2)\\
&=(-2)\left[\left(\sum_{i=1}^{c_1}n(i)\right)\log\left(\sum_{i=1}^{c_1}\frac{n(i)}{c_1n}\right)\right.\\
&+\sum_{j=2}^{(c-c_1-c_2/r)+1}\left\{\left(\sum_{i=c_1+(j-2)r+1}^{c_1+(j-1)r}n(i)\right)\log\left(\sum_{i=c_1+(j-2)r+1}^{c_1+(j-1)r}\frac{n(i)}{rn}\right)\right\}\\
&\left.+\left(\sum_{i=c-c_2+1}^{c}n(i)\right)\log\left(\sum_{i=c-c_2+1}^{c}\frac{n(i)}{c_2n}\right)\right]\\
&+2\left(\frac{c-c_1-c_2}{r}+1\right) \ . \qquad (5.59)
\end{aligned}$$

If we denote by $n'(j)$ and c' respectively the frequency in the j-th interval and the number of intervals in the resulting frequency histogram obtained by pooling some of the initial classes, (5. 59) reduces to

$$\begin{aligned}
AIC(c_1, r, c_2)=(-2)&\left\{n'(1)\log\frac{n'(1)}{c_1n}+\sum_{j=2}^{c'-1}n'(j)\log\frac{n'(j)}{rn}\right.\\
&\left.+n'(c')\log\frac{n'(c')}{c_2n}\right\}+2(c'-1) \ . \qquad (5.60)
\end{aligned}$$

In calculating AIC it is convenient to substitute e^{-1} for $n'(j)$ if $n'(j)=0$. This is based on the consideration that $x \log x$ in (5.59) and in (5.60) takes its minimum at $x=e^{-1}$ and AIC takes its largest value at that point. This modification is intended to penalize a model with too many classes, that is, too many free parameters.

We shall illustrate the AIC in this case by considering [Example 5.7]. The smallest and largest value in the data are respectively 0.02, 101.75 and the precision of measurement is 0.01. These values give the interval $[0.015, 101.755]$ and an initial categorization is given by dividing this interval into $[2 \cdot 100^{1/2} - 1]$ equal parts. Then each class interval length is equal to $(101.755-0.015)/19=5.35474$ and the 19 cell frequencies arranged in increasing order of cell midpoints are as follows:

16 22 11 6 7 9 11 4 3 2 3 2 1 1 0 0 0 0 2.

If we apply (5.60) to the frequency tables obtained for various values of c_1, r and c_2 and then search for those values that give $MAICE$, we can determine an optimal frequency table. The optimal frequency histogram for the example is shown in Figure 5.4. As an example the calculation by (5.60) for the model corresponding to this figure is as follows:

$$AIC(2,5,7) = (-2)\left\{38 \log \frac{38}{2\times 100} + 44 \log \frac{44}{5\times 100} + 14 \log \frac{14}{5\times 100} + 4 \log \frac{4}{7\times 100}\right\} + 2(4-1)$$

$$= 487.53 \ .$$

INITIAL CATEGORIZATION :

### MIN OF DATA	INTERVAL LENGTH	MAX OF DATA	NO. OF INTIAL CATEGORIES ###
0.15000D-01	0.53547D+01	0.10175D+03	19

OPTIMAL FREQUENCY HISTOGRAM

NO.	CLASS BOUNDARIES	FREQUENCIES	ESTIMATES
1	0.015 ~ 5.370	16	0.19000
2	5.370 ~ 10.724	22	0.19000
3	10.724 ~ 16.079	11	0.08800
4	16.079 ~ 21.434	6	0.08800
5	21.434 ~ 26.789	7	0.08800
6	26.789 ~ 32.143	9	0.08800
7	32.143 ~ 37.498	11	0.08800
8	37.498 ~ 42.853	4	0.02800
9	42.853 ~ 48.208	3	0.02800
10	48.208 ~ 53.562	2	0.02800
11	53.562 ~ 58.917	3	0.02800
12	58.917 ~ 64.272	2	0.02800
13	64.272 ~ 69.627	1	0.00571
14	69.627 ~ 74.981	1	0.00571
15	74.981 ~ 80.336	0	0.00571
16	80.336 ~ 85.691	0	0.00571
17	85.691 ~ 91.046	0	0.00571
18	91.046 ~ 96.400	0	0.00571
19	96.400 ~ 101.755	2	0.00571
	TOTAL	100	

(Histogram scale: 0.0 0.0200 0.0400 0.0600 0.0800 0.1000 0.1200 0.1400 0.1600 0.1800 0.2000)

MAICE 487.53

Figure 5.4 Optimal frequency histogram for Example 5.7

Table 5.7 AIC values for some models for Example 5.7

Rank	*MODEL*	c'	*AIC*	*AIC-min AIC*
<1>	(2, 5, 7)	4	487.53	-
2	(2, 5, 2)	5	488.71	1.18
3	(3, 5, 1)	5	490.29	2.76
<4>	(2, 6, 5)	4	490.66	3.13
<5>	(3, 4, 8)	4	491.11	3.58
6	(3, 5, 6)	4	492.00	4.47
7	(3, 4, 4)	5	492.43	4.90
8	(2, 6, 11)	3	494.88	7.35
9	(2, 5, 12)	3	495.28	7.75
<10>	(2, 7, 10)	3	496.60	9.07
<18>	(4, 3, 6)	5	498.14	10.61
<23>	(3, 8, 8)	3	499.30	11.77
<29>	(2, 9, 8)	3	499.96	12.43
<40>	(2, 10, 7)	3	501.82	14.29
<73>	(7, 11, 1)	3	506.70	19.17

Similarly we can easily compute *AIC* values for other models. Only a part of the results are shown in Table 5.7 in which models are arranged in ascending order of *AIC* value. In Table 5.7 the model whose model number is enclosed with the symbol < > has the minimum *AIC* among models in each of which the same number of classes are pooled in the central part. Hence this table illustrates that frequency tables obtained by pooling five classes in the central part are better than the others. Further, from the point of view of the number of free parameters in the model, although more frequency tables with three or four parameters rank higher, a few with the same number of parameters rank lower. Finally, the value of *AIC* depends not only on the number of free parameters but also on the way of choosing class boundaries.

In fact, the data in [Example 5.7] were randomly drawn from an exponential distribution with $\lambda=0.04$ having probability density function

$$f(x|\lambda)=\lambda e^{-\lambda x} \quad x>0 \quad . \tag{5.61}$$

The long right tail which is a feature of the exponential distribution can be seen in the optimal frequency histogram shown in Figure 5.4.

Generally we have to check many models to find out the optimal frequency histogram and it is difficult to check the *AIC* values for all possible models by hand calculation. For this reason it may be convenient to use the FORTRAN program CATDAP-11 given in part III. Of course, the comparison of a few models by hand calculation is possible since the basic statistic (5.60) is very simple.

In classical books of statistics, the number of classes c' of the frequency histogram is determined, for example, by Sturges' formula

$$c' = 1 + \frac{\log_{10} n}{\log_{10} 2} = 1 + 3.32 \log_{10} n \ .$$

To clarify the behavior of the present procedure from this aspect, we shall consider the following experiment. Three probability distributions are defined as

$1°$ $\quad f_1(x) = 0.04e^{-0.04x} \qquad x > 0$

$2°$ $\quad f_2(x) = \frac{1}{\sqrt{2\pi}} e^{-x^2/2} \qquad -\infty < x < +\infty$

$3°$ $\quad f_3(x) = 0.5 \times \frac{1}{\sqrt{2\pi}} e^{-x^2/2} + 0.5 \times \frac{1}{\sqrt{2\pi}} e^{-(x-4.0)^2/2} \qquad -\infty < x < +\infty$

These are examples of an exponential distribution, a normal distribution and a mixed normal distribution, respectively. For each distribution random samples of size *20*, *50*, *100*, *200*, *500* and *1000* were generated and the value of c' was determined for the optimal frequency histogram selected by the present procedure. Such experiments were repeated *100* times for each case and the mean values of c' were obtained. The results are summarized in Table 5.8. From the table we see that: 1) the mean value of c' increases with the sample size; 2) the mean value of c' depends on the type of probability distribution even when the sample size are the same; 3) the mean value at each sample size almost always increases in the order, exponential distribution, normal distribution, mixed normal distribution; 4) the rate of increase of the mean value is larger with the present procedure than that given by Sturges' formula and this tendency is greater in large samples. Figure 5.5 shows an example of the optimal frequency histogram obtained by CATDAP-11 in the case of a sample size of *500* taken from the mixed normal distribution which has a more complicated shape involving two peaks. The third point above seems to suggest

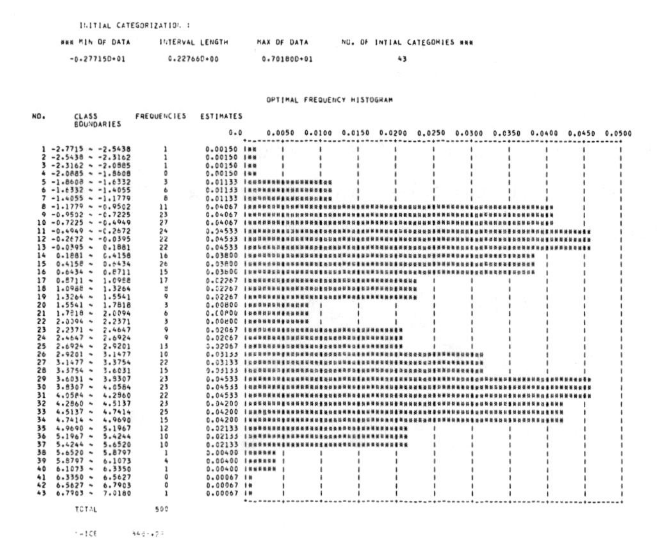

Figure 5.5 An optimal frequency histogram for the mixed normal distribution

Table 5.8 Comparison of the number of classes in the optimal frequency histogram

Sample Size	Exponential Distribution	Normal Distribution	Mixed Normal Distribution	Sturges' Formula
50	3.23	3.15	3.60	6.64
100	4.05	4.25	5.44	7.64
200	5.06	6.30	7.99	8.64
500	7.26	10.13	13.39	9.96
1000	8.96	13.25	16.88	10.96

that the choice of the optimal number of classes by the present procedure depends on the complexity of the shape of the distribution.

Finally we should point out that the present procedure is applicable to distributions of any shape and has a wide range of applications. However, if the sample size is small, the optimal number of classes is less as is seen from Table 5.8. This may cause a misleading inferences to be drawn. This is mainly due to the fact that a continuous variate has been discretized, a procedure which fails to take account of all the information contained in the continuous variate. A procedure adapting this modification is discussed in Ishiguro & Sakamoto (1984).

PROBLEMS

5.1 A random sample of *400* persons were asked which they like better, an apple or a banana. *220* persons liked an apple, whereas *180* persons liked a banana. Is there a significant difference of opinion here?

5.2 Suppose that, to the similar question, the number of respondents choosing "apple", "banana" and "others" were respectively *215*, *175* and *10*. Is there a significant difference between the proportions choosing "apple" and "banana"?

5.3 Suppose that a survey concerning wages gave the results that *60* out of *100* workers get twenty thousand dollars and over and the remainder received less. Can you conclude that the median wage is twenty thousand dollars? Here, when x is a continuous random variable having a probability density function $f(x)$, the median of x is defined by the value x_0 that satisfies

$$\int_{-\infty}^{x_0} f(x)\,dx = \frac{1}{2}.$$

Such a problem corresponds to the classical median test.

5.4 For the data in [Example 5.4], compare the two hypotheses shown in section 5.2.2 with the hypothesis that the distributions of voters' concern about a general election are equal only within the two cities, Tokyo and Machida.

5.5 Compare the following two frequency tables, taking note of their class boundaries.

Class Boundaries	Frequency
0. 00-2. 00	20
2. 00-4. 00	60
4. 00-6. 00	20
Total	100

Class Boundaries	Frequency
0. 00-3. 00	49
3. 00-6. 00	51
Total	100

5. 6 Find the maximum likelihood estimator of the parameter λ on the basis of n observation $(x_1, \cdots, x_n)$ taken from the exponential distribution defined by (5. 61). And then find the distribution function corresponding to (5. 61).

5. 7 From the class boundaries given in Figure 5. 4, the following frequency table for [Example 5. 7] was generated. Check from the table whether the data follow an exponential distribution? (The maximum likelihood estimator found in the preceding problem is available and the sample mean is known to be *23. 2*.)

Class Boundaries	Frequency
0. 015 - 10. 724	38
10. 742 - 37. 498	44
37. 498 - 64. 272	14
64. 272 -	4
Total	100

CHAPTER 6

CONTINGENCY TABLE ANALYSIS MODELS

6. 1 Check of Independence

[*Example 6. 1*] Table 6. 1 is an example from the survey results obtained by the 1978 nation-wide survey of the Japanese national character which was conducted by The Institute of Statistical Mathematics. Can we conclude from these data that the responses to the question "If you could be born again, would you rather be a man or a woman?" and sex of each respondent are independent?

Table 6. 1 "If you could be born again, would you rather be a man or a woman?"

Sex (I_2)	Men or Women ? (I_1) 1. Men	2. Women	Total
1. Male	749	83	832
2. Female	445	636	1081
Total	1194	719	1913

Table 6. 1, in which the frequencies of every classification for a given set of data are recorded, is a typical example of a contingency table. A contingency table is usually made for the purpose of studying the relationship between the classification variables. In particular, as we shall do in this example, one may wish to check the hypothesis that the two variables are

independent.

Suppose we denote the two variables by I_1 and I_2 and suppose that they have c_1 and c_2 categories, respectively. We assume that each variable I_j takes a value $i_j(i_j=1, \cdots, c_j)$, where i_j denotes the i_j-th category of variable I_j. It is clear that $c_1=c_2=2$ in [Example 6. 1]. Let $p(i_1, i_2)$ be the probability that the variables I_1 and I_2 take the values i_1 and i_2 respectively and let $n(i_1, i_2)$ be the corresponding cell frequency. If we denote the sample size by n, then

$$\sum_{i_1=1}^{c_1}\sum_{i_2=1}^{c_2} p(i_1, i_2)=1 \tag{6. 1}$$

$$\sum_{i_1=1}^{c_1}\sum_{i_2=1}^{c_2} n(i_1, i_2)=n \tag{6. 2}$$

and the probability $p(\{n(i_1, i_2)\}|\{p(i_1, i_2)\})$ of getting the cell frequencies $\{n(i_1, i_2)\}$ $(i_1=1, \cdots, c_1, i_2=1, \cdots, c_2)$ is obtained from the multinomial distribution as

$$P(\{n(i_1, i_2)\}|\{p(i_1, i_2)\})$$
$$=\frac{n!}{\prod_{i_1=1}^{c_1}\prod_{i_2=1}^{c_2} n(i_1, i_2)!}\prod_{i_1=1}^{c_1}\prod_{i_2=1}^{c_2} p(i_1, i_2)^{n(i_1, i_2)}. \tag{6. 3}$$

By putting

$$K_5= \log\{n!/\prod_{i_1}\prod_{i_2} n(i_1, i_2)!\}, \tag{6. 4}$$

the log likelihood $l(\{p(i_1, i_2)\})$ is given by

$$l(\{p(i_1, i_2)\})= K_5+\sum_{i_1=1}^{c_1}\sum_{i_2=1}^{c_2} n(i_1, i_2) \log p(i_1, i_2). \tag{6. 5}$$

Now, the hypothesis that the two variables I_1 and I_2 are independent is equivalent to

$$MODEL(0): \ p(i_1, i_2)=\theta(i_1, \cdot)\theta(\cdot, i_2) \ ,$$

where

$$\theta(i_1, \cdot)=\sum_{i_2=1}^{c_2} p(i_1, i_2) \tag{6.6}$$

$$\theta(\cdot, i_2)=\sum_{i_1=1}^{c_1} p(i_1, i_2). \tag{6.7}$$

Substituting (6.6) and (6.7) in (6.1), it is seen that *MODEL* (*0*) has the constraint

$$\sum_{i_1=1}^{c_1} \theta(i_1, \cdot)=\sum_{i_2=1}^{c_2} \theta(\cdot, i_2)=1. \tag{6.8}$$

The alternative with no restrictions is represented as

$$MODEL(1): \ p(i_1, i_2)=\theta(i_1, i_2).$$

From (6.1), this model has only one constraint

$$\sum_{i_1=1}^{c_1} \sum_{i_2=1}^{c_2} \theta(i_1, i_2)=1. \tag{6.9}$$

Under *MODEL*(*0*), the maximum likelihood estimators $\hat{\theta}(i_1, \cdot)$ and $\hat{\theta}(\cdot, i_2)$ of the parameters $\theta(i_1, \cdot)$ and $\theta(\cdot, i_2)$ are given by $\theta(i_1, \cdot)$ and $\theta(\cdot, i_2)$ which maximize the corresponding log likelihood

$$l(\{\theta(i_1, \cdot)\theta(\cdot, i_2)\})$$
$$=K_5+\sum_{i_1=1}^{c_1} \sum_{i_2=1}^{c_2} n(i_1, i_2) \log \theta(i_1, \cdot)\theta(\cdot, i_2). \tag{6.10}$$

Solving $\partial l(\{\theta(i_1, \cdot)\theta(\cdot, i_2)\})/\partial\theta(i_1, \cdot)=0$ and $\partial l(\{\theta(i_1, \cdot)\theta(\cdot, i_2)\})/\partial\theta(\cdot, i_2)=0$ under (6.8), we have the following maximum likelihood estimators

$$\hat{\theta}(i_1, \cdot)=\frac{n(i_1, \cdot)}{n} =\frac{\sum_{i_2=1}^{c_2} n(i_1, i_2)}{n} \qquad (6.11)$$

$$\hat{\theta}(\cdot, i_2)=\frac{n(\cdot, i_2)}{n} =\frac{\sum_{i_1=1}^{c_1} n(i_1, i_2)}{n}. \qquad (6.12)$$

Then, from the two constraints shown in (6.8), the number of free parameters is equal to $(c_1-1)+(c_2-1)$ and the corresponding AIC is given by

$$AIC^*(0)=(-2)\left[K_5+\sum_{i_1}\sum_{i_2} n(i_1, i_2) \log \frac{n(i_1, \cdot)n(\cdot, i_2)}{n^2}\right] +2(c_1+c_2-2). \qquad (6.13)$$

The maximum likelihood estimator for $MODEL(1)$ is found by a similar calculation to that given for the multinomial distribution in Chapter 3, and its AIC is

$$AIC^*(1)=(-2)\left[K_5+\sum_{i_1}\sum_{i_2} n(i_1, i_2) \log \frac{n(i_1, i_2)}{n}\right] +2(c_1c_2-1). \qquad (6.14)$$

Ignoring the common constant $(-2)K_5$ as before, the AIC's are reduced to

$$AIC(0)=(-2)\left[\sum_{i_1=1}^{c_1} n(i_1) \log n(i_1)+\sum_{i_2=1}^{c_2} n(i_2) \log n(i_2)-2n \log n\right] +2(c_1+c_2-2) \qquad (6.15)$$

$$AIC(1)=(-2)\left[\sum_{i_1=1}^{c_1}\sum_{i_2=1}^{c_2} n(i_1, i_2) \log n(i_1, i_2)-n \log n\right] +2(c_1c_2-1), \qquad (6.16)$$

where $n(i_1)=n(i_1, \cdot)$ and $n(i_2)=n(\cdot, i_2)$.

If we apply (6.15) and (6.16) to the data in [Example 6.1], we have

$$AIC(0)=(-2)[1194 \log 1194+\cdots+1081 \log 1081-2\times 1913 \log 1913]$$
$$+2(2+2-2)=5156.27$$
$$AIC(1)=(-2)[749 \log 749+\cdots+636 \log 636-1913 \log 1913]$$
$$+2(2\times 2-1)=4630.20 \ .$$

It is obvious that $AIC(1) < AIC(0)$ and that $MODEL(1)$ is a better model. From this result we should conclude that the responses to the question depend on sex. This coincides with the observations from Table 6. 1 which show that a clear majority of men (*90%*) preferred to be born men, but only *59%* of women chose to be born women.

This type of problem is traditionally handled by a test for independence based on the statistic

$$\chi^2=\sum_{i_1}\sum_{i_2}\{n(i_1, i_2)-n(i_1)n(i_2)/n\}^2/\{n(i_1)n(i_2)/n\}, \qquad (6.17)$$

which is approximately distributed as a chi-square distribution with $(c_1-1)(c_2-1)$ degrees of freedom when $MODEL(0)$ is true. The relation between this conventional test procedure and the procedure proposed in this section can be seen from the following consideration. Following the *MAICE* procedure we adopt the independence model if the value of (6. 15) is smaller than that of (6. 16), otherwise we adopt the dependence one. This means that we choose the model by the sign of $AIC(1)-AIC(0)$. Now,

$$AIC(1)-AIC(0)$$
$$=(-2)\sum_{i_1}\sum_{i_2}n(i_1, i_2) \log [n\cdot n(i_1, i_2)/\{n(i_1)n(i_2)\}]$$
$$+2(c_1-1)(c_2-1) \qquad (6.18)$$
$$\cong -\chi^2+2(c_1-1)(c_2-1), \qquad (6.19)$$

where the symbol $\cong$ means that the difference of both sides converges to 0 in probability. Taking into account the fact that the degrees of freedom of the χ^2 statistic are $(c_1-1)(c_2-1)$ and the expectation of χ^2 equals its degrees of freedom, checking the sign of (6.18) is asymptotically equivalent to the comparison of χ^2 with twice its expectation. The actual probability that a realization of a chi-square random variable exceeds twice its expectation varies with increasing degrees of freedom from *1* to *10* as follows: *0.1573, 0.1353, 0.1116, 0.0916, 0.0752, 0.0620, 0.0512, 0.0424, 0.0352, 0.0293*. From these probabilities we see that if the two variables are independent, the probability of adopting the independence model increases with the degrees of freedom, that is, with the number of independent cells. Only in the particular case where the degrees of freedom equal *7* does the present procedure give almost the same result as the conventional test procedure at the *5%* level of significance. In all other cases, the present procedure automatically adjusts the level of significance in the way illustrated above.

6.2 Comparison of Contingency Tables
—— Selection of Optimal Explanatory Variables

[*Example 6.2*] Table 6.2 is a three-way table pertaining to the same question as in [Example 6.1], but with respondents categorized by sex and age. Check which of the two tables, Tables 6.1 and 6.2, is more effective for predicting the distribution of responses to the question.

Table 6.2 "If you could be born again, would you rather be a man or a woman?"

Sex (I_2)	Age (I_3)	Men or Wemen ? (I_1) 1. Men	2. Women	Total
1. Male	1. Under 30 yrs.	165	22	187
	2. 30yrs. and over	584	61	645
2. Female	1. Under 30 yrs.	113	125	238
	2. 30yrs. and over	332	511	843
	Total	1194	719	1913

A table recording the cell frequencies concerning two variables, as in Table 6.1, is called a two-way table and a table recording the cell frequencies concerning three or more variables, as in Table 6.2, is called a multi-way table, or multidimensional contingency table.

In general, more detailed information on the distribution of a response variable can be obtained using more explanatory variables. However the sample size is finite and the estimated structure is subject to greater sampling variation as the number of explanatory variables increases. In such circumstances we might wish to determine an optimal dimension of explanatory variables to effectively estimate the distribution of the response variable. In this section we shall consider such a problem.

We denote by I_1, I_2 and I_3 the response variable, sex and age, respectively. We assume that I_j $(j=1, 2, 3)$ takes values i_j $(i_j=1, \cdots, c_j)$ as before. For this example, $c_1=c_2=c_3=2$. We further denote the joint probabilities by $p(i_1, i_2, i_3)$ and the cell frequencies by $n(i_1, i_2, i_3)$. In this chapter we use notation such as the following;

$$p(i_1, i_2)=\sum_{i_3=1}^{c_3} p(i_1, i_2, i_3)$$

$$n(i_1, i_3)=\sum_{i_2=1}^{c_2} n(i_1, i_2, i_3).$$

It is obvious that

$$\sum_{i_1, i_2, i_3} p(i_1, i_2, i_3)=\sum_{i_1, i_2} p(i_1, i_2)=\cdots=\sum_{i_3} p(i_3)=1 \qquad (6.20)$$

$$\sum_{i_1, i_2, i_3} n(i_1, i_2, i_3)=\sum_{i_1, i_2} n(i_1, i_2)=\cdots=\sum_{i_3} n(i_3)=n, \qquad (6.21)$$

where $\sum_{i_1, i_2, i_3}$ means $\sum_{i_1}\sum_{i_2}\sum_{i_3}$. Now the probability $p(\{n(i_1, i_2, i_3)\}|\{p(i_1, i_2, i_3)\})$ of observing the cell frequencies $\{n(i_1, i_2, i_3)\}$ $(i_j=1, \cdots, c_j, \ j=1, 2, 3)$ is expressed by the multinomial distribution

$$P(\{n(i_1, i_2, i_3)\} | \{p(i_1, i_2, i_3)\})$$
$$= \frac{n!}{\prod_{i_1, i_2, i_3} n(i_1, i_2, i_3)!} \prod_{i_1, i_2, i_3} p(i_1, i_2, i_3)^{n(i_1, i_2, i_3)}. \quad (6.22)$$

In cases such as [Example 6.2], we concentrate our attention on the relation between the response variable I_1 and the explanatory variables I_2 and I_3. If we denote by $p(i_1 | i_2, i_3)$ the conditional probability that I_1 takes a value i_1 given a set of values (i_2, i_3) of I_2 and I_3, then

$$p(i_1, i_2, i_3) = p(i_1 | i_2, i_3) p(i_2, i_3) \quad (6.23)$$

and the righthand side of (6.22) can be written as

$$\frac{n!}{\prod_{i_1, i_2, i_3} n(i_1, i_2, i_3)!} \prod_{i_1, i_2, i_3} \{p(i_1 | i_2, i_3) p(i_2, i_3)\}^{n(i_1, i_2, i_3)}.$$

Taking into account the fact that

$$\prod_{i_2, i_3} \{\prod_{i_1} p(i_2, i_3)^{n(i_1, i_2, i_3)}\} = \prod_{i_2, i_3} p(i_2, i_3)^{\sum_{i_1} n(i_1, i_2, i_3)}$$
$$= \prod_{i_2, i_3} p(i_2, i_3)^{n(i_2, i_3)},$$

the probability (6.22) is now given by

$$\left[\prod_{i_2, i_3} \left\{\frac{n(i_2, i_3)!}{\prod_{i_1} n(i_1, i_2, i_3)!} \prod_{i_1} p(i_1 | i_2, i_3)^{n(i_1, i_2, i_3)}\right\}\right]$$
$$\times \left\{\frac{n!}{\prod_{i_2, i_3} n(i_2, i_3)!} \prod_{i_2, i_3} p(i_2, i_3)^{n(i_2, i_3)}\right\}. \quad (6.24)$$

In (6.24), the first bracketed expression corresponds to the conditional probability of I_1 given a value (i_2, i_3) of I_2 and I_3, and the second bracketed expression corresponds to the probability of observing a value (i_2, i_3) of I_2 and I_3.

The probability (5.30) in section 5.2.2 is an example where the relevant probability is obtained only from the first bracketed expression in (6.24). This is due to the fact that in [Example 5.4] the conditioning variable is not a random variable. Contrary to this, in [Example 6.2], the conditioning or explanatory variable is a random variable whose probability is given by the second bracketed expression in (6.24). However, the problem in question is not the relation between I_2 and I_3, but the dependence of I_1 on I_2 and I_3. If we concentrate our attention on the first bracketed expression in (6.24), the conditional log likelihood $l(\{p(i_1 \mid i_2, i_3)\})$ with respect to the parameters $\{p(i_1 \mid i_2, i_3)\}$ is given by

$$l(\{p(i_1 \mid i_2, i_3)\})=K_6+\sum_{i_1, i_2, i_3} n(i_1, i_2, i_3) \log p(i_1 \mid i_2, i_3), \tag{6.25}$$

where

$$K_6=\log \{\prod_{i_2, i_3} n(i_2, i_3)! / \prod_{i_1, i_2, i_3} n(i_1, i_2, i_3)!\}$$

and

$$\sum_{i_1} p(i_1 \mid i_2, i_3)=1, \quad i_2=1, \cdots, c_2, \quad i_3=1, \cdots, c_3. \tag{6.26}$$

To simplify the comparison of contingency tables, we shall review the two-way table in the preceding section. If we denote by $p(i_1 \mid i_2)$ the conditional probability that the variable I_1 takes a value i_1 given that the variable I_2 has the value i_2, the probability $p(i_1 \mid i_2)$ is obtained from

$$p(i_1 \mid i_2)=p(i_1, i_2)/p(i_2). \tag{6.27}$$

The dependence of I_1 on I_2 is measured by how the conditional probability $p(i_1 \mid i_2)$ for each i_1 varies with i_2. If the two variables are not related, $p(i_1 \mid i_2)$ takes a constant value independently of i_2 and

$$p(i_1 \mid i_2)=p(i_1). \tag{6.28}$$

From (6. 27), it generally holds that

$$p(i_1, i_2)=p(i_1 \mid i_2)p(i_2) \tag{6.29}$$

and, substituting (6. 28) in this relation, we have

$$p(i_1, i_2)=p(i_1)p(i_2). \tag{6.30}$$

This is nothing but the independence model in the preceding section. Thus this shows that the independence model assumes that $p(i_1 \mid i_2)$ has the simplified structure given by (6. 28). In such a case, even if we knew the category of I_2 to which a response belongs, this does not supply any information on the category of I_1 to which it belongs.

We shall apply a similar idea to the problem concerning the comparison of contingency tables. Table 6. 1 may be regarded as a two-way table between I_1 and I_2. However, this table may also be derived from a three-way table provided the distribution of I_1, I_2 and I_3 satisfies certain assumptions. Suppose that, in the three-way table consisting of I_1, I_2 and I_3, the conditional probability $p(i_1 \mid i_2, i_3)$ can be determined by the value i_2 of sex I_2 independently of the value i_3 of age I_3. Then the information on age I_3 can be ignored in the determination of the relation of the question I_1 on sex I_2 and Table 6. 1 retains the full information that is contained in the three-way table which corresponds to Table 6. 2. Imposing such a restriction on $p(i_1 \mid i_2, i_3)$ in (6. 25) yields the model corresponding to Table 6. 1, namely

$$MODEL(I_1;I_2):\ p(i_1 \mid i_2, i_3)=\theta(i_1 \mid i_2),$$
$$i_j=1, \cdots, c_j, \quad j=1, 2, 3.$$

In Table 6. 2, the conditional probability $p(i_1 | i_2, i_3)$ cannot be determined without the set of values (i_2, i_3) and this yields the model

$$MODEL(I_1; I_2, I_3):\ p(i_1 | i_2, i_3) = \theta(i_1 | i_2, i_3),$$
$$i_j = 1, \cdots, c_j, \quad j = 1, 2, 3 .$$

$MODEL(I_1; I_2)$ implies that the corresponding conditional log likelihood is given by the product of c_2 multinomial distributions and $MODEL(I_1; I_2, I_3)$ implies that the corresponding log likelihood is obtained from the product of $c_2 c_3$ multinomial distributions. Hence the necessary maximum likelihood estimators are found in the same manner as in section 5. 2. 2 and are given by

$$\hat{\theta}(i_1 | i_2) = \frac{n(i_1, i_2)}{n(i_2)}$$

and

$$\hat{\theta}(i_1 | i_2, i_3) = \frac{n(i_1, i_2, i_3)}{n(i_2, i_3)} .$$

Now $MODEL(I_1; I_2)$ and $MODEL(I_1; I_2, I_3)$ have $c_1 c_2$ and $c_1 c_2 c_3$ parameters, respectively, but the number of free parameters are, from (6. 26), $(c_1 - 1) c_2$ and $(c_1 - 1) c_2 c_3$. Thus we have

$$AIC^*(I_1; I_2) = (-2) \left[K_6 + \sum_{i_1, i_2, i_3} n(i_1, i_2, i_3) \log \frac{n(i_1, i_2)}{n(i_2)} \right] + 2c_2(c_1 - 1) \quad (6.\ 31)$$

$$AIC^*(I_1; I_2, I_3) = (-2) \left[K_6 + \sum_{i_1, i_2, i_3} n(i_1, i_2, i_3) \log \frac{n(i_1, i_2, i_3)}{n(i_2, i_3)} \right] + 2c_2 c_3 (c_1 - 1). \quad (6.\ 32)$$

Taking into account the relations

$$\sum_{i_1, i_2, i_3} n(i_1, i_2, i_3) \log \frac{n(i_1, i_2)}{n(i_2)}$$

$$= \sum_{i_1, i_2} \left\{ \log \frac{n(i_1, i_2)}{n(i_2)} \right\} \sum_{i_3} n(i_1, i_2, i_3)$$

$$= \sum_{i_1, i_2} n(i_1, i_2) \log \frac{n(i_1, i_2)}{n(i_2)}$$

and ignoring the common constant as before, we have, after adding another common constant $-2\sum_{i_1} n(i_1) \log n/n(i_1) - 2(c_1 - 1)$, the AIC's

$$AIC(I_1 ; I_2) = (-2) \sum_{i_1, i_2} n(i_1, i_2) \log \frac{n \cdot n(i_1, i_2)}{n(i_1) n(i_2)} + 2(c_1 - 1)(c_2 - 1) \tag{6. 33}$$

$$AIC(I_1 ; I_2, I_3) = (-2) \sum_{i_1, i_2, i_3} n(i_1, i_2, i_3) \log \frac{n \cdot n(i_1, i_2, i_3)}{n(i_1) n(i_2, i_3)} + 2(c_1 - 1)(c_2 c_3 - 1). \tag{6. 34}$$

This modification brings the AIC's into correspondence with the ordinary likelihood ratio statistic.

We shall solve [Example 6. 2]. Substituting the values in Tables 6. 1 and 6. 2 into (6. 33) and (6. 34), respectively, we obtain

$$AIC(I_1 ; I_2) = (-2) \left\{ 749 \log \frac{1913 \times 749}{1194 \times 832} + \cdots + 636 \log \frac{1913 \times 636}{719 \times 1081} \right\} + 2(2-1)(2-1) = -526. 08$$

$$AIC(I_1 ; I_2, I_3) = (-2) \left\{ 165 \log \frac{1913 \times 165}{1194 \times 187} + \cdots + 511 \log \frac{1913 \times 511}{719 \times 843} \right\} + 2(2-1)(2 \times 2 - 1) = -527. 89.$$

Since that $AIC(I_1 ; I_2, I_3) < AIC(I_1 ; I_2)$, we see that $MODEL(I_1 ; I_2, I_3)$ is better than $MODEL(I_1 ; I_2)$. This shows that the combination of sex and age is more effective than the single predictor, sex, to estimate the distribution of the responses to the question. This result means that the responses to the question vary with the sex of a respondent and that they also depend on age. In fact, in Table 6.2, we see that the percentages of females that preferred to be born men differ according to age.

Suppose that we wish to compare Table 6.1 and the two-way table shown in Table 6.3. Assuming the model

$$MODEL(I_1;I_3):\ p(i_1 \mid i_2, i_3)=\theta(i_1 \mid i_3),$$

we have the corresponding AIC

$$AIC(I_1;I_3)=(-2)\Sigma n(i_1, i_3)\ log\frac{n\cdot n(i_1, i_3)}{n(i_1)n(i_3)}+2(c_1-1)(c_3-1). \tag{6.35}$$

This problem is easily solved by comparing the value of (6.35) with that of $AIC(I_1;I_2)$. Actual computational results obtained by substituting the values of Table 6.3 into (6.35) give $AIC(I_1;I_3)$ as -0.11. This value is much larger than that of $AIC(I_1;I_2)$. From this calculation we reach the conclusion that, for responses to this question, sex provides more effective predictive information than age. This is identical to our judgement that followed from the inspection of these tables.

Table 6.3 "If you could be born again, would you rather be a man or a woman?"

Age (I_3)	Men or Women ? (I_1) 1. Men	2. Women	Total
1. Under 30yrs.	278	147	425
2. 30yrs. and over	916	572	1488
Total	1194	719	1913

Generally, when a response variable I_1 and $k-1$ explanatory variables $I_2, \cdots, I_k$ are given, the evaluation of possible combination of explanatory variables can be realized by comparing the following $2^{k-1} (={}_{k-1}C_{k-1}+\cdots+{}_{k-1}C_0)$ models:

$$\begin{cases} MODEL(I_1; I_2, \cdots, I_k): p(i_1 \mid i_2, \cdots, i_k)=\theta(i_1 \mid i_2, \cdots i_k) \\ MODEL(I_1; I_2, \cdots, I_{k-1}): p(i_1 \mid i_2, \cdots, i_k)=\theta(i_1 \mid i_2, \cdots, i_{k-1}) \\ \quad \cdots\cdots \quad \cdots\cdots \\ MODEL(I_1; I_3, \cdots, I_k): p(i_1 \mid i_2, \cdots, i_k)=\theta(i_1 \mid i_3, \cdots, i_k) \\ \quad \vdots \quad\quad \vdots \\ MODEL(I_1; I_k): p(i_1 \mid i_2, \cdots, i_k)=\theta(i_1 \mid i_k) \\ \quad \cdots\cdots \quad \cdots\cdots \\ MODEL(I_1; I_2): p(i_1 \mid i_2, \cdots, i_k)=\theta(i_1 \mid i_2) \\ MODEL(I_1; \phi): p(i_1 \mid i_2, \cdots, i_k)=\theta(i_1) \end{cases} \tag{6.36}$$

If we denote any subset of the set of explanatory variables $I=\{I_2, \cdots, I_k\}$ by J and denote respective realization of I and J by i and j, each model in (6.36) is represented by

$$MODEL(I_1; J): \ p(i_1 \mid i)=\theta(i_1 \mid j). \tag{6.37}$$

As is seen from the preceding discussions, AIC corresponding to (6.37) is given by

$$AIC(I_1; J)=(-2) \sum_{i_1, j} n(i_1, j) \log \frac{n \cdot n(i_1, j)}{n(i_1) n(j)} + 2(c_1-1)(c_J-1), \tag{6.38}$$

where $n(j)$ and c_J denote the marginal frequency of a combination of explanatory variables J and the number of categories concerning J, respectively. In this case the symbol Σ denotes summation over all cells in a given table. We assume here that $n(\phi)=n$ and $c_\phi=1$. As an example, if we put $J=\{I_2, I_3\}$ in

(6. 38), we have $n(j)=n(i_2, i_3)$, $n(i_1, j)=n(i_1, i_2, i_3)$, $c_J=c_2c_3$ and (6. 34) results.

It is sometimes effective to use supplementary model

$$MODEL(I_1\,;\,\phi_{-1}) : p(i_1 \mid i_2, \cdots, i_k)=\frac{1}{c_1}.$$

This model assumes that the conditional probabilities uniformly take a constant value, and is referred to as a minus one dimensional model. Corresponding to the form of (6. 38) the necessary AIC is expressed as

$$AIC(I_1\,;\,\phi_{-1})=(-2)\Sigma n(i_1)\ log\frac{n}{n(i_1)c}-2(c_1-1).$$

The problem of checking for independence in the preceding section is interpreted as that of comparing $MODEL(I_1\,;I_2)$ and $MODEL(I_1\,;\,\phi)$ which are obtained by putting $I=J=\{I_2\}$ in (6. 37). Since $AIC(I_1\,;\,\phi)=0$, we have

$$AIC(I_1\,;I_2)-AIC(I_1\,;\,\phi) = AIC(I_1\,;I_2)\ .$$

The righthand side of this statistic is equivalent to (6. 33) and equals (6. 18) in the preceding section. Therefore the sign of the statistic (6. 38) indicates whether the relevant model is better than the model that assumes that I_1 and J are independent. Thus the statistic (6. 38) gives information on the independence between I_1 and J as well as the amount of information which J retains on I_1. As an example, since all AIC's for those models concerning [Example 6. 2] are negative, we know that all possible combinations of I_2 and I_3 are independent of I_1 and that the two-dimensional combination of I_2 and I_3 has the largest amount of information on I_1.

We should note here that in calculating the value of (6. 38) we do not always need the highest dimensional table given by a set of variables. For example, we do not need the three dimensional table, Table 6.2, for the comparison between Table 6.1 and Table 6.3. This means that the comparison of low dimensional tables is possible using only those given tables. Further, increasing the dimension of the explanatory variables means an increase in the value of c_J, and this in turn produces an increase in the value of (6.38) which results in an increased *AIC* value. Finally, we can easily find an effective set of explanatory variables by applying a stepwise selection procedure with respect to J as in the case of subset regression analysis even though we must check the effect over many explanatory variables. A FORTRAN program CATDAP-01 was developed to carry out this procedure. To save space, only a part of it, the program for the comparison of two-way tables, is given in the part III.

The efficiency of the estimate of the probability of a response variable depends not only on the combination of explanatory variables but also on how we categorize each one. However, the conclusion is that to search for an optimal pooling of categories of explanatory variables, we have only to apply the statistic (6.38) to each table with a different categorization and then pick out the categorization with the minimum *AIC*. We note that both the problems in section 5.1.2 and in section 5.2.2 can be solved using (6.38).

Finally, the present procedure can be used to search for an optimal combination of explanatory variables with an optimal categorization provided only that the response variable is categorical. It will even apply in the case where explanatory variables are continuous variables. For the latter case we developed the program CATDAP-02 which has not been listed here for want of space. For details of that procedure, see Katsura & Sakamoto(1980) and Sakamoto(1982) and for the further refinements of it, see Ishiguro & Sakamoto (1983) and Sakamoto & Ishiguro(1985).

PROBLEMS

6.1 Show that the statistic obtained by ignoring K_6 in (6.31) reduces to (6.33).

6.2 Table 6.4 is the two-way table relating to the question considered in [Example 6.1] and another question, "Which do you think gets the greater pleasure out of life, men or women ?" Out of three explanatory variables, sex, age and this last question, determine the most effective model for predicting responses to the original question considered in [Example 6.1].

Table 6.4

Greater Pleasure (I_4)	Men or Women ? (I_1) 1. Men	2. Women	Total
1. Men	875	391	1266
2. Women	319	328	647
Total	1194	719	1913

6.3 Suppose that in [Example 5.4] we denote concern about a general election and cities by I_1 and J respectively and then apply (6.38) to that example. Show that the value of $AIC(I_1;J)$ for this case is equal to the difference in value between (5.37) and (5.38), that is, *4456.98-4487.11=-30.13*.

6.4 Rearrange (6.38) in a convenient form for hand calculation.

CHAPTER 7

NORMAL DISTRIBUTION MODELS

7.1 Fitting a Normal Distribution

Figure 7.1 shows a graph of the standard normal density function. Short vertical lines at the bottom of the figure are the following 20 independent realizations from this distribution:

-1.76	-1.09	-0.87	-0.78	-0.49
-0.40	-0.36	-0.19	-0.06	0.11
0.46	0.54	0.54	0.91	1.24
1.31	1.31	1.64	1.76	1.97

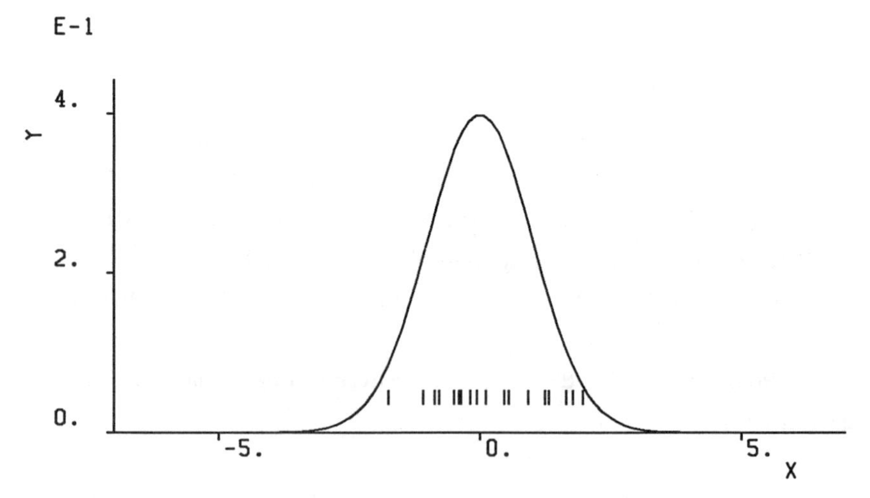

Figure 7.1 Normal density function and the first set of data

As is typically seen in this figure, the normal distribution is suitable to represent the distribution of data which scatter symmetrically around a point.

We have two other sets of data

-3. 64	-3. 44	-2. 78	-2. 20	-2. 02
-1. 81	-1. 68	-1. 55	-1. 37	-1. 09
-1. 05	-0. 82	-0. 81	-0. 79	-0. 78
-0. 78	-0. 72	-0. 21	0. 04	0. 30

and

0. 17	1. 60	1. 64	1. 65	2. 20
2. 35	2. 36	2. 44	2. 46	2. 62
2. 71	2. 78	2. 81	2. 89	2. 91
3. 00	3. 03	3. 72	3. 87	4. 53

which are shown in Figures 7. 2 and 7. 3. The appearance of the second set of data shown in Figure 7. 2 resembles that of the first set shown in Figure 7. 1. On the other hand, the third set of data shown in Figure 7. 3 whose minimum value is rather detached from the others, looks slightly different.

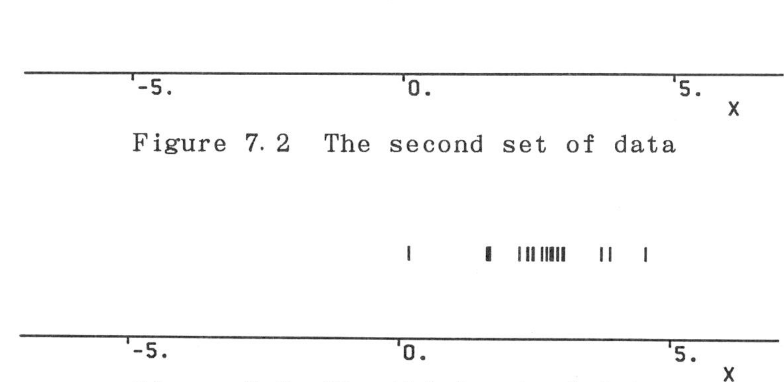

Figure 7. 2 The second set of data

Figure 7. 3 The third set of data

Let us fit the normal distributions to these three sets of data.

The density function of the normal distribution is given by

$$f(y \mid \mu, \sigma^2) = \frac{1}{\sqrt{2\pi\sigma^2}} e^{-(y-\mu)^2/2\sigma^2}, \tag{7.1}$$

where μ and σ^2 are parameters to be estimated. When data y_i $(i=1, \ldots, n)$ are obtained, the log likelihood is given by

$$\begin{aligned} l(\mu, \sigma^2) &= \sum_{i=1}^{n} \log \left\{ \frac{1}{\sqrt{2\pi\sigma^2}} e^{-(y_i-\mu)^2/2\sigma^2} \right\} \\ &= -\frac{n}{2} \log 2\pi - \frac{n}{2} \log \sigma^2 - \frac{1}{2\sigma^2} \left(\sum_{i=1}^{n} y_i^2 - 2\mu \sum_{i=1}^{n} y_i + n\mu^2 \right) \\ &= -\frac{n}{2} \log 2\pi - \frac{n}{2} \log \sigma^2 - \frac{n}{2\sigma^2} \left\{ \left(\mu - \frac{1}{n}\Sigma y_i \right)^2 + \frac{1}{n}\Sigma y_i^2 - \left(\frac{1}{n}\Sigma y_i \right)^2 \right\}. \end{aligned} \tag{7.2}$$

As derived in Chapter 3, the maximum likelihood estimators of μ and σ^2 are

$$\hat{\mu} = \frac{1}{n}\Sigma y_i \tag{7.3}$$

and

$$\hat{\sigma}^2 = \frac{1}{n}\Sigma y_i^2 - \left(\frac{1}{n}\Sigma y_i \right)^2, \tag{7.4}$$

respectively. Using these quantities, the log likelihood (7.2) can be rewritten into the form :

$$l(\mu, \sigma^2) = -\frac{n}{2} \log 2\pi - \frac{n}{2} \log \sigma^2 - \frac{n}{2\sigma^2} \left\{ (\mu - \hat{\mu})^2 + \hat{\sigma}^2 \right\}. \tag{7.5}$$

When μ is fixed, the maximum likelihood estimator $\tilde{\sigma}^2$ of σ^2 is given by

$$\tilde{\sigma}^2 = (\mu - \hat{\mu})^2 + \hat{\sigma}^2. \tag{7.6}$$

The maximum likelihood estimates of parameters obtained from the three sets of data are as follows :

	$\hat{\mu}$	$\hat{\sigma}^2$
Data of Figure 7.1	0.29	1.02^2
Data of Figure 7.2	-1.36	1.03^2
Data of Figure 7.3	2.59	0.90^2

The graphs of normal distributions with these values of the parameters are shown by ++++ in Figures 7.4 to 7.6 . The real density functions of each set of data are also shown in these Figures (solid line). In Figures 7.4 and 7.5 the fitted normal distributions are reproducing the real densities fairly well. In Figure 7.6, however, the discrepancy between the fitted distribution and the real distribution is large. The normal distribution model is not appropriate for this set of data. In

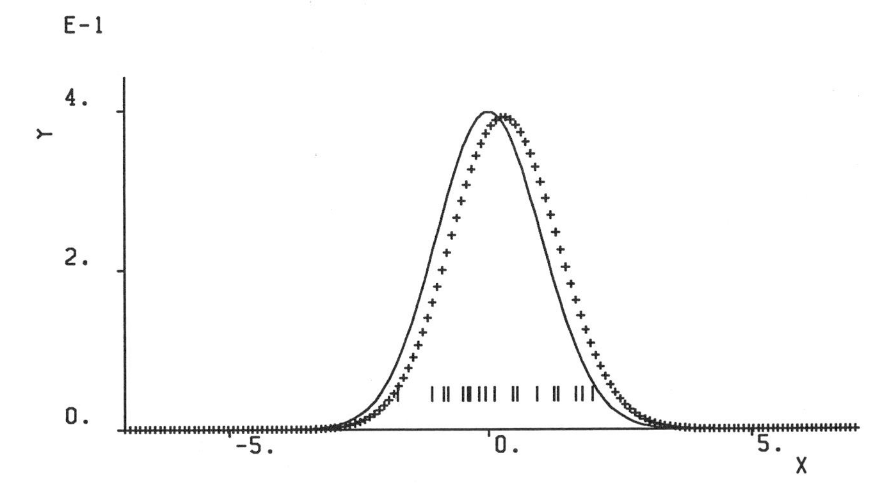

Figure 7.4 The normal distribution fitted to the first set of data

fact, the real density shown in Figure 7.5 is not normal, either. In this case, however, the approximation given by the normal density may well be good enough in practice.

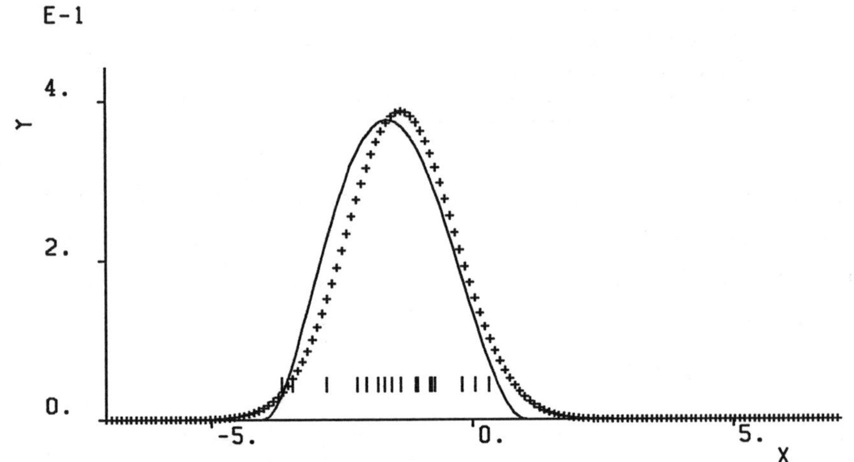

Figure 7.5 The normal distribution fitted to the second set of data

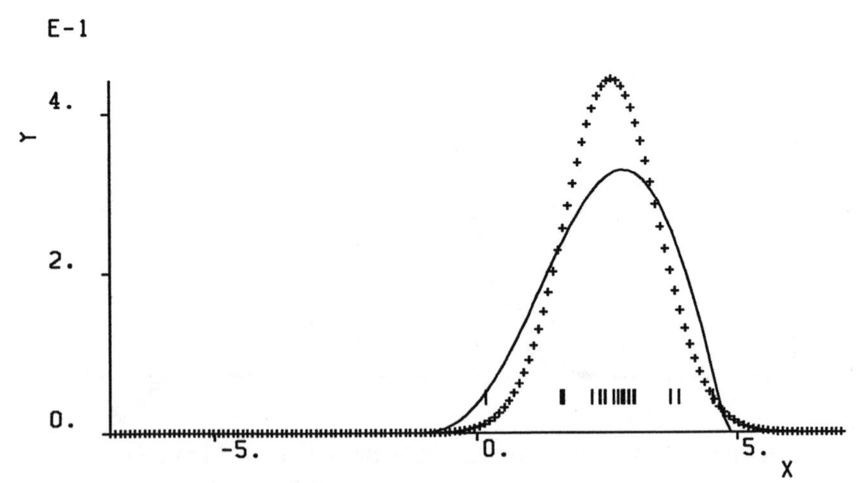

Figure 7.6 The normal distribution fitted to the third set of data

7.2 *Restricted Normal Distribution Models*

[*Example 7.1*] A certain machine produces ball bearings. It is known that the diameters of the products are distributed as a normal distribution with mean 1 cm and the standard deviation 0.01 cm when the machine is operating normally. One day 20 ball bearings were randomly chosen and their diameters measured. The results were as follows.

Table 7.1 Diameters of ball bearings

0.999	1.013	0.974	0.993	0.989
1.001	1.008	1.003	0.989	1.009
1.001	0.977	1.023	0.994	0.988
1.005	1.006	0.995	1.003	1.027

Can we conclude from this set of data that the machine is still operating normally?

If the machine had been operating normally, this set of data would be independently distributed as $N(1.0, 0.01^2)$ with density function given by

$$f(y \mid 1.0, 0.01^2) = \frac{1}{\sqrt{2\pi \times 0.01^2}} e^{-(y-1.0)^2/2(0.01)^2}. \qquad (7.7)$$

Let us examine whether this model or the un-restricted model

$$MODEL(\mu, \sigma^2) : f(y \mid \mu, \sigma^2) = \frac{1}{\sqrt{2\pi\sigma^2}} e^{-(y-\mu)^2/2\sigma^2} \qquad (7.8)$$

fits the data better.

Generalizing the problem we will compare the above model

with the restricted model

$$MODEL(\mu_0, \sigma_0^2) : f(y \mid \mu_0, \sigma_0^2) = \frac{1}{\sqrt{2\pi\sigma_0^2}} e^{-(y-\mu_0)^2/2\sigma_0^2}, \tag{7. 9}$$

where μ_0 and σ_0^2 are pre-assigned constants. (7. 7) is a special case of $\mu_0=1.0$, $\sigma_0^2=0.01^2$.

When the data y_i, $i=1, \ldots, n$ are given, the maximum log likelihood, or the log likelihood itself in this case, of $MODEL(\mu_0, \sigma_0^2)$ is given from (7. 5) by

$$l(\mu_0, \sigma_0^2) = -\frac{n}{2}\log 2\pi - \frac{n}{2}\log\sigma_0^2 - \frac{n}{2\sigma_0^2}\{(\mu_0-\hat{\mu})^2+\hat{\sigma}^2\}. \tag{7. 10}$$

The maximum log likelihood of $MODEL(\mu, \sigma^2)$ is

$$l(\hat{\mu}, \hat{\sigma}^2) = -\frac{n}{2}\log 2\pi - \frac{n}{2}\log\hat{\sigma}^2 - \frac{n}{2}, \tag{7. 11}$$

which is obtained by substituting $\hat{\mu}$ and $\hat{\sigma}^2$ for μ and σ^2 in (7. 5).

The number of free parameters of $MODEL(\mu_0, \sigma_0^2)$ and $MODEL(\mu, \sigma^2)$ are zero and two, respectively, and hence AIC's for these models are given by

$$\begin{aligned} AIC(\mu_0, \sigma_0^2) &= -2l(\mu_0, \sigma_0^2)+2\times 0 \\ &= n\left[\log 2\pi + \log\sigma_0^2 + \frac{1}{\sigma_0^2}\left\{(\mu_0-\hat{\mu})^2+\hat{\sigma}^2\right\}\right] \end{aligned} \tag{7. 12}$$

and

$$\begin{aligned} AIC(\mu, \sigma^2) &= -2l(\hat{\mu}, \hat{\sigma}^2)+2\times 2 \\ &= n\{\log 2\pi + \log\hat{\sigma}^2+1\}+4. \end{aligned} \tag{7. 13}$$

For [Example 7.1], we have

$$\Sigma y_i^2 = 19.997355, \quad \Sigma y_i = 19.997000, \quad n=20$$

$$\hat{\mu} = (19.997000/20) = 0.99985,$$

$$\hat{\sigma}^2 = (19.997355/20) - 0.99985^2 = 0.000168$$

and

$$(\mu_0 - \hat{\mu})^2 = 0.00015^2 = 2.25 \times 10^{-8}. \tag{7.14}$$

Thus the AIC's are calculated to be

$$AIC(\mu_0, \sigma_0^2) = 20\left\{\log 2\pi + \log 0.01^2 + \frac{1}{0.01^2}\{0.00015^2 + 0.000168\}\right\}$$

$$= -113.8$$

$$AIC(\mu, \sigma^2) = 20\{\log 2\pi + \log 0.000168 + 1\} + 4 = -113.1.$$

The smaller value of $AIC(\mu_0, \sigma_0^2)$ supports the conclusion that the ball bearing machine has been operating normally. The difference between AIC's, however, is merely 0.7.

We will further check two different models

$$MODEL(\mu_0, \sigma^2) : f(y \mid \mu_0, \sigma^2) = \frac{1}{\sqrt{2\pi\sigma^2}} e^{-(y-\mu_0)^2/2\sigma^2}$$

$$MODEL(\mu, \sigma_0^2) : f(y \mid \mu, \sigma_0^2) = \frac{1}{\sqrt{2\pi\sigma_0^2}} e^{-(y-\mu)^2/2\sigma_0^2},$$

where the first one has the restriction $\mu = \mu_0$ and the second one has $\sigma^2 = \sigma_0^2$.

When $MODEL(\mu_0, \sigma^2)$ is fitted, substituting μ_0 for μ in (7.6), the maximum likelihood estimator $\hat{\sigma}_1^2$ of σ^2 is given by

$$\hat{\sigma}_1^2 = (\mu_0 - \hat{\mu})^2 + \hat{\sigma}^2. \tag{7. 15}$$

The maximum log likelihood of $MODEL(\mu_0, \sigma^2)$ is obtained by replacing μ and σ^2 in (7. 5) by μ_0 and $\hat{\sigma}_1^2$, respectively. Thus we have

$$l(\mu_0, \hat{\sigma}_1^2) = -\frac{n}{2}\log 2\pi - \frac{n}{2}\log \hat{\sigma}_1^2 - \frac{n}{2}. \tag{7. 16}$$

Since the number of free parameters of this model is one, AIC is given by

$$AIC(\mu_0, \sigma^2) = n\{\log 2\pi + \log \hat{\sigma}_1^2 + 1\} + 2 \quad . \tag{7. 17}$$

The maximum likelihood estimator $\hat{\mu}_1$ of the free parameter μ of $MODEL(\mu, \sigma_0^2)$ is

$$\hat{\mu}_1 = \frac{\Sigma y_i}{n} = \hat{\mu}$$

and hence the maximum log likelihood of this model is, substituting $\hat{\mu}_1$ and $\hat{\sigma}_0^2$ for μ and σ^2 in (7. 5),

$$l(\hat{\mu}_1, \sigma_0^2) = -\frac{n}{2}\log 2\pi - \frac{n}{2}\log \sigma_0^2 - \frac{n\hat{\sigma}^2}{2\sigma_0^2}. \tag{7. 18}$$

Since the number of free parameters of this model is also one, we have

$$AIC(\mu, \sigma_0^2) = n\{\log 2\pi + \log \sigma_0^2 + \frac{\hat{\sigma}^2}{\sigma_0^2}\} + 2. \tag{7. 19}$$

For [Example 7. 1], values of AIC's, (7. 17) and (7. 19), are calculated, using values of (7. 14), to yield

$$AIC(\mu_0, \sigma^2) = 20\{ \log 2\pi + \log (0.00015^2 + 0.000168) + 1\} + 2$$
$$= -115.1$$
$$AIC(\mu, \sigma_0^2) = 20\left\{ \log 2\pi + \log 0.01^2 + \frac{0.000168}{0.01^2} \right\} + 2$$
$$= -111.8.$$

Summarizing all these results we have

MODEL	Restrictions	(Number of free parameters)	Estimate of μ	Estimate of σ	*AIC*	Diff. in *AIC*
(μ_0, σ^2)	$\mu=\mu_0$	(1)	1.0	0.013^2	-115.1	
						1.3
(μ_0, σ_0^2)	$\mu=\mu_0, \sigma^2=\sigma_0^2$	(0)	1.0	0.01^2	-113.8	
						0.7
(μ, σ^2)	*none*	(2)	0.9999	0.013^2	-113.1	
						1.3
(μ, σ_0^2)	$\sigma^2=\sigma_0^2$	(1)	0.9999	0.01^2	-111.8	

$AIC(\mu_0, \sigma^2)$ is less than $AIC(\mu_0, \sigma_0^2)$ by 1.3. This difference is considered to be significant. When $MODEL(\mu_0, \sigma^2)$ is adopted the estimate of σ^2 is 0.013^2. This means that the variance of the diameter of ball bearings had been greater than its normal value. The machine seems to have been out of control.

7.3 *Conditional Normal Distribution Models*

7.3.1 *Check of Identity of Two Distributions*

[*Example 7.2*] Two sets of data, Data set (*1*) and (*2*), are given. Can we conclude that these two sets of data are identically normally distributed?

DATA SET (1)

0.73	-0.06	1.04	2.29	0.51	-0.45
1.03	0.44	0.02	0.11	-2.42	

DATA SET (2)

0.10	0.56	-1.11	-0.48	3.46	-2.39
0.36	4.56				

The basic model is given by

$$MODEL(\mu_1, \mu_2, \sigma_1^2, \sigma_2^2) : f(y \mid \mu, \sigma^2) = \frac{1}{\sqrt{2\pi\sigma^2}} e^{-(y-\mu)^2/2\sigma^2} \tag{7.20}$$

where

$$(\mu, \sigma^2) = \begin{cases} (\mu_1, \sigma_1^2) & \text{if } y \text{ belongs to data set (1)} \\ (\mu_2, \sigma_2^2) & \text{if } y \text{ belongs to data set (2)} \end{cases}$$

This model implies that the two sets of data are differently distributed. This model is a conditional model with four free parameters which take values in the parameter space

$$\Theta = \{ (\mu_1, \sigma_1^2, \mu_2, \sigma_2^2) \mid \sigma_1^2 > 0, \sigma_2^2 > 0 \}. \tag{7.21}$$

If two sets of data are identically noramlly distributed,

$$MODEL(\mu, \mu, \sigma^2, \sigma^2) : f(y \mid \mu, \sigma^2) = \frac{1}{\sqrt{2\pi\sigma^2}} e^{-(y-\mu)^2/2\sigma^2} \tag{7.22}$$

with constraints $\mu_1 = \mu_2 = \mu$ and $\sigma_1^2 = \sigma_2^2 = \sigma^2$ should apply to both data sets (1) and (2). The number of free parameters of this model is two.

Let us examine which of the two models fits better. Denote

the observations of the k-th data set by $y_k(i)$, $i=1, 2, \ldots, n(k)$, $k=1, 2$. Then the conditional log likelihood of $MODEL(\mu_1, \mu_2, \sigma_1^2, \sigma_2^2)$ is given by

$$\begin{aligned}
l(\mu_1, \mu_2, \sigma_1^2, \sigma_2^2) &= \sum_{i=1}^{n(1)} \log f(y_1(i) \mid \mu_1, \sigma_1^2) + \sum_{i=1}^{n(2)} \log f(y_2(i) \mid \mu_2, \sigma_2^2) \\
&= -\frac{n(1)}{2} \log 2\pi - \frac{n(1)}{2} \log \sigma_1^2 \\
&\quad -\frac{1}{2\sigma_1^2}\Big(\sum_{i=1}^{n(1)} y_1(i)^2 - 2\mu_1 \sum_{i=1}^{n(1)} y_1(i) + n(1)\mu_1^2\Big) \\
&\quad -\frac{n(2)}{2} \log 2\pi - \frac{n(2)}{2} \log \sigma_2^2 \\
&\quad -\frac{1}{2\sigma_2^2}\Big(\sum_{i=1}^{n(2)} y_2(i)^2 - 2\mu_2 \sum_{i=1}^{n(2)} y_2(i) + n(2)\mu_2^2\Big) \\
&= -\frac{n}{2} \log 2\pi \\
&\qquad -\frac{n(1)}{2} \log \sigma_1^2 - \frac{1}{2\sigma_1^2}\Big\{\sum_{i=1}^{n(1)} y_1(i)^2 - 2\mu_1 y_1 + n(1)\mu_1^2\Big\} \\
&\qquad -\frac{n(2)}{2} \log \sigma_2^2 - \frac{1}{2\sigma_2^2}\Big\{\sum_{i=1}^{n(2)} y_2(i)^2 - 2\mu_2 y_2 + n(2)\mu_2^2\Big\},
\end{aligned} \tag{7. 23}$$

where

$$\begin{aligned}
n &= n(1) + n(2) \\
y_k &= \sum_{i=1}^{n(k)} y_k(i) \qquad k=1, 2.
\end{aligned} \tag{7. 24}$$

To maximize the log likelihood with respects to (μ_1, σ_1^2), we have only to maximize the first line of the rightmost side of (7. 23). Thus the maximum likelihood estimator of μ_1 and σ_1^2 are given respectively by

$$\hat{\mu}_1 = \frac{y_1}{n(1)} \qquad \text{and} \qquad \hat{\sigma}_1^2 = \frac{1}{n(1)} \sum_{i=1}^{n(1)} y_1(i)^2 - \hat{\mu}_1^2. \tag{7. 25}$$

The maximum likelihood estimator of (μ_2, σ_2^2) are likewise obtained, from the second line, to be

$$\hat{\mu}_2 = \frac{y_2}{n(2)} \quad \text{and} \quad \hat{\sigma}_2^2 = \frac{1}{n(2)} \sum_{i=1}^{n(2)} y_2(i)^2 - \hat{\mu}_2^2. \tag{7.26}$$

Hence, the maximum log likelihood is

$$l(\hat{\mu}_1, \hat{\mu}_2, \hat{\sigma}_1^2, \hat{\sigma}_2^2) = -\frac{n}{2}\log 2\pi - \frac{n(1)}{2}\log \hat{\sigma}_1^2 - \frac{n(2)}{2}\log \hat{\sigma}_2^2 - \frac{n}{2}. \tag{7.27}$$

When the restricted model, $MODEL(\mu, \mu, \sigma^2, \sigma^2)$, is assumed, the log likelihood reduces to

$$l(\mu, \sigma^2) = -\frac{n}{2}\log 2\pi - \frac{n}{2}\log \sigma^2 - \frac{1}{2\sigma^2}\left\{\sum_{i=1}^{n(1)} y_1^2(i) + \sum_{i=1}^{n(2)} y_2^2(i) - 2\mu(y_1+y_2) + n\mu^2\right\} \tag{7.28}$$

and the maximum likelihood estimators and the maximum log likelihood are given by

$$\hat{\mu} = \frac{y_1+y_2}{n}, \qquad \hat{\sigma}^2 = \frac{1}{n}\left\{\sum_{i=1}^{n(1)} y_1^2(i) + \sum_{i=1}^{n(2)} y_2^2(i)\right\} - \hat{\mu}^2 \tag{7.29}$$

and

$$l(\hat{\mu}, \hat{\sigma}^2) = -\frac{n}{2}\log 2\pi - \frac{n}{2}\log \hat{\sigma}^2 - \frac{n}{2}. \tag{7.30}$$

Then, the AIC's for $MODEL(\mu_1, \mu_2, \sigma_1^2, \sigma_2^2)$ and $MODEL(\mu, \mu, \sigma^2, \sigma^2)$ are

$$AIC(\mu_1, \mu_2, \sigma_1^2, \sigma_2^2) = n\log 2\pi + n(1)\log \hat{\sigma}_1^2 + n(2)\log \hat{\sigma}_2^2 + n + 2\times 4 \tag{7.31}$$

and

$$AIC(\mu, \mu, \sigma^2, \sigma^2) = n\log 2\pi + n\log \hat{\sigma}^2 + n + 2\times 2, \tag{7.32}$$

respectively. Note that (7.31) is equivalent to the sum of the *AIC* of $MODEL(\mu_1, \sigma_1^2)$ fitted to the data set (1),

$$AIC(\mu_1, \sigma_1^2) = n(1) \log 2\pi + n(1) \log \hat{\sigma}_1^2 + n(1) + 2\times 2$$

and the *AIC* of $MODEL(\mu_2, \sigma_2^2)$ fitted to the data set (2),

$$AIC(\mu_2, \sigma_2^2) = n(2) \log 2\pi + n(2) \log \hat{\sigma}_2^2 + n(2) + 2\times 2.$$

It follows that the *AIC* of $MODEL(\mu_1, \mu_2, \sigma_1^2, \sigma_2^2)$ is given by

$$AIC(\mu_1, \mu_2, \sigma_1^2, \sigma_2^2) = AIC(\mu_1, \sigma_1^2) + AIC(\mu_2, \sigma_2^2). \tag{7.33}$$

For [Example 7.2], we have

$$n(1) = 11, \qquad n(2) = 8$$

$$\sum_{i=1}^{n(1)} y_1(i)^2 = 14.4482, \qquad \sum_{i=1}^{n(2)} y_2(i)^2 = 40.3930$$

$$y_1 = \sum_{i=1}^{n(1)} y_1(i) = 3.24, \qquad y_2 = \sum_{i=1}^{n(2)} y_2(i) = 5.06$$

and

$$\hat{\mu}_1 = \frac{3.24}{11} = 0.2945$$

$$\hat{\sigma}_1^2 = \frac{14.4482}{11} - 0.2945^2 = 1.2267$$

$$\hat{\mu}_2 = \frac{5.06}{8} = 0.6325$$

$$\hat{\sigma}_2^2 = \frac{40.3930}{8} - 0.6325^2 = 4.6491$$

$$\hat{\mu} = \frac{3.24+5.06}{8+11} = 0.4368$$

$$\hat{\sigma}^2 = \frac{14.4482+40.3930}{8+11} - 0.4368^2 = 2.6956$$

and hence

$$AIC(\mu_1, \sigma_1^2) = 11(\log 2\pi + \log 1.2267 + 1) + 4 = 37.5$$

$$AIC(\mu_2, \sigma_2^2) = 8(\log 2\pi + \log 4.6491 + 1) + 4 = 39.0$$

$$AIC(\mu_1, \mu_2, \sigma_1^2, \sigma_2^2) = 37.5 + 39.0 = 76.5$$

$$AIC(\mu, \mu, \sigma^2, \sigma^2) = 19(\log 2\pi + \log 2.6956 + 1) + 4 = 76.8.$$

These results show that $MODEL(\mu_1, \mu_2, \sigma_1^2, \sigma_2^2)$ is better than the alternative which implies that the two distributions are identical. It suggests that the two sets of data are differently distributed. The difference between *AIC*'s, however, is only 0.3.

7.3.2. Comparison of Variances

Imposing a restriction $\sigma_1^2 = \sigma_2^2 = \sigma^2$ on $MODEL(\mu_1, \mu_2, \sigma_1^2, \sigma_2^2)$ we have the new model

$$MODEL(\mu_1, \mu_2, \sigma^2, \sigma^2) : f(y \mid \mu, \sigma^2) = \frac{1}{\sqrt{2\pi\sigma^2}} e^{-(y-\mu)^2/2\sigma^2}, \tag{7.34}$$

where

$$\mu = \begin{cases} \mu_1 & \text{if } y \text{ belongs to data set (1)} \\ \mu_2 & \text{if } y \text{ belongs to data set (2)} \end{cases}.$$

Comparing this model with $MODEL(\mu_1, \mu_2, \sigma_1^2, \sigma_2^2)$ we can check if two sets of data have the same variance. The conditional log likelihood of this model is given by

$$l(\mu_1, \mu_2, \sigma^2) = -\frac{n}{2}\log 2\pi - \frac{n}{2}\log \sigma^2$$
$$-\frac{1}{2\sigma^2}\{\sum_{i=1}^{n(1)} y_1(i)^2 - 2\mu_1 y_1 + n(1)\mu_1^2$$
$$+\sum_{i=1}^{n(2)} y_2(i)^2 - 2\mu_2 y_2 + n(2)\mu_2^2\}. \qquad (7.35)$$

The maximum likelihood estimators of the free parameters of this model are easily derived as

$$\hat{\mu}_1 = \frac{y_1}{n(1)}, \qquad \hat{\mu}_2 = \frac{y_2}{n(2)}$$
$$\hat{\sigma}_3^2 = \frac{1}{n}\{\sum_{i=1}^{n(1)} y_1(i)^2 - n(1)\hat{\mu}_1^2 + \sum_{i=1}^{n(2)} y_2(i)^2 - n(2)\hat{\mu}_2^2\}$$
$$= \frac{n(1)}{n}\hat{\sigma}_1^2 + \frac{n(2)}{n}\hat{\sigma}_2^2, \qquad (7.36)$$

where $\hat{\sigma}_1^2$ and $\hat{\sigma}_2^2$ are defined by (7.25) and (7.26), respectively. The maximum log likelihood and *AIC* are

$$l(\hat{\mu}_1, \hat{\mu}_2, \hat{\sigma}_3^2) = -\frac{n}{2}\log 2\pi - \frac{n}{2}\log \hat{\sigma}_3^2 - \frac{n}{2} \qquad (7.37)$$

and

$$AIC(\mu_1, \mu_2, \sigma^2, \sigma^2) = n\log 2\pi + n\log \hat{\sigma}_3^2 + n + 2\times 3, \qquad (7.38)$$

respectively.

For [Example 7.2], we have

$$\hat{\sigma}_3^2 = \frac{11}{19}\times 1.2267 + \frac{8}{19}\times 4.6491 = 2.6677$$

and hence

$$AIC(\mu_1, \mu_2, \sigma^2, \sigma^2) = 19(\log 2\pi + \log 2.6677 + 1) + 6 = 78.6.$$

This model is the worst of the three models considered so far. Thus we can conclude that the two sets of data have different variances.

7.3.3 *Comparison of Means*

Imposing the constraint $\mu_1=\mu_2=\mu$ on $MODEL(\mu_1, \mu_2, \sigma_1^2, \sigma_2^2)$, we have the model

$$MODEL(\mu, \mu, \sigma_1^2, \sigma_2^2) : f(y\,|\,\mu, \sigma^2) = \frac{1}{\sqrt{2\pi\sigma^2}} e^{-(y-\mu)^2/2\sigma^2}, \tag{7.39}$$

where

$$\sigma^2 = \begin{cases} \sigma_1^2 & \text{if } y \text{ belongs to data set (1)} \\ \sigma_2^2 & \text{if } y \text{ belongs to data set (2)} \end{cases}.$$

By comparing this model with $MODEL(\mu_1, \mu_2, \sigma_1^2, \sigma_2^2)$, we can check if the two sets of data have the same mean. The conditional log likelihood of this model is given by

$$\begin{aligned} l(\mu, \sigma_1^2, \sigma_2^2) &= -\frac{n}{2}\log 2\pi \\ &\quad -\frac{n(1)}{2}\log\sigma_1^2 - \frac{1}{2\sigma_1^2}\left\{\sum_{i=1}^{n(1)} y_1(i)^2 - 2\mu y_1 + n(1)\mu^2\right\} \\ &\quad -\frac{n(2)}{2}\log\sigma_2^2 - \frac{1}{2\sigma_2^2}\left\{\sum_{i=1}^{n(2)} y_2(i)^2 - 2\mu y_2 + n(2)\mu^2\right\} \\ &= -\frac{n}{2}\log 2\pi \\ &\quad -\frac{n(1)}{2}\log\sigma_1^2 - \frac{n(1)}{2\sigma_1^2}\{(\mu-\hat{\mu}_1)^2 + \hat{\sigma}_1^2\} \\ &\quad -\frac{n(2)}{2}\log\sigma_2^2 - \frac{n(2)}{2\sigma_2^2}\{(\mu-\hat{\mu}_2)^2 + \hat{\sigma}_2^2\}, \end{aligned} \tag{7.40}$$

where $\hat{\mu}_1$, $\hat{\mu}_2$, $\hat{\sigma}_1^2$ and $\hat{\sigma}_2^2$ are those defined by (7.25) and (7.26). From the three necessary conditions

$$\frac{\partial l}{\partial \mu} = 0, \quad \frac{\partial l}{\partial \sigma_1^2} = 0, \quad \frac{\partial l}{\partial \sigma_2^2} = 0$$

for the maximum likelihood estimates, we have the maximum likelihood equations :

$$\begin{aligned} &\mu^3 - A\mu^2 + B\mu - C = 0 \\ &\sigma_1^2 = (\mu - \hat{\mu}_1)^2 + \hat{\sigma}_1^2 \\ &\sigma_2^2 = (\mu - \hat{\mu}_2)^2 + \hat{\sigma}_2^2 . \end{aligned} \tag{7.41}$$

Here, defining p_1, p_2, D_1 and D_2 by

$$\begin{aligned} &p_1 = \frac{n(1)}{n}, \qquad p_2 = \frac{n(2)}{n} \\ &D_1 = p_1 \frac{\sum_{i=1}^{n(2)} y_2(i)^2}{n(2)}, \qquad D_2 = p_2 \frac{\sum_{i=1}^{n(1)} y_1(i)^2}{n(1)}, \end{aligned} \tag{7.42}$$

A, B and C are given by

$$\begin{aligned} &A = (1+p_1)\hat{\mu}_2 + (1+p_2)\hat{\mu}_1 \\ &B = 2\hat{\mu}_1\hat{\mu}_2 + D_1 + D_2 \\ &C = \hat{\mu}_1 D_1 + \hat{\mu}_2 D_2 \quad . \end{aligned} \tag{7.43}$$

The maximum likelihood estimator $\tilde{\mu}$ of μ is obtained as a solution of the equation (7.41). Once $\tilde{\mu}$ is obtained, the maximum likelihood estimators of σ_1^2 and σ_2^2 and the maximum log likelihood are calculated to be

$$\tilde{\sigma}_1^2 = (\tilde{\mu}-\hat{\mu}_1)^2+\hat{\sigma}_1^2$$

$$\tilde{\sigma}_2^2 = (\tilde{\mu}-\hat{\mu}_2)^2+\hat{\sigma}_2^2$$

$$l(\tilde{\mu}, \tilde{\sigma}_1^2, \tilde{\sigma}_2^2) = -\frac{n}{2}\log 2\pi-\frac{n(1)}{2}\log\tilde{\sigma}_1^2-\frac{n(2)}{2}\log\tilde{\sigma}_2^2-\frac{n}{2} \tag{7.44}$$

and hence the AIC of this model is

$$AIC(\mu, \mu, \sigma_1^2, \sigma_2^2) = n\log 2\pi+n(1)\log\tilde{\sigma}_1^2+n(2)\log\tilde{\sigma}_2^2+n+2\times 3. \tag{7.45}$$

[Newton's method]

Although there is a way to solve (7.41) directly, we will introduce here the more practical Newton procedure. Define

$$g(\mu) = \mu^3-A\mu^2+B\mu-C$$

$$g'(\mu) = 3\mu^2-2A\mu+B$$

and let ε be a properly chosen small positive constant. Then an approximate solution of (7.41) is obtained by the following algorithm:

1° $k=0;\ \mu_0 = \frac{n(1)}{n}\hat{\mu}_1+\frac{n(2)}{n}\hat{\mu}_2$
2° $g_k = g(\mu_k)$
3° $g'_k = g'(\mu_k)$
4° $\mu_{k+1} = \mu_k-g_k/g'_k$
5° Go to 7° if $|g_k/g'_k| \leqq \varepsilon$
6° Add 1 to k and go back 2°
7° Put $\tilde{\mu} = \mu_{k+1}$

In the case of [Example 7.2], we have

$p_1 = \frac{11}{19} = 0.5789$ $\qquad$ $p_2 = \frac{8}{19} = 0.4211$

$D_1 = 0.5789 \times \frac{40.3930}{8} = 2.9229$ $\quad$ $D_2 = 0.4211 \times \frac{14.4482}{11} = 0.5531$

$A = (1+0.5789)\times 0.6325 + (1+0.4211)\times 0.2945 = 1.4172$

$B = 2\times 0.6325\times 0.2945 + 2.9229 + 0.5531 = 3.8485$

$C = 0.2945\times 2.9229 + 0.6325\times 0.5531 = 1.2106$

and

$$g(\mu) = ((\mu - 1.4172)\mu + 3.8485)\mu - 1.2106$$

$$g'(\mu) = (3\mu - 2.8344)\mu + 3.8485$$

$$\mu_0 = \frac{11}{19}\times 0.2945 + \frac{8}{19}\times 0.6325 = 0.437.$$

When we set $\varepsilon = 0.0001$, Newton's method proceeds as shown in the following table:

k	μ_k	g_k	g'_k	g_k/g'_k
0	0.4370	0.2840	3.1828	0.0892
1	0.3478	-0.0015	3.2256	-0.0004
2	0.3482	-0.0002	3.2253	-0.0001
3	0.3483			

Thus, we have

$$\tilde{\mu} = 0.3483$$

$$\tilde{\sigma}_1^2 = (0.3483 - 0.2945)^2 + 1.2267 = 1.2296$$

$$\tilde{\sigma}_2^2 = (0.3483 - 0.6325)^2 + 4.6491 = 4.7299$$

and

$$AIC(\mu, \mu, \sigma_1^2, \sigma_2^2) = 19 \log 2\pi + 11 \log 1.2296 + 8 \log 4.7299 + 19 + 6$$
$$= 74.6.$$

The results for the four models fitted so far are summarized as follows:

MODEL	Restrictions (Number of free parameters)		AIC	Difference in AIC	Estimates of μ_1	μ_2	σ_1^2	σ_2^2
$(\mu, \mu, \sigma_1^2, \sigma_2^2)$	$\mu_1=\mu_2$	(3)	74.6		0.3483	0.3483	1.2296	4.7299
				1.9				
$(\mu_1, \mu_2, \sigma_1^2, \sigma_2^2)$	none	(4)	76.5		0.2945	0.6325	1.2267	4.6491
				0.3				
$(\mu, \mu, \sigma^2, \sigma^2)$	$\mu_1=\mu_2$, $\sigma_1^2=\sigma_2^2$	(2)	76.8		0.4368	0.4368	2.6956	2.6956
				1.8				
$(\mu_1, \mu_2, \sigma^2, \sigma^2)$	$\sigma_1^2=\sigma_2^2$	(3)	78.6		0.2945	0.6325	2.6677	2.6677

The best fit is $MODEL(\mu, \mu, \sigma_1^2, \sigma_2^2)$. We would conclude that the two sets of data have the same mean, but the variances are not equal.

7.4. *Correlation of Two-dimensional Data*

[*Example* 7.3] Suppose that 15 pairs of observations are obtained. It is known that (x, y) is distributed as a two-dimensional normal distribution. Can we regard the correlation coefficient as being 0?

(1.847, -1.700) (1.531, 0.226) (-0.498, 1.688)
(1.405, -0.312) (0.635, 0.274) (1.817, 0.749)
(0.498, 0.644) (2.266, 1.828) (-1.048, -1.370)
(0.926, -1.071) (0.509, -0.734) (0.237, 0.564)
(0.613, -0.819) (1.202, 1.783) (-0.819, 1.154)

The generic form of the two dimensional normal distribution model is given by

$$MODEL(1): f(\boldsymbol{z} \mid \mu, \Sigma) = \frac{1}{2\pi\sqrt{\det\Sigma}} e^{-(\boldsymbol{z}-\mu)^T\Sigma^{-1}(\boldsymbol{z}-\mu)/2}, \qquad (7.46)$$

where $z=(x, y)^T$ and $\mu=(\mu_x, \mu_y)^T$. Denote the correlation coefficient by ρ then

$$\Sigma = \begin{bmatrix} \sigma_x^2, & \rho\sqrt{\sigma_x^2\sigma_y^2} \\ \rho\sqrt{\sigma_x^2\sigma_y^2}, & \sigma_y^2 \end{bmatrix}.$$

The parameter space of the model (7.46) is

$$\Theta=\{(\mu_1, \mu_2, \sigma_x^2, \sigma_y^2, \rho) \mid \sigma_x^2>0, \sigma_y^2>0, -1<\rho<1\}.$$

The problem is which of the two models, *MODEL(1)* or the model with a constraint $\rho=0$, better fits data.

Generalizing the problem, let us denote the model with the constraint $\rho=\rho_0$ by *MODEL(0)*. When data z_i, $i=1, \ldots, n$ are given, the log likelihood of *MODEL(1)* is given by

$$\begin{aligned} l(\mu_1, \mu_2, \sigma_x^2, \sigma_y^2, \rho) &= -n\log 2\pi - \frac{n}{2}\log(\det\Sigma) - \frac{1}{2}\sum_{i=1}^{n}(z_i-\mu)^T\Sigma^{-1}(z_i-\mu) \\ &= -n\log 2\pi - \frac{n}{2}\log(\det\Sigma) - \frac{1}{2}\mathrm{tr}\left\{\Sigma^{-1}\sum_{i=1}^{n}(z_i-\mu)(z_i-\mu)^T\right\}. \end{aligned} \tag{7.47}$$

Introducing the following notation

$$Z = \sum_{i=1}^{n} z_i z_i^T, \quad \xi = \sum_{i=1}^{n} z_i, \quad R = Z - \frac{\xi\xi^T}{n} = \begin{bmatrix} R_{11} & R_{12} \\ R_{12} & R_{22} \end{bmatrix},$$

(7.47) is transformed as follows:

$$\begin{aligned} l(\mu_1, \mu_2, \sigma_x^2, \sigma_y^2, \rho) = &-n\log 2\pi - \frac{n}{2}\log(\det\Sigma) \\ &- \frac{1}{2}\mathrm{tr}[\Sigma^{-1}\{Z - \xi\mu^T - \mu\xi^T + n\mu\mu^T\}] \end{aligned}$$

$$= -n \log 2\pi - \frac{n}{2} \log (\det\Sigma)$$

$$-\frac{1}{2}\mathrm{tr}\left[\Sigma^{-1}\left\{\frac{(n\mu-\xi)(n\mu-\xi)^T}{n}+Z-\frac{\xi\xi^T}{n}\right\}\right]$$

$$= -n \log 2\pi - \frac{n}{2} \log (\det\Sigma)$$

$$-\frac{1}{2}\left[\left(\mu-\frac{\xi}{n}\right)^T\Sigma^{-1}\left(\mu-\frac{\xi}{n}\right)\right]-\frac{1}{2}\mathrm{tr}\Sigma^{-1}R. \qquad (7.48)$$

Thus the maximum likelihood estimator of μ is given by

$$\hat{\mu} = \frac{\xi}{n} = \frac{1}{n}\sum_{i=1}^{n} \boldsymbol{z}_i \qquad (7.49)$$

and putting $\mu=\hat{\mu}=(\hat{\mu}_x, \hat{\mu}_y)^T$ the log likelihood is reduced to

$$l(\hat{\mu}_1, \hat{\mu}_2, \sigma_x^2, \sigma_y^2, \rho)=-n \log 2\pi - \frac{n}{2} \log (\det\Sigma) - \frac{1}{2}\mathrm{tr}(\Sigma^{-1}R). \qquad (7.50)$$

Defining U by

$$U = \begin{bmatrix} u_{11} & u_{12} \\ u_{12} & u_{22} \end{bmatrix} \equiv \Sigma^{-1},$$

then putting each of partial derivatives of (7.50) with respects to u_{11}, u_{12} and u_{22} equal to zero, we have the maximum likelihood equations

$$n\frac{u_{11}}{\det U} = R_{11}, \quad -n\frac{u_{12}}{\det U} = R_{12} \text{ and } n\frac{u_{22}}{\det U} = R_2.$$

These equations imply

$$nU^{-1} = R.$$

Consequently, the maximum likelihood estimator of $\Sigma=U^{-1}$ is given by

$$\hat{\Sigma} = \frac{1}{n}R \ . \tag{7.51}$$

Substituting this value into (7. 50) we have the maximum log likelihood

$$l(\hat{\mu}_1, \hat{\mu}_2, \hat{\sigma}_x^2, \hat{\sigma}_y^2, \hat{\rho}) = -n \log 2\pi - \frac{n}{2} \log (\det\hat{\Sigma}) - n \tag{7.52}$$

of *MODEL(1)*.

The maximum likelihood estimator of μ is still given by (7. 49) even if we assumed *MODEL(0)*. In this case, however, we have

$$\Sigma = \begin{bmatrix} \sigma_x^2, & \rho_0\sqrt{\sigma_x^2\sigma_y^2} \\ \rho_0\sqrt{\sigma_x^2\sigma_y^2}, & \sigma_y^2 \end{bmatrix}$$

and

$$\det\Sigma = \sigma_x^2\sigma_y^2(1-\rho_0^2)$$

$$\Sigma^{-1} = \frac{1}{\det\Sigma}\begin{bmatrix} \sigma_y^2, & -\rho_0\sqrt{\sigma_x^2\sigma_y^2} \\ -\rho_0\sqrt{\sigma_x^2\sigma_y^2}, & \sigma_x^2 \end{bmatrix}.$$

By equating each of the partial derivatives of (7. 50) with respect to σ_x^2 and σ_y^2 to zero, the maximum likelihood equations are given by

$$\frac{R_{11}}{\sigma_x^2} - \frac{\rho_0 R_{12}}{\sqrt{\sigma_x^2\sigma_y^2}} = n(1-\rho_0^2)$$

$$\frac{R_{22}}{\sigma_y^2}-\frac{\rho_0 R_{12}}{\sqrt{\sigma_x^2\sigma_y^2}} = n(1-\rho_0^2).$$

these equations are solved to obtain the maximum likelihood estimators

$$\tilde{\sigma}_x^2 = \frac{\sqrt{R_{11}R_{22}}-\rho_0 R_{12}}{n(1-\rho_0^2)}\sqrt{\frac{R_{11}}{R_{22}}}$$

$$\tilde{\sigma}_y^2 = \frac{\sqrt{R_{11}R_{22}}-\rho_0 R_{12}}{n(1-\rho_0^2)}\sqrt{\frac{R_{22}}{R_{11}}}.$$

Thus the maximum log likelihood of *MODEL(0)* is given by

$$l(\hat{\mu}_1, \hat{\mu}_2, \tilde{\sigma}_x^2, \tilde{\sigma}_y^2) = -\frac{n}{2}\log \tilde{\sigma}_x^2\tilde{\sigma}_y^2-\frac{n}{2}\log(1-\rho_0^2)-n \quad . \qquad (7.53)$$

AIC's of *MODEL(1)* and *MODEL(0)* are obtained from (7.52) and (7.53) as

$$AIC(1)=2n\log 2\pi+n\log(\det\hat{\Sigma})+2n+2\times 5 \qquad (7.54)$$

$$AIC(0)=2n\log 2\pi+n\log(\tilde{\sigma}_x^2\tilde{\sigma}_y^2)+n\log(1-\rho_0^2)+2n+2\times 4 \qquad (7.55)$$

For [Example 7.3], we have

$$\xi=\begin{pmatrix}11.122\\ 2.904\end{pmatrix}$$

$$R=\begin{pmatrix}21.825600 & 2.826228\\ 2.826228 & 19.342520\end{pmatrix}-\frac{1}{15}\begin{pmatrix}11.122\\ 2.904\end{pmatrix}(11.122, 2.904)$$
$$=\begin{pmatrix}13.579008 & 0.673009\\ 0.673009 & 18.780306\end{pmatrix}.$$

From these values and $\rho_0 = 0$, (7.54) and (7.55) give

$$AIC(1) = 30 \log 2\pi + 15 \log \left(\frac{R_{11}R_{22}-R_{12}^2}{15^2}\right) + 30 + 10 = 97.0$$

$$AIC(0) = 30 \log 2\pi = 15 \log \left(\frac{R_{11}R_{22}}{15^2}\right) + 30 + 8 = 95.0$$

We would conclude that the correlation coefficient is equal to zero.

Note that when $\rho=0$, $MODEL(1)$ is equivalent to $MODEL(\mu_1, \mu_2, \sigma_1^2, \sigma_2^2)$ in the previous section. When ρ_0 in (7.55) is set equal to zero, and n, $n(1)$ and $n(2)$ in (7.31) are replaced by $2n$, n and n, respectively, (7.55) and (7.31) are equivalent.

PROBLEMS

7.1 Can we regard that the following set of 20 data are independent realizations from the normal distribution $N(0.0, 1.0)$?

1.347	0.264	-0.360	0.599	0.271
0.366	-0.972	-0.584	-0.643	0.960
-1.062	1.103	1.454	1.296	-0.024
0.721	-0.004	0.713	0.318	0.227

7.2 a) Derive estimator for σ_1^2 and σ_2^2 ; and AIC for the model obtained by imposing the additional constraint $\mu=0$ on $MODEL(\mu, \mu, \sigma_1^2, \sigma_2^2)$.

b) Fit this model to the data of [Example 7.2].

CHAPTER 8

REGRESSION MODELS

In this chapter, we introduce a model that describes the stochastic structure between several continuous variables. We will pay attention to one particular variable called the objective variable, and will assume that the conditional distribution of the objective variable given the other variables (explanatory variables) is represented by a normal distribution whose mean value is a function of the explanatory variables.

Specifically, we will introduce the polynomial regression model, the multiple regression model and the autoregressive model that is frequently used in time series analysis. We will see that the polynomial regression model is a special case of a multiple regression model and there is no essential reason why we should consider these models separately. The difference is that, in the case of the polynomial regression, the model is uniquely specified by the number of parameters; whereas in the multiple regression there are many possible combinations of explanatory variables for each number of parameters. We will use *AIC* as a tool for model selection. Inspite of the seeming repetition, we will first develop the polynomial regression in details so that reader can obtain a good grasp of this chapter.

The expressions for the maximum likelihood estimators of the regression models presented in this chapter are only for the theoretical understanding and for calculation by a hand calculator. We suggest that readers with access to a computer use the program REGRES given in Part III. This is based on an orthogonal transformation given in section 8. 5.

8. 1 *Polynomial Regression Model*

[*Example 8. 1*] Table 8. 1 shows the 21 pairs of x_i and y_i. Figure 8. 1 shows the plot of these observations. From this data set find the relation between x and y.

Table 8. 1

i	x_i	y_i
1	0. 00	0. 125
2	0. 05	0. 156
3	0. 10	0. 193
4	0. 15	-0. 032
5	0. 20	-0. 075
6	0. 25	-0. 064
7	0. 30	0. 006
8	0. 35	-0. 135
9	0. 40	0. 105
10	0. 45	0. 131
11	0. 50	0. 154
12	0. 55	0. 114
13	0. 60	-0. 094
14	0. 65	0. 215
15	0. 70	0. 035
16	0. 75	0. 327
17	0. 80	0. 061
18	0. 85	0. 383
19	0. 90	0. 357
20	0. 95	0. 605
21	1. 00	0. 499

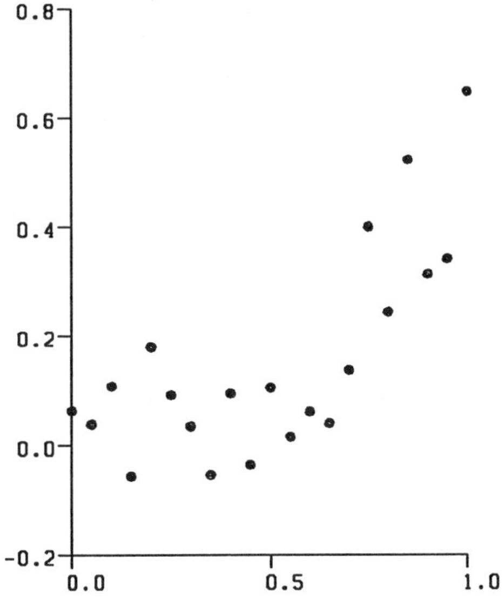

Figure 8. 1 Plot of (x_i, y_i), $i=1, \ldots, 21$.

In this situation, a common approach is to fit a function of the form

$$y = f(x), \tag{8. 1}$$

to the data. In particular, for data similar to that shown in Figure 8.1, we frequently fit a polynomial

$$y = a_0 + a_1x + a_2x^2 + \ldots + a_mx^m \qquad (8.2)$$

Here we introduce the polynomial regression model

$$y_i = a_0 + a_1x_i + \ldots + a_mx_i^m + \varepsilon_i, \qquad (8.3)$$

where ε_i is independent normal random variable that follows $N(0, \sigma^2)$. The variables x_i and y_i are called the explanatory variable and objective variable, respectively. The integer m is called the order of the polynomial regression model. This model explains the variation of y_i as the sum of a polynomial in the deterministic variable x_i and the random error ε_i.

The polynomial regression model of order m is a conditional distribution model of which, given the explanatory variable x_i, the distribution of the objective variable y is a normal distribution with the mean $a_0 + a_1x_i + \ldots. + a_mx_i^m$ and the variance σ^2; i.e.

$$f(y_i \mid a_0, \cdots, a_m, \sigma^2)$$
$$= \frac{1}{\sqrt{2\pi\sigma^2}} exp\left\{-\frac{1}{2\sigma^2}(y_i - a_0 - a_1x_i - \cdots - a_mx_i^m)^2\right\}. \qquad (8.4)$$

Hereafter we will denote the polynomal regression model of order m by *MODEL(m)*. *MODEL(0)* is the constant regression model of which the probability distribution of the objective variable is specified by the mean a_0 and the variance σ^2 irrespective of the explanatory variable x (Figure 8.2.1). This model has two parameters, a_0 and σ^2. *MODEL(1)* defines a probability distribution with the mean value $a_0 + a_1x_i$ that is the linear function of x_i (Figure 8.2.2). This model has 3 parameters, a_0, a_1 and σ^2. Similarly *MODEL(2)* is the quadratic regression model (Figure 8.2.3). For convenience, *MODEL(-1)* will denote the

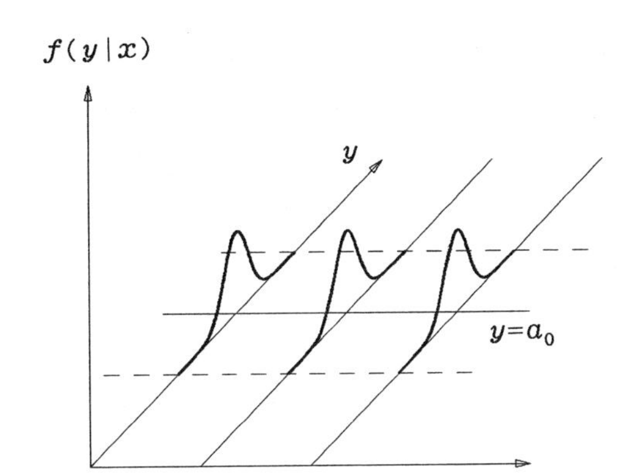

Figure 8.2.1

MODEL(0)

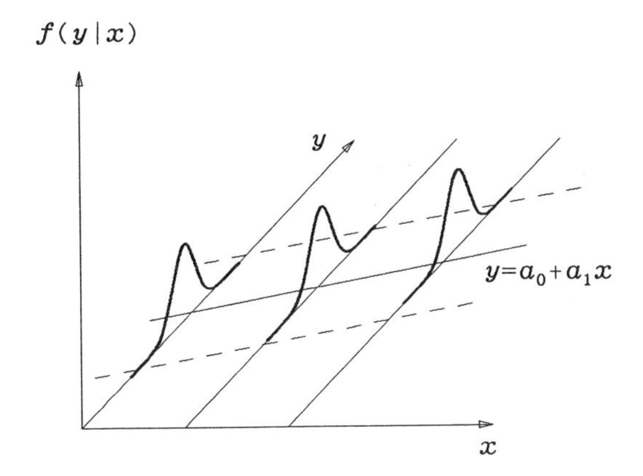

Figure 8.2.2

MODEL(1)

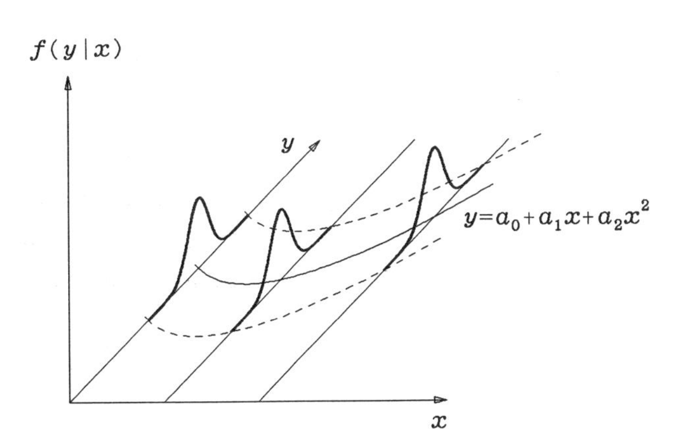

Figure 8.2.3

MODEL(2)

model for which the distribution of y is the normal distribution with mean zero and the variance σ^2 independent of the value of x. This model has only one parameter, σ^2. Fitting a polynomial to the data is equivalent to estimating a probability distribution for the objective variable y as a function of the objective variable x. The fitting of the polynomial regression model with order m, $MODEL(m)$, means that we are restricting this probability distribution to a particular type with $m+2$ parameters.

8.1.1 *Likelihood of a Polynomial Regression Model*

Given n pairs of observations $(x_1, y_1), \ldots, (x_n, y_n)$, we will fit

$$MODEL(m): \qquad y_i = a_0 + a_1 x_i + \cdots + a_m x_i^m + \varepsilon_i. \qquad (8.5)$$

As shown in the previous section, the probability density function of y_i that follows $MODEL(m)$ is given by (8.4). Therefore, given a set of n data points $\{(x_i, y_i); i=1, \ldots, n\}$, the likelihood is given by

$$\begin{aligned} &L(y_1, \ldots, y_n \mid a_0, \cdots, a_m, \sigma^2) \\ &= \prod_{i=1}^{n} f(y_i \mid a_0, \cdots, a_m, \sigma^2) \\ &= \prod_{i=1}^{n} \frac{1}{\sqrt{2\pi\sigma^2}} \exp\left\{ -\frac{1}{2\sigma^2} \sum_{i=1}^{n} (y_i - a_0 \cdots - a_m x_i^m)^2 \right\}. \\ &= \left(\frac{1}{2\pi\sigma^2}\right)^{\frac{n}{2}} \exp\left\{ -\frac{1}{2\sigma^2} \sum_{i=1}^{n} (y_i - a_0 - a_1 x_i - \cdots - a_m x_i^m)^2 \right\}. \end{aligned} \qquad (8.6)$$

Given data $(x_1, y_1), \ldots, (x_n, y_n)$, the values of $a_0, a_1, \ldots, a_m$ and σ^2 that maximize the likelihood constitute the maximum likelihood estimates of the regression coefficients and the residual variance of the polynomial regression model, $MODEL(m)$, respectively.

8.1.2 Maximum Likelihood Estimates of the Parameters

By taking the natural logarithm of the likelihood function we obtain the log likelihood

$$\begin{aligned} l(y \mid a_0, \cdots, a_m, \sigma^2) &= \log L(y_1, \cdots, y_n \mid a_0, \cdots, a_m, \sigma^2) \\ &= -\frac{n}{2} \log 2\pi - \frac{n}{2} \log \sigma^2 - \frac{1}{2\sigma^2} \sum_{i=1}^{n} (y_i - a_0 - \cdots - a_m x_i^m)^2. \end{aligned} \qquad (8.7)$$

To obtain the maximum likelihood estimates we must find the values of $a_0, a_1, \ldots, a_m$ and σ^2 that maximize the log likelihood. The log likelihood is maximized with respect to $a_0, a_1, \ldots, a_m$ when

$$S = \sum_{i=1}^{n} (y_i - a_0 - a_1 x_i - \cdots - a_m x_i^m)^2 \qquad (8.8)$$

is minimized. Thus in the case of the polynomial model fitting, the maximum likelihood method is equivalent to the ordinary least squares method.

The necessary conditions that $a_0, \ldots, a_m$ maximize S are

$$\begin{aligned} \frac{\partial S}{\partial a_0} &= -2 \sum_{i=1}^{n} (y_i - a_0 - a_1 x_i - \cdots - a_m x_i^m) = 0 \\ \frac{\partial S}{\partial a_1} &= -2 \sum_{i=1}^{n} x_i (y_i - a_0 - a_1 x_i - \cdots - a_m x_i^m) = 0 \\ &\vdots \\ \frac{\partial S}{\partial a_m} &= -2 \sum_{i=1}^{n} x_i^m (y_i - a_0 - a_1 x_i - \cdots - a_m x_i^m) = 0. \end{aligned} \qquad (8.9)$$

Thus the maximum likelihood estimates $\hat{a}_0, \ldots, \hat{a}_m$ are obtained by solving the system of linear equations (normal equations),

$$\begin{bmatrix} n & \Sigma x_i & \cdots & \Sigma x_i^m \\ \Sigma x_i & \Sigma x_i^2 & \cdots & \Sigma x_i^{m+1} \\ \cdot & \cdot & & \cdot \\ \cdot & \cdot & & \cdot \\ \cdot & \cdot & & \cdot \\ \Sigma x_i^m & \Sigma x_i^{m+1} & \cdots & \Sigma x_i^{2m} \end{bmatrix} \begin{bmatrix} a_0 \\ a_1 \\ \cdot \\ \cdot \\ \cdot \\ a_m \end{bmatrix} = \begin{bmatrix} \Sigma y_i \\ \Sigma x_i y_i \\ \cdot \\ \cdot \\ \cdot \\ \Sigma x_i^m y_i \end{bmatrix}, \qquad (8.10)$$

where Σ means $\sum_{i=1}^{n}$.

The necessary condition that σ^2 maximizes the log likelihood (8.7) is

$$\left.\frac{\partial l}{\partial \sigma^2}\right|_{\sigma^2=\hat{\sigma}^2} = -\frac{n}{2\hat{\sigma}^2} + \frac{1}{2(\hat{\sigma}^2)^2}\sum_{i=1}^{n}(y_i - \hat{a}_0 - \cdots - \hat{a}_m x_i^m)^2 = 0. \qquad (8.11)$$

Therefore the maximum likelihood estimate of the residual variance is obtained from

$$\begin{aligned} \hat{\sigma}^2 &= \frac{1}{n}\sum_{i=1}^{n}(y_i - \hat{a}_0 - \hat{a}_1 x_i - \cdots - \hat{a}_m x_i^m)^2 \\ &= \frac{1}{n}\left\{\sum_{i=1}^{n} y_i^2 - 2\sum_{i=0}^{m}\hat{a}_i \sum_{j=1}^{n} x_j^i y_j + \sum_{i=0}^{m}\hat{a}_i \sum_{j=0}^{m}\hat{a}_j \sum_{k=1}^{n} x_k^{i+j}\right\} \qquad (8.12) \\ &= \frac{1}{n}\left\{\sum_{i=1}^{n} y_i^2 - \sum_{i=0}^{m}\hat{a}_i \sum_{j=1}^{n} x_j^i y_j\right\}. \end{aligned}$$

Hereafter the residual variance σ^2 for the *MODEL*(m) will be denoted by $d(m)$.

From (8.7) and (8.12), the maximum likelihood is given by

$$l(y \mid \hat{a}_0, \ldots, \hat{a}_m, \hat{\sigma}^2) = -\frac{n}{2}\log 2\pi - \frac{n}{2}\log d(m) - \frac{n}{2}. \qquad (8.13)$$

Example of the estimates for some simple models are given below.

A. MODEL(-1)

$$d(-1) = \frac{1}{n}\sum_{i=1}^{n} y_i^2. \tag{8.14}$$

B. MODEL(0) (Constant regression model)

From the normal equation (8.10),

$$\hat{a}_0 = \frac{1}{n}\sum_{i=1}^{n} y_i,$$

$$d(0) = \frac{1}{n}\Big(\sum_{i=1}^{n} y_i^2 - \hat{a}_0 y_i\Big) \tag{8.15}$$

$$= \frac{1}{n}\sum_{i=1}^{n} y_i^2 - \Big(\frac{1}{n}\sum_{i=1}^{n} y_i\Big)^2.$$

C. MODEL(1) (Linear regression model)

From the normal equation

$$\begin{bmatrix} n & \Sigma x_i \\ \Sigma x_i & \Sigma x_i^2 \end{bmatrix} \begin{bmatrix} \hat{a}_0 \\ \hat{a}_1 \end{bmatrix} = \begin{bmatrix} \Sigma y_i \\ \Sigma x_i y_i \end{bmatrix}, \tag{8.16}$$

we obtain

$$\hat{a}_0 = \frac{\Sigma x_i^2 \Sigma y_i - \Sigma x_i \Sigma x_i y_i}{n\Sigma x_i^2 - (\Sigma x_i)^2}$$

$$\hat{a}_1 = \frac{n\Sigma x_i y_i - \Sigma x_i \Sigma y_i}{n\Sigma x_i^2 - (\Sigma x_i)^2} \tag{8.17}$$

$$d(1) = \frac{1}{n}\Big(\sum_{i=1}^{n} y_i^2 - \hat{a}_0\sum_{i=1}^{n} y_i - \hat{a}_1\sum_{i=1}^{n} x_i y_i\Big).$$

8.1.3 AIC

$MODEL(m)$ has $m+2$ free parameters, namely the regression coefficients $a_0, \ldots, a_m$ and the variance σ^2. Therefore, on substituting $k=m+2$ and (8.13) into

$$AIC = -2(\text{maximum log likelihood}) + 2k \qquad (8.18)$$

we obtain the AIC of $MODEL(m)$ as

$$AIC(m) = n \log 2\pi + n \log d(m) + n + 2(m+2). \qquad (8.19)$$

8.1.4 Numerical Example

Let us fit polynomial regression models to the data in [Example 8.1]. As preliminary computations, we obtain

$$\Sigma x_i = 10.5, \quad \Sigma x_i^2 = 7.175, \quad \Sigma x_i^3 = 5.5125,$$
$$\Sigma x_i^4 = 4.51666, \quad \Sigma y_i = 3.295, \quad \Sigma x_i y_i = 2.4904,$$
$$\Sigma x_i^2 y_i = 2.12248, \quad \Sigma y_i^2 = 1.23667.$$

The parameter estimates and AIC values for various models are given below.

A. $MODEL(-1)$ (*Normal distribution with mean 0*)

$$d(-1) = \frac{1}{n}\Sigma y_i^2 = \frac{1.23667}{21} = 0.0589$$
$$AIC(-1) = 11(\log 2\pi + 1 + \log 0.0589) + 2\times 1 = 2.12.$$

B. $MODEL(0)$ (*Normal distribution with mean a_0*)

From (8.15),

$$\hat{a}_0 = \frac{1}{n}\Sigma y_i = \frac{3.295}{21} = 0.1569$$

$$\begin{aligned} d(0) &= \frac{1}{n}(\Sigma y_i^2 - \hat{a}_m \Sigma y_i) \\ &= (1.23667 - 0.1569 \times 3.295)/21 = 0.0343 \\ AIC(0) &= 21(\log 2\pi + 1 + \log 0.0343) + 2\times 2 = -7.25. \end{aligned}$$

C. *MODEL(1)* (*Linear regression model*)

From (8.17), we obtain

$$\begin{aligned} \hat{a}_0 &= \frac{7.175\times 3.295 - 10.5\times 2.4904}{21\times 7.175 - 10.5\times 10.5} = -0.0620 \\ \hat{a}_1 &= \frac{21\times 2.4904 - 10.5\times 3.295}{21\times 7.175 - 10.5\times 10.5} = 0.4379 \\ d(1) &= \{1.23667-(-0.0620)(3.295) \\ &\qquad -(0.4379)(2.4904)\}/21 = 0.0167 \\ AIC(1) &= 21(\log 2\pi + 1 + \log 0.0167) + 2\times 3 = -20.35. \end{aligned}$$

D. *MODEL(2)* (*Quadratic regression model*)

The normal equations (8.10) become

$$\begin{aligned} 21a_0 + 10.5a_1 + 7.175a_2 &= 3.295 \\ 10.5a_0 + 7.175a_1 + 5.5125a_2 &= 2.4904 \\ 7.175a_0 + 5.5125a_1 + 3.85416a_2 &= 2.12248 \end{aligned}$$

Solving these, we obtain

$$\hat{a}_0 = 0.11164, \quad \hat{a}_1 = -0.65902, \quad \hat{a}_2 = 1.09689.$$

From (8.12), the residual variance is obtained as

$$d(2) = \frac{1}{n}\left\{\Sigma y_i^2 - \hat{a}_0\Sigma y_i - \hat{a}_1\Sigma x_i y_i - \hat{a}_2\Sigma x_i^2 y_i\right\}$$

$$= \{1.23667-(0.11164)(3.295) -(-0.65902)(2.3904) -(1.09689)(2.12248)\}/21$$

$$= 0.00866$$

and the *AIC* is given by

$$AIC(2) = -32.13.$$

Similarly, we can fit polynomial regression models for orders greater than two. Table 8.2 summarizes the values of the residual variance and *AIC* for each order.

Table 8.2 Summary of Fitted Models

Order	Number of free parameters	Residual variance	AIC
-1	1	0.05889	2.12
0	2	0.03427	-7.25
1	3	0.01669	-20.35
2	4	0.00866	-32.13
3	5	0.00839	-30.80
4	6	0.00800	-29.79
5	7	0.00798	-27.86

According to this table, the residual variance decreases as the order increases but that reduction is not so significant for orders greater than 2. $AIC(m)$ takes its minimum value -32.13 at $m = 2$ and then increases for $m > 2$. From this we conclude that for [Example 8.1]

$$y_t = 0.11164 - 0.65902x_t + 1.09689x_t^2 + \varepsilon_t \qquad (8.21)$$

is the best model.

In reality, the data given in [Example 8.1] were obtained by assuming the following regression model as the true structure

$$y_i = e^{(x_i-0.3)^2} - 1 + \varepsilon_i, \qquad (8.22)$$

where ε_i's are normal random numbers with mean zero and variance 0.01. Obviously this model cannot be expressed by finite order polynomial regression model. Figure 8.3 shows the 21 pairs of observations, fitted polynomials for orders 0 through 5 and the true regression curve

$$y_i = e^{(x_i-0.3)^2} - 1. \qquad (8.23)$$

We can see that the fitted curve varies with increasing order.

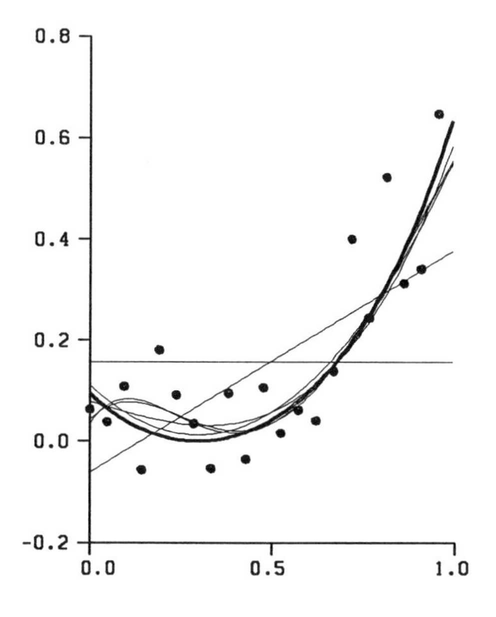

Figure 8.3

To obtain a better feel for the role of *AIC* in model selection, we performed the following numerical experiment. As in the case of [Example 8.1] we assumed that

$$y = e^{(x-0.3)^2} - 1 + \varepsilon$$
$$\sigma^2 = E\varepsilon^2 = 0.01 \qquad (8.24)$$

is the true model and obtained the realizations y_i for 21 different values of x, $x_i=0.05(i-1)$, $i=1, 2, \ldots, 21$, by Monte Carlo methods. We then fitted polynomial models of order 0 through 5. Figure 8.4.1 to Figure 8.4.6 show the fitted polynomials when we repeated this process 10 times. Figure 8.4.1 shows the fitted 0th order polynomials $y=a_0$. We see that the fitted 10 lines are far from the true regression curve $y = exp\{(x-0.3)^2\} - 1$. Figure 8.4.2 shows the fitted first order polynomials which are closer to the true regression curve. Figure 8.4.3 shows that the fitted second order polynomials reasonably approximate the true curve. Figure 8.4.4 and Figure 8.4.6 show that, with increased model order, the variation of the estimated polynomials becomes large and the estimates become unstable.

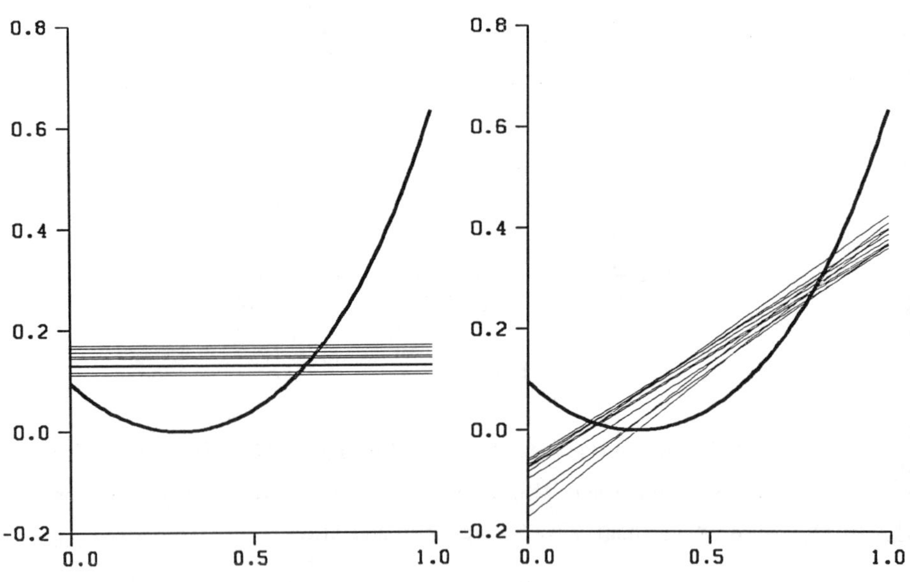

Fig. 8.4.1 0th order model Fig. 8.4.2 1st order model

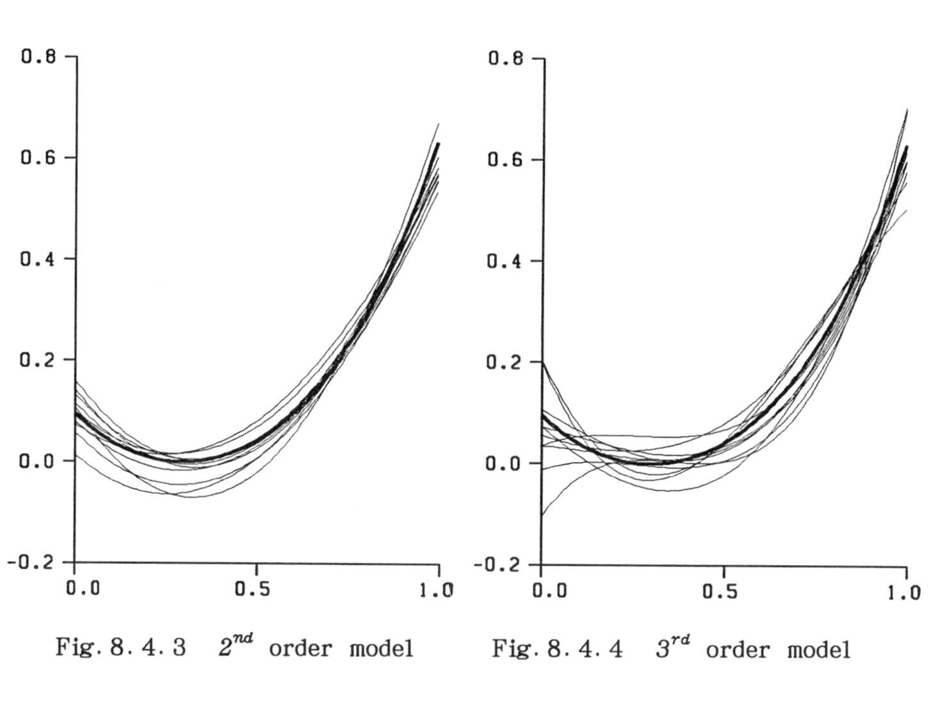

Fig. 8. 4. 3 2nd order model

Fig. 8. 4. 4 3rd order model

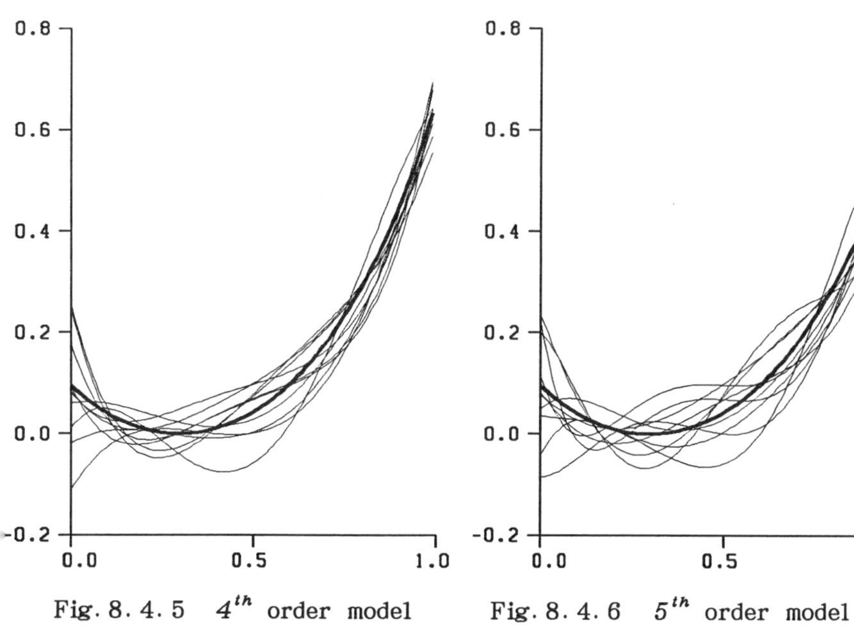

Fig. 8. 4. 5 4th order model

Fig. 8. 4. 6 5th order model

From this simulation study we conclude that with increasing model order, the bias from the true curve becomes small but the variance of the estimator becomes increasingly large. Thus it is necessary to make a compromise between two contradicting demands, the reduction of bias and the reduction of variance. As shown before, we resolve this using

$$AIC = -2(maximum\ log\ likelihood) + 2k. \tag{8.25}$$

Here the first term plus k expresses the bias of the best approximating model from the true distribution and the remaining k in the second term expresses the variance of the estimated model.

8.1.5 Restricted Model

There are various ways of improving the general polynomial regression model introduced above. Consider for example, the model

$$y_i = a(x_i + b)^2 + \varepsilon_i \tag{8.26}$$

for [Example 8.1]. This model is a restricted polynomial regression model and has only three parameters, a, b and σ^2. Whereas the general second order polynomial regression model $MODEL(2)$ has four parameters. If there is some reasons to believe that the minimum of the regression curve is 0, then we may get a better fit by using (8.26) since it has fewer parameters.

The least squares method yields

$$\hat{a} = 1.1515 \qquad \hat{b} = -0.3035 \qquad \hat{\sigma}^2 = 0.00875.$$

Therefore we have

$$AIC = 21(\log 2\pi + 1 + \log 0.00875) + 2\times 3$$
$$= -33.92$$

and we conclude that this model is better than the second order polynomial regression model shown in Table 8.2. In fact, the true curve given in (8.23) has its minimum value 0 at $x = 0.30$. Figure 8.5 shows the regression curve fitted to the same data used for fitting the models in Figures 8.4.1 to 8.4.6. It is obvious that the estimated curves are much closer to the true one than the ones in Figure 8.4.3.

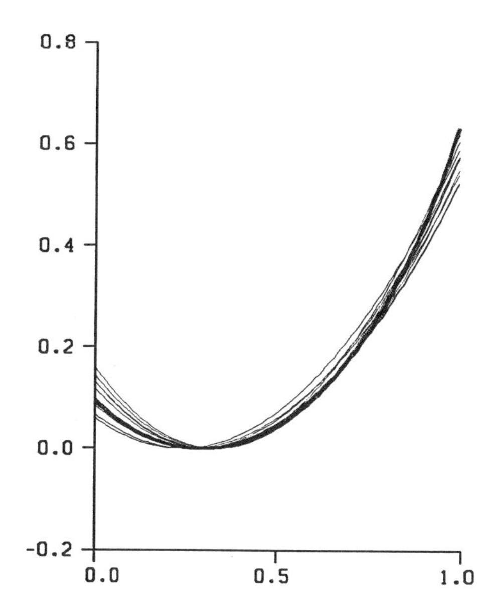

Figure 8.5 Estimated regression curves by the model (8.26)

8.2 *Multiple Regression Model*

[*Example 8.2*] The table below gives the daily minimum temperatures in January averaged from 1941 through 1970, the latitude, longitude and altitude of twenty cities in Japan.

(1) Find a model to estimate the average daily minimum temperature in January.

(2) Estimate the average minimum temperature in January for Fukushima ($x_1 = 37.73$, $x_2 = 140.9$ and $x_3 = 38.9$) and Kobe ($x_1 = 34.68$, $x_2 = 135.18$ and $x_3 = 59.3$).

No.	Cities	Temperature (y)	Latitude (x_1)	Longitude (x_2)	Altitude (x_3)
1	Wakkanai	-8.0	45.42	141.68	2.8
2	Asahikawa	-13.6	43.77	142.37	111.9
3	Sapporo	-9.5	43.05	141.33	17.2
4	Aomori	-5.4	40.82	140.78	3.0
5	Morioka	-6.7	39.70	141.17	155.2
6	Sendai	-3.2	38.27	140.90	38.9
7	Kanazawa	-0.1	36.55	136.65	26.1
8	Nagano	-5.5	36.67	138.20	418.2
9	Takayama	-7.6	36.15	137.25	560.2
10	Karuizawa	-10.0	36.33	138.55	999.1
11	Nagoya	-0.9	35.17	136.97	51.1
12	Iida	-4.7	35.52	137.83	481.8
13	Tokyo	-0.4	35.68	139.77	5.3
14	Tottori	0.5	35.48	134.23	7.1
15	Kyoto	-0.6	35.02	135.73	41.4
16	Hiroshima	0.2	34.37	132.43	29.3
17	Fukuoka	1.5	33.58	130.38	2.5
18	Kagoshima	2.0	31.57	130.55	4.3
19	Kochi	0.1	33.55	133.53	1.9
20	Naha	13.5	26.23	127.68	34.9

(From Chronological Table of Physics)

Suppose we have m explanatory variables $x_1, \ldots, x_m$. The model that expresses the value of the objective variable y as

$$MODEL(x_1, \cdots, x_m): \quad y = a_0 + a_1 x_1 + \cdots + a_m x_m + \varepsilon, \qquad (8.27)$$

is called the multiple regression model. The variable ε is called the residual and assumed to be a normal random variable with mean zero and variance σ^2.

Given a set of n independent observations $\{(y_i, x_{1i}, \cdots, x_{mi});\ i=1, \cdots, n\}$, the likelihood of the multiple regression model is

$$L(a_0, a_1, \cdots, a_m, \sigma^2)$$
$$= \left(\frac{1}{2\pi\sigma^2}\right)^{\frac{n}{2}} exp\left\{-\frac{1}{2\sigma^2}\sum_{i=1}^{n}\left(y_i - a_0 - \sum_{j=1}^{m} a_j x_{ji}\right)^2\right\}. \qquad (8.28)$$

Thus the log likelihood is given by

$$l(a_0, a_1, \cdots, a_m, \sigma^2)$$
$$= -\frac{n}{2} \log 2\pi\sigma^2 - \frac{1}{2\sigma^2}\sum_{i=1}^{n}\left(y_i - a_0 - \sum_{j=1}^{m} a_j x_{ji}\right)^2, \qquad (8.29)$$

and the maximum likelihood estimators $\hat{a}_0, \hat{a}_1, \ldots, \hat{a}_m$ of the regression coefficients $a_0, a_1, \ldots, a_m$ are obtained as the solution to the system of linear equations

$$\begin{bmatrix} n & \Sigma x_{1i} & \cdots & \Sigma x_{mi} \\ \Sigma x_{1i} & \Sigma x_{1i}^2 & \cdots & \Sigma x_{1i} x_{mi} \\ \cdot & \cdot & \cdot & \cdot \\ \cdot & \cdot & \cdot & \cdot \\ \cdot & \cdot & \cdot & \cdot \\ \Sigma x_{mi} & \Sigma x_{mi} x_{1i} & \cdots & \Sigma x_{mi}^2 \end{bmatrix} \begin{bmatrix} a_0 \\ a_1 \\ \cdot \\ \cdot \\ \cdot \\ a_m \end{bmatrix} = \begin{bmatrix} \Sigma y_i \\ \Sigma x_{1i} y_i \\ \cdot \\ \cdot \\ \cdot \\ \Sigma x_{mi} y_i \end{bmatrix}. \qquad (8.30)$$

The maximum likelihood estimate $\hat{\sigma}^2$ is

$$\hat{\sigma}^2 = \frac{1}{n}\left\{\sum_{i=1}^{n} y_i^2 - \hat{a}_0 \sum_{i=1}^{n} y_i - \sum_{j=1}^{m} \hat{a}_j \sum_{i=1}^{n} x_{ji} y_i\right\}. \qquad (8.31)$$

On substituting this into (8.29), the maximum log likelihood is given by

$$l(\hat{a}_0, \hat{a}_1, \cdots, \hat{a}_m, \hat{\sigma}^2) = -\frac{n}{2}\log 2\pi - \frac{n}{2}\log d(x_1, \cdots, x_m) - \frac{n}{2}, \quad (8.32)$$

where $d(x_1, \ldots, x_m)$ is the residual variance, σ^2, of the model with the explanatory variables $x_1, \ldots, x_m$ given in (8.31).

Since the number of free parameters contained in the multiple regression model is $m+2$, AIC for this model is

$$AIC(x_1, \cdots, x_m) = n(\log 2\pi + 1) + n\log d(x_1, \cdots, x_m) + 2(m+2). \quad (8.33)$$

In multiple regression analysis, all of the given explanatory variables are not necessarily effective for predicting the objective variable. An estimated model with an unnecessarily large number of explanatory variables may be unstable. By selecting the model with minimum AIC among possible combinations of the explanatory variables, we can get a reasonable model.

Consider the problem of [Example 8.2]. For the full order model with all the explanatory variables, the normal equations (8.30) are

$$\begin{bmatrix} 20 & 732.90 & 2737.98 & 2992.20 \\ & 27222.42 & 100647.61 & 109383.46 \\ & & 375169.76 & 413442.23 \\ & & & 1764671.60 \end{bmatrix} \begin{bmatrix} a_0 \\ a_1 \\ a_2 \\ a_3 \end{bmatrix} = \begin{bmatrix} -58.40 \\ -2558.17 \\ -8392.72 \\ -21283.70 \end{bmatrix}.$$

Solving these equations yields the full order model. Here, since the matrix is symmetric, only the elements above the diagonal are shown. From (8.31) and $\Sigma y_i^2 = 823.74$, we can get $\hat{\sigma}^2$. The lower order models are also obtained by solving linear equations which are obtained by eliminating appropriate rows and columns from the above system of equations. Table 8.3 summarizes the obtained results. From this table we see that the AIC for the

MODEL 1 which has latitude x_1, altitude x_3 and the constant term as explanatory variables is the minimum. *MODEL 2* which includes all explanatory variables is the second best. However, since the reduction of the residual variance is minuscule compared to that of the *AIC* best model, it is evident that information on the latitude x_2 is almost useless if we have information about x_1 and x_3. The *AIC* of *MODEL 7* which includes only x_3 is very large, whereas the *AIC* values of the models that include x_1 and x_3 or x_2 and x_3 (*MODEL 1* and *MODEL 3*) are less than that of the *MODEL 4* which includes x_1 and x_2. From this we deduce that x_1 and x_2 have similar information whereas x_3 has independent information. This is consistent with the geography of Japan. Since it runs from northeast to southwest, x_1 and x_2 have strong positive correlation. Thus they might be expected to have similar effect on temperature.

Table 8. 3

MODEL No.	*Explanatory variables*	*Order*	*Residual variance*	*AIC*	*Regression coefficients* a_0	a_1	a_2	a_3
1	1, x_1, x_3	3	2. 46	82. 8	40. 7	-1. 15	-	-0. 0098
2	1, x_1, x_2, x_3	4	2. 46	84. 8	38. 3	-1. 17	0. 02	-0. 0098
3	1, x_2, x_3	3	7. 01	103. 7	147. 1	-	-1. 09	-0. 0064
4	1, x_1, x_2	3	7. 74	105. 7	94. 3	-0. 69	-0. 52	-
5	1, x_1	2	8. 73	106. 1	39. 0	-1. 14	-	-
6	1, x_2	2	9. 60	108. 0	155. 8	-	-1. 16	-
7	1, x_3	2	26. 68	128. 4	-1. 5	-	-	-0. 0095
8	1	1	32. 66	130. 5	-2. 9	-	-	-
9	*None*	0	41. 19	133. 1	-	-	-	-

The minimum AIC model is

$$y = 40.7 - 1.15x_1 - 0.0098x_3 + \varepsilon$$
$$\varepsilon \sim N(0, 2.46).$$

The regression coefficient on the altitude x_3, -0.0098, is about 50% larger than the one given by the convensional wisdom in meteorology which suggests that the temperature should drop by 6℃ with a rise in altitude of 1000 meters. We shall follow this up in section 8.4.

Next let us estimate the minimum temperature at Fukushima and Kobe. For Fukushima, $x_1 = 37.75$ and $x_3 = 67.4$, we have

$$\hat{y} = 40.7 - (1.15)(37.75) - (0.0098)(67.4) = -3.37.$$

For Kobe, $x_1 = 34.68$ and $x_3 = 59.3$,

$$\hat{y} = 40.7 - (1.15)(34.68) - (0.0098)(59.3) = 0.24.$$

These estimates are in good accordance with the actual average values of -3.1 for Fukushima and 1.2 for Kobe. Incidentally, the estimates from the second best model which includes all explanatory variables are -3.43 for Fukushima and -0.15 for Kobe. Both have larger prediction errors than that for *MODEL 1*.

8.3 *Autoregressive Model*

A sequence of observations from a stochastic phenomenon that fluctuates irregularly with time is called a time series. Measurements such as roll of a ship cruising under wave and wind, inputs and outputs of a chemical plant and annual or monthly economic statistics are typical examples of time series. Time series analysis is used to analyse the structure of the stochastic dynamic systems which generate these time series and eventually to predict or control the future observations.

[*Example 8.3*] The records of the daily average temperature in Tokyo over the period July 1, 1979 to July 31, 1979 are given as follows.

24.4 27.3 25.5 22.0 24.0 26.0 25.5 24.2 24.8 25.4
26.1 27.2 23.4 20.9 21.6 20.6 22.3 20.4 22.0 21.6
24.4 26.9 28.1 28.8 28.2 27.8 26.4 27.6 28.8 29.1
29.9

A plot of the data against date is given in Figure 8.6.

(a) Find a model to predict the average temperature for a certain day based on the record up to the previous day.
(b) The records of the average temperature in July, 1980 are given as follows. Predict the temperatures for each day and compare these with the actual values.

20.0 17.2 21.7 23.9 22.7 24.5 24.8 23.6 21.1 23.7
22.8 24.4 23.4 24.8 26.6 21.4 20.8 20.2 24.4 26.8
27.5 27.3 29.0 28.6 26.2 25.6 22.5 23.9 23.1 22.7
24.2

(From a meteorological table for Tokyo)

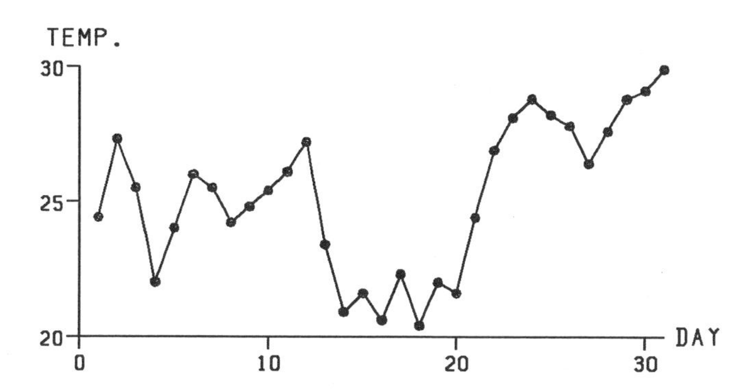

Figure 8.6 Record of daily average temperature.

The most fundamental and useful model in time series analysis is the autoregressive model or AR model for short. For simplicity, we will consider a univariate time series $\{x_t;\ t=1, \ldots, n\}$ in which t is a parameter that indicates time. The AR model expresses the present value of a time series as a linear combination of past values and a random component,

$$MODEL(AR(m)): \qquad x_t = \sum_{i=1}^{m} a_i x_{t-i} + \varepsilon_t. \tag{8.34}$$

This model is a kind of regression model that regresses the time series on past values of itself; hence the name autoregressive model. The integer m is called the order of the model, and the a_i are called the autoregressive coefficients. The random variable ε_t is assumed to be a normal random variable with mean 0 and variance σ^2. In other words, given the past values, $x_{t-m}, \ldots, x_{t-1}$, the x_t is distributed as a normal distribution with mean $a_1 x_{t-1} + \cdots + a_m x_{t-m}$ and variance σ^2.

For simplicity, assuming that $x_{1-m}, \ldots, x_0$ are known, the likelihood of the model given data $x_1, \ldots, x_n$ is obtained by

$$\begin{aligned} L(a_1, \ldots, a_m, \sigma^2) &= f(x_1, \ldots, x_n | x_{1-m}, \ldots, x_0) \\ &= \prod_{t=1}^{n} f(x_t | x_{t-m}, \ldots, x_{t-1}). \end{aligned} \tag{8.35}$$

Here $f(x_t | x_{t-m}, \ldots, x_{t-1})$ is the conditional density of x_t given $x_{t-m}, \ldots, x_{t-1}$ and this is a normal density with mean $a_1 x_{t-1} + \cdots + a_m x_{n-m}$ and variance σ^2, i.e.

$$f(x_t | x_{t-m}, \ldots, x_{t-1}) = \frac{1}{\sqrt{2\pi\sigma^2}} \exp\left\{ -\frac{1}{2\sigma^2} \left(x_t - \sum_{i=1}^{m} a_i x_{t-i} \right)^2 \right\}. \tag{8.36}$$

Thus the likelihood of $MODEL(AR(m))$ assuming that $x_{1-m}, \ldots, x_0$ are known can be written as

$$L(a_1, \ldots, a_m, \sigma^2) = \left(\frac{1}{2\pi\sigma^2}\right)^{\frac{n}{2}} exp\left\{-\frac{1}{2\sigma^2}\sum_{t=1}^{n}\left(x_t - \sum_{i=1}^{m} a_i x_{t-i}\right)^2\right\}. \quad (8.37)$$

By taking logarithm, the likelihood of the model is given by

$$l(a_1, \ldots, a_m, \sigma^2) = -\frac{n}{2} \log 2\pi\sigma^2 - \frac{1}{2\sigma^2}\sum_{t=1}^{n}\left(x_t - \sum_{i=1}^{m} a_i x_{t-i}\right)^2. \quad (8.38)$$

To obtain the maximum likelihood estimators of $a_1, \ldots, a_m$ and σ^2, we must solve

$$\begin{aligned}
\frac{\partial l}{\partial a_1} &= \frac{1}{\sigma^2}\sum_{t=1}^{n} x_{t-1}\left(x_t - \sum_{i=1}^{m} a_i x_{t-i}\right) &= 0 \\
&\vdots &\vdots \\
\frac{\partial l}{\partial a_m} &= \frac{1}{\sigma^2}\sum_{t=1}^{n} x_{t-m}\left(x_t - \sum_{i=1}^{m} a_i x_{t-i}\right) &= 0 \\
\frac{\partial l}{\partial \sigma^2} &= -\frac{n}{2\sigma^2} + \frac{1}{2\sigma^4}\sum_{t=1}^{n}\left(x_t - \sum_{i=1}^{m} a_i x_{t-i}\right)^2 &= 0.
\end{aligned} \quad (8.39)$$

Thus, like other regression models, the maximum likelihood estimators $\hat{a}_1, \ldots, \hat{a}_m$ are obtained as the solution to the normal equation

$$\begin{bmatrix} C(1,1) & \cdot & \cdot & \cdot & C(1,m) \\ \cdot & \cdot & & & \cdot \\ \cdot & & \cdot & & \cdot \\ \cdot & & & \cdot & \cdot \\ C(m,1) & \cdot & \cdot & \cdot & C(m,m) \end{bmatrix} \begin{bmatrix} a_1 \\ \cdot \\ \cdot \\ \cdot \\ a_m \end{bmatrix} = \begin{bmatrix} C(1,0) \\ \cdot \\ \cdot \\ \cdot \\ C(m,0) \end{bmatrix}, \quad (8.40)$$

where

$$C(i, j) = \sum_{t=1}^{n} x_{t-i} x_{t-j}. \quad (8.41)$$

The maximum likelihood estimator σ^2 is

$$\hat{\sigma}^2 = \frac{1}{n}\sum_{t=1}^{n}\left(x_t - \sum_{i=1}^{m}\hat{a}_i x_{t-i}\right)^2 \\ = \frac{1}{n}\left(C(0, 0) - \sum_{i=1}^{m}\hat{a}_i C(i, 0)\right). \tag{8.42}$$

Substitution of this into (8.38) yields the maximum log likelihood

$$l(\hat{a}_1, \ldots, \hat{a}_m, \hat{\sigma}^2) = -\frac{n}{2}\log 2\pi\hat{\sigma}^2 - \frac{n}{2}.$$

Since the autoregressive model with order m has $m+1$ free parameters, the corresponding AIC is given by

$$AIC(m) = -2\, l(\hat{a}_1, \ldots, \hat{a}_m, \hat{\sigma}^2) + 2(m+1) \\ = n(\log 2\pi + 1) + n\log\hat{\sigma}^2 + 2(m+1). \tag{8.43}$$

Here, the first term in the second equation is sometimes neglected since it is a common constant which does not depend on the order.

A procedure for obtaining the best AR model from a given set of data $\{z_1, \ldots, z_n\}$ is as follows:

(1) Select an appropriate maximum order for the autoregression, M. The order M is at most $n/3$, but it is better to choose M less than $2n^{1/2}$.

(2) Obtain the sample mean of the original series and generate the mean corrected series z_t.

(3) Put $x_t = z_{t+M}$ $(t=1-M, \ldots, n-M)$ and replace the number of observations n by $n-M$.

(4) For each order m $(m=0, 1, \ldots, M)$, estimate the AR coefficients by (8.40) and compute $AIC(m)$ by (8.43).

(5) Obtain the order m^* which attains the minimum of

$AIC(m)$ $(m=0, 1, \ldots, M)$. The model with order m^* defines the MAICE model.

The model estimated by this procedure can be used immediately for prediction. When the time series x_n is given up to the present time, the one-step-ahead predicted value is given by

$$\hat{x}_{n+1} = \sum_{i=1}^{m^*} \hat{a}_i x_{n+1-i}. \tag{8.44}$$

More precisely, x_{n+1} is distributed normally with estimated mean $\hat{x}_{n+1}$ and estimated variance $\hat{\sigma}^2$. By using $\hat{x}_{n+1}$, the predicted value of x_{n+2} is

$$\hat{x}_{n+2} = \hat{a}_1 \hat{x}_{n+1} + \sum_{i=2}^{m^*} \hat{a}_i x_{n+2-i}.$$

Repeating this procedure, we can successively obtain predictions for future values.

Let us consider [Example 8.3]. Here we put $M=3$ and the sample mean of the time series is 25.2. The cross covariance matrix $C(i, j)$ is shown in Table 8.4.

Table 8.4 Covariance Matrix

i j	1	2	3	0
1	205.04	159.28	105.62	176.98
2		194.24	143.56	122.78
3			181.92	84.88
0				227.04

By substituting this into (8.40), we obtain the estimates of the autoregressive coefficients. The AIC are obtained by substituting these into (8.42) and (8.43). Table 8.5 summarizes the results.

Table 8.5

Order	Residual variance	AIC
0	8.11	140.06
1	2.65	110.78
2	2.54	111.60
3	2.52	113.37

From this table, we see that the AR model of order 1 is the best fitting model. It is given by

$$x_n = 0.863 x_{n-1} + \varepsilon_n$$
$$\varepsilon_n \sim N(0, 2.65).$$

Next, we will predict the temperatures during July, 1980 by using this estimated model and the actual record of previous days. Since the temperature of July 1 is $x_1=20.0$, the predicted value of x_2 is

$$\hat{x}_2 = (0.863)(20.0-25.2)+25.2 = 20.7.$$

Since the actual value is $x_2=17.2$, the prediction error is

$$x_2-\hat{x}_2 = -3.5.$$

The predicted value of x_3 is

$$\hat{x}_3 = (0.863)(17.2-25.2)+25.2 = 18.3.$$

Similarly, we can compute the one-step-ahead predicted value and prediction error for each day. Figure 8.7 shows these values. The mark ○ shows the predicted value, ● shows the actual value and the difference of these is the prediction error.

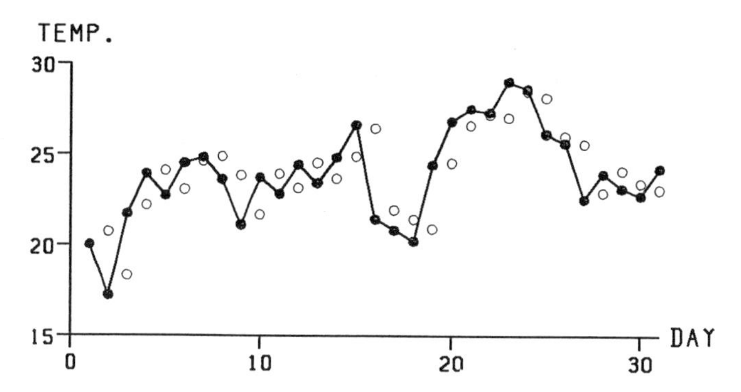

Figure 8.7 Predicted and Actual Temperatures

8.4 Prediction Errors

[*Example 8.4*] For various locations in Japan the average daily minimum temperatures in January were predicted and prediction errors were computed using the model

$$\hat{y} = 40.7 - 1.15x_1 - 0.0098x_3 \tag{8.45}$$

obtained from [Example 8.2] of section 8.2. Examine the sampling distribution of the prediction errors.

	x_1	x_3	y	$\hat{y}$	$\Delta y = y - \hat{y}$
Haboro	44.37	7.9	-9.7	-10.43	0.73
Hakodate	41.82	33.2	-7.8	-7.74	-0.06
Miyako	39.65	42.5	-4.3	-5.34	1.04
Sakata	38.90	3.1	-1.8	-4.09	2.29
Toyama	36.70	8.6	-1.1	-1.61	0.51
Gifu	35.40	12.7	-0.6	-0.15	-0.45
Hamamatsu	34.70	31.7	1.5	0.47	1.03
Yokohama	35.43	39.1	0.0	-0.44	0.44
Osaka	34.68	23.1	0.6	0.58	0.02
Nagasaki	32.73	26.9	2.9	2.79	0.11

Miyazaki	31. 92	6. 5	1. 5	3. 92	-2. 42
Yamagata	38. 25	152. 5	-4. 7	-4. 80	0. 10
Matsumoto	36. 25	610. 0	-6. 4	-6. 96	0. 56
Kofu	35. 67	271. 9	-4. 1	-2. 99	-1. 11
Nara	34. 68	104. 7	-0. 7	-0. 29	-0. 48
Murotomisaki	33. 25	185. 0	4. 1	0. 64	3. 46
Utsunomiya	36. 55	118. 9	-4. 9	-2. 51	-2. 39
Maebashi	36. 40	111. 2	-1. 9	-2. 26	0. 36
Oshima	34. 77	190. 2	3. 4	-1. 16	4. 56

By fitting a normal distribution model to the prediction errors Δy we obtain the following maximum likelihood estimates and AIC value;

$$\hat{\mu} = 0.4368, \qquad \hat{\sigma}^2 = 1.6418^2$$
$$AIC = 76.8 .$$

The two sets of data used in [Example 7. 2] were obtained by classifying the prediction errors into two groups, the group with altitude lower than 100m (*data set 1*) and the group with altitude higher than 100m (*data set 2*). The results in [Example 7. 2] indicated that the two sets of data, *data set 1* and *data set 2*, are distributed normally with identical means but different variances. Since the variance of *data set 2* is larger than that of *data set 1*, the predictions will not be so precise for x_3 greater than 100.

Figure 8. 8 illustrates the (x_1, x_3) pair for the data in [Example 8. 2]. We see that the observed points are sparse for $x_3 > 100$. As explained in the introduction to the conditional log likelihood, when a conditional distribution model is fitted, it is evaluated with a weight proportional to the sample distribution of the explanatory variables. Hence the prediction capability of the fitted model may not be very good in a region where the explanatory variables are rarely observed in the data.

For example, the actual value of the temperature at Mt. Fuji ($x_1 = 35.35$, $x_3 = 3775.0$) is -22.3℃ whereas the predicted value is $\hat{y} = -36.81$ yielding a large prediction error of 14.51. We should be aware of these anomalies when predicting with a regression model.

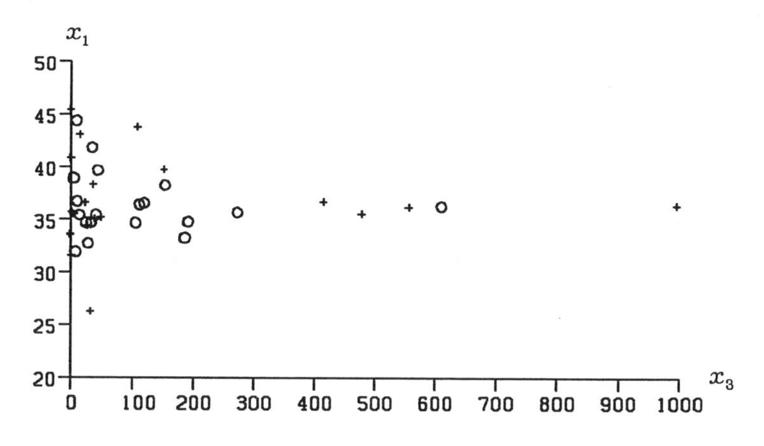

Fig. 8.8 Distribution of Altitude (x_3) against Latitude (x_1)

+: Locations used for fitting model
○: Locations used for prediction

Incidentally, if we fit a regression model to the set of data in [Example 8.2] augmented with the one in [Example 8.44], we obtain

$$\hat{y} = 40.04 - 1.138x_1 - 0.0063x_3.$$

We note that this model is in accordance with the convensional wisdom that temperature decreases by approximately 6℃ with an increase in altitude of x_3 by 1000m.

[*Prediction error variance and AIC*]

The quantity FPE (Final Prediction Error);

$$FPE = \frac{(n+m)}{(n-m)}\hat{\sigma}^2 \tag{8. 46}$$

is an estimate of the prediction error variance when we use a regression model estimated by maximum likelihood for prediction. Here n is the data length (or, sample size) and m is the number of explanatory variables. FPE is derived under the assumption that the distributions of the predictors of both data sets, one used for fitting the model and the other used for prediction, are identical (Akaike 1969).

From FPE and the best fitting model to the data in [Example 8. 2] the expected value of the squared prediction error is obtained to be

$$\frac{20+3}{20-3}\times 2.46 = 3.3.$$

The actual mean square of the prediction errors for the data in [Example 8. 4] is 2. 9. Taking account of the fact that the distributions of the explanatory variables in [Example 8. 2] and [Example 8. 4] are different, we see that FPE has provided a good estimate of the actual prediction error variance.

For large n, since

$$\begin{aligned} n \log FPE &= n \log \hat{\sigma}^2 + n \log \frac{n+m}{n-m} \\ &= n \log \hat{\sigma}^2 + n \log \left(1+\frac{2m}{n-m}\right) \qquad (8.\ 47) \\ &\fallingdotseq n \log \hat{\sigma}^2 + 2m, \end{aligned}$$

AIC given in (8. 33) and $n \log FPE$ are approximately identical if we neglect a common constant term. Thus, when fitting a regression model, minimizing the AIC is equivalent to finding a model with least mean square prediction error.

8.5* *Least Squares Method via Householder Transformation*

The regression coefficients for the polynomial regression model, multiple regression model and autoregressive model are obtained by minimizing the residual variance $\hat{\sigma}^2$. In other words, the maximum likelihood estimates of the regression coefficients are obtained by the method of least squares. In this section, we give an algorithm for the solution to the least squares problem which is based on an orthogonal transformation and is known to be numerically efficient and convenient in modeling (Golub 1965).

First, we define the vector y of objective variable and the matrix Z of explanatory variables by

$$y = \begin{bmatrix} y_1 \\ y_2 \\ \cdot \\ \cdot \\ \cdot \\ y_n \end{bmatrix}, \qquad Z = \begin{bmatrix} z_{11} & \cdots & z_{1m} \\ z_{21} & \cdots & z_{2m} \\ \cdot & & \cdot \\ \cdot & & \cdot \\ \cdot & & \cdot \\ z_{n1} & \cdots & z_{nm} \end{bmatrix}, \tag{8.48}$$

where m is the order of the model, and n is the number of observations. By appropreately defining the explanatory variable z_{ij}, the three models shown in this chapter are expressible in the generic form

$$y = Za + \varepsilon, \tag{8.49}$$

where $a = (a_1, \ldots, a_m)^T$ is the vector of regression coefficients and $\varepsilon = (\varepsilon_1, \ldots, \varepsilon_n)^T$ is the vector of residuals.

The least squares estimate of a is obtained by minimizing the sum of squares of the residuals

$$\| \varepsilon \|^2 = \| y - Za \|^2. \tag{8.50}$$

Note that we can get the normal equation by partially differenciating the above with respect to the a_i and then setting the partial derivative to zero.

For any $n \times n$ orthogonal matrix U, we have

$$\|y-Za\|^2 = \|U(y-Za)\|^2 = \|Uy-UZa\|^2 \tag{8.51}$$

and it follows that the vector a which minimizes $\|Uy-UZa\|^2$ also minimizes $\|y-Za\|^2$. Thus the least squares estimate can be obtained by finding the a that minimizes

$$\|\tilde{y}-\tilde{Z}a\|^2 = \|Uy-UZa\|^2 \tag{8.52}$$

when U is chosen so that UZ becomes a convenient form for subsequent computations.

We next show, that any matrix X can be transformed to upper triangular form by an orthogonal transformation. The transformation we consider here is called the Householder transformation. We can also reduce the matrix X to upper triangular form by a modified Gram-Schmidt method. Consider the following three steps.

1. Given a unit vector w, we define a plane M_w which is orthogonal to w and includes the origin. The mirror image b of a vector a with respect to this plane M_w, is given (see Figure 8.9) by $a - 2w(w^Ta)$. This is obtained by forming Ua where

$$U = I - 2ww^T. \tag{8.53}$$

The transformation U is called the mirror transformation, and is orthogonal since

$$UU^T = U^TU = I - 4ww^T + 4ww^Tww^T = I.$$

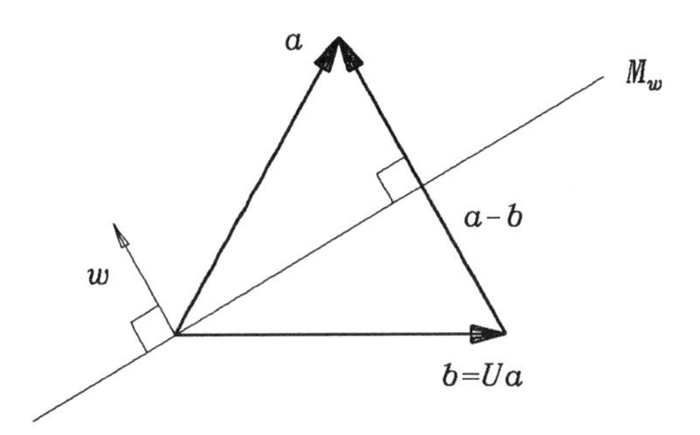

Figure 8.9 Mirror Transformation

2. If a and b are vectors with identical length. Then there is a mirror transformation U which transforms a to b. For this purpose it is intuitively obvious from Figure 8.9 that we should take the w in step 1. to be

$$w = \frac{a-b}{\|a-b\|}.$$

In fact for this w,

$$\begin{aligned} Ua &= (I-2ww^T)a \\ &= a - \frac{(a-b)(a-b)^T}{\|a-b\|^2}2a \\ &= a - \frac{(a-b)(a-b)^T}{\|a-b\|^2}\left\{(a-b)+(a+b)\right\} \qquad (8.54) \\ &= a - (a-b) - \frac{(a-b)(\|a\|^2-\|b\|^2)}{\|a-b\|^2} \\ &= b. \end{aligned}$$

3. We can reduce any $n \times k$ matrix X $(n \geq k)$ to upper triangular form by repeating the mirror transformation derived in step 2. Assume that X is of the form

$$X = \begin{bmatrix} x_{11} & x_{12} & \cdot & \cdot & \cdot & x_{1k} \\ x_{12} & x_{22} & \cdot & \cdot & \cdot & x_{2k} \\ \cdot & \cdot & & & & \cdot \\ \cdot & \cdot & & & & \cdot \\ \cdot & \cdot & & & & \cdot \\ x_{n1} & x_{n2} & \cdot & \cdot & \cdot & x_{nk} \end{bmatrix}. \tag{8.55}$$

If we define $a_1=(x_{11}, x_{21}, \ldots, x_{n1})^T$, $b_1=(x_{11}^{(1)}, 0, \ldots, 0)^T$, $x_{11}^{(1)}=\mp\|a_1\|$ (the sign is determined so that $x_{11}^{(1)}x_{11}$ becomes negative), there is an orthogonal transformation U_1 which transforms a_1 to b_1. Thus

$$U_1X = \begin{bmatrix} x_{11}^{(1)} & x_{12}^{(1)} & \cdot & \cdot & \cdot & x_{1k}^{(1)} \\ 0 & x_{22}^{(1)} & \cdot & \cdot & \cdot & x_{2k}^{(1)} \\ \cdot & \cdot & & & & \cdot \\ \cdot & \cdot & & & & \cdot \\ \cdot & \cdot & & & & \cdot \\ 0 & x_{n2}^{(1)} & \cdot & \cdot & \cdot & x_{nk}^{(1)} \end{bmatrix}.$$

Next, for $a_2=(x_{12}^{(1)}, x_{22}^{(1)}, \ldots, x_{n2}^{(1)})^T$, $b_2=(x_{12}^{(1)}, x_{22}^{(2)}, 0, \ldots, 0)^T$ where $x_{22}^{(2)}$ is the square root of the sum of squares of $x_{22}^{(1)}$ through $x_{n2}^{(1)}$, there is an orthogonal transformation U_2 which transforms a_2 to b_2 and we have

$$U_2U_1X = \begin{bmatrix} x_{11}^{(1)} & x_{12}^{(1)} & x_{13}^{(1)} & \cdot & \cdot & \cdot & x_{1k}^{(1)} \\ 0 & x_{22}^{(2)} & x_{23}^{(2)} & \cdot & \cdot & \cdot & x_{2k}^{(2)} \\ 0 & 0 & x_{33}^{(2)} & \cdot & \cdot & \cdot & x_{3k}^{(2)} \\ \cdot & \cdot & \cdot & & & & \cdot \\ \cdot & \cdot & \cdot & & & & \cdot \\ \cdot & \cdot & \cdot & & & & \cdot \\ 0 & 0 & x_{n3}^{(2)} & \cdot & \cdot & \cdot & x_{nk}^{(2)} \end{bmatrix}.$$

Note here that by this transformation the first row and the first column are unchanged. Repeating this procedure k times $UX = U_k \cdots U_1 X$ reduces to upper triangular form. The transformation U is called the Householder transformation.

We next show how we can obtain the least squares estimates using the Householder transformation. We first augment the matrix Z of the explanatory as

$$X = [Z \vdots y]. \tag{8.56}$$

Then, by the Householder transformation, we obtain

$$UX = S = \begin{bmatrix} s_{11} & \cdot & \cdot & \cdot & s_{1\,m+1} \\ & \cdot & & & \cdot \\ & & \cdot & & \cdot \\ & & & \cdot & \cdot \\ & & & & s_{m+1\,m+1} \\ & & 0 & & \end{bmatrix}. \tag{8.57}$$

For this U,

$$\|Uy - UZa\|^2 = \left\| \begin{bmatrix} s_{1\,m+1} \\ \cdot \\ \cdot \\ \cdot \\ s_{m\,m+1} \\ s_{m+1\,m+1} \\ 0 \\ \cdot \\ \cdot \\ \cdot \\ 0 \end{bmatrix} - \begin{bmatrix} s_{11} & \cdot & \cdot & \cdot & s_{1m} \\ 0 & \cdot & & & \cdot \\ \cdot & \cdot & \cdot & & \cdot \\ \cdot & & \cdot & \cdot & \cdot \\ \cdot & & & \cdot & s_{mm} \\ \cdot & & & & 0 \\ \cdot & & & & \cdot \\ \cdot & & & & \cdot \\ \cdot & & & & \cdot \\ \cdot & & & & \cdot \\ 0 & \cdot & \cdot & \cdot & 0 \end{bmatrix} \begin{bmatrix} a_1 \\ \cdot \\ \cdot \\ \cdot \\ a_m \end{bmatrix} \right\|^2 \tag{8.58}$$

$$= \left\| \begin{bmatrix} s_{1\,m+1} \\ \cdot \\ \cdot \\ \cdot \\ s_{m\,m+1} \end{bmatrix} - \begin{bmatrix} s_{11} & \cdot & \cdot & \cdot & s_{1m} \\ & \cdot & & & \cdot \\ & & \cdot & & \cdot \\ & & & \cdot & \cdot \\ 0 & & & & s_{mm} \end{bmatrix} \begin{bmatrix} a_1 \\ \cdot \\ \cdot \\ \cdot \\ a_m \end{bmatrix} \right\|^2 + s_{m+1\,m+1}^2 .$$

Since the second term on the right hand side of (8.58) is a constant which is independent of a, $\|Uy-UZa\|^2$ is minimized by equating the first term to be zero, namely

$$\begin{bmatrix} s_{11} & \cdot & \cdot & \cdot & s_{1m} \\ & \cdot & & & \cdot \\ & & \cdot & & \cdot \\ & & & \cdot & \cdot \\ 0 & & & & s_{mm} \end{bmatrix} \begin{bmatrix} a_1 \\ \cdot \\ \cdot \\ \cdot \\ a_m \end{bmatrix} = \begin{bmatrix} s_{1\,m+1} \\ \cdot \\ \cdot \\ \cdot \\ s_{m\,m+1} \end{bmatrix} . \tag{8.59}$$

Thus by solving (8.59), we obtain the least squares estimate of a. Since the matrix in (8.59) is of triangular form this system of equations is easy to solve. The least squares estimate of the residual variance is given by $s_{m+1,\,m+1}^2/n$.

By using this upper triangular matrix, we can also get the regression coefficients for the lower order models. For $k<m$, the regression coefficients of the kth order model are obtained as the solution to

$$\begin{bmatrix} s_{11} & \cdot & \cdot & \cdot & s_{1k} \\ & \cdot & & & \cdot \\ & & \cdot & & \cdot \\ & & & \cdot & \cdot \\ 0 & & & & s_{kk} \end{bmatrix} \begin{bmatrix} a_1 \\ \cdot \\ \cdot \\ \cdot \\ a_k \end{bmatrix} = \begin{bmatrix} s_{1\,m+1} \\ \cdot \\ \cdot \\ \cdot \\ s_{k\,m+1} \end{bmatrix} . \tag{8.60}$$

The residual variance for this reduced order model is given by $(s_{k+1\,m+1}^2 + \cdots + s_{m+1\,m+1}^2)/n$.

PROBLEMS

8.1 In the examination of water quality the following measurements giving the amount of contained iron (x) and the colour level (y) were observed;

x	.1	.05	.02	.05	.6	.4	.45	.27	.3	.2
y	2	1	0	2	13	8	10	7	6	5

(i) Compute $\sum_{t=1}^{10} x_t$, $\sum_{t=1}^{10} x_t^2$, $\sum_{t=1}^{10} y_t$, $\sum_{t=1}^{10} x_t Y_t$ and $\sum_{t=1}^{10} y_t^2$.

(ii) Fit an appropriate regression model.

(iii) Subjective opinion states that the measured colour level is 20 times the level of the contained iron. Examine this formula.

8.2 The annual Wolfer sunspot numbers from 1950 to 1979 are given as follows;

83.9	69.4	31.5	13.9	4.4	38.0	141.7	190.2
184.8	159.0	112.3	53.9	37.5	27.9	10.2	15.1
47.0	93.8	105.5	105.5	104.5	66.6	68.9	38.2
15.5	12.6	27.5	92.5	155.4			

(i) Take common logarithms of the sunspot numbers and fit an AR model to the resulting series.

(ii) Discuss the reason why we should fit a time series model to the log transformed data.

(iii) Discuss whether we can compare the goodness of fit of the model fitted to the original series and the model fitted after taking logarithms?

CHAPTER 9

ANALYSIS OF VARIANCE MODELS

Models for the analysis of statistical structure between a discrete objective variable and discrete explanatory variables are given in Chapter 6. The cases where both variables are continuous are dealt with in chapter 8. In this chapter we introduce models, analysis of variance models, for the analysis of data comprising a continuous objective variable and discrete explanatory variables.

The most important problem here is again the selection of explanatory variables which best explain the variation of the objective variable. The conventional method to judge whether a particular explanatory variable is necessary or not is called analysis of variance. Hence we call the models dealt with in this chapter analysis of variance models. The basic strategy of analysis of variance is to analyse/decompose the *sample variance** of the objective variable into terms contributable to each explanatory variable and other unknown factors. If the contribution of some explanatory variable is 'small' then this explanatory variable is judged to be negligible. The difficulty with this approach is the lack of an objective criterion for judging 'smallness'.

From our point of view, to neglect some particular explanatory variable is to impose a constraint on the model. The

---

* Let y_i, $i=1,\ldots,n$ be data, then the sample variance is given by $\frac{1}{n}\sum_{i=1}^{n}(y_i-y)^2$.

variable selection problem is equivalent to the model selection problem and can be solved by the minimum AIC procedure.

9. 1 Analysis of Variance Model

[*Example 9. 1*] A series of experiments to study the wear resistance of four kinds of material, 1, 2, 3 and 4, was performed. A sample of each material was attached to one of four positions, 1, 2, 3 and 4, of a testing machine and after some fixed time of rubbing the reduction of the weight of each sample was measured in mg unit. To avoid the inequality of four positions of the testing machine, four rounds of experiment were done changing the position of each material. The results are summarized in Table 9. 1. Can we find the difference of the wear resistance between four materials from this set of data? (Data from *Design and Analysis of Industrial Experiment* by Davies, O. L., 1954, Oliver and Boyd)

Table 9. 1 Results of experiments

		Position (X_2)			
	$k_1 \backslash k_2$	1	2	3	4
Material(X_1)	1	268	233	254	281
	2	240	249	231	314
	3	256	250	280	291
	4	265	250	248	271

Generalizing the problem, we consider the case where m characteristics (e. g. factors, or explanatory variables) X_j, $j=1, \ldots, m$ are measured along with an objective variable Y. The j-th characteristic X_j takes its value in the set $[c_j]$ =

$\{1, 2, \ldots, c_j\}$ of c_j categories (e.g. levels). Assume that Y is distributed as a normal distribution with mean dependent on the realization of the characteristics. Let $\lambda(k_1, \ldots, k_m)$ denote the mean when the realization of $(X_1, \ldots, X_m)$ belongs to the category $(k_1, \ldots, k_m)$, $k_j \in [c_j]$, $j=1, 2, \ldots, m$, then the basic model is given by

$$f(y \mid \lambda(k_1, \cdots, k_m), \sigma^2) = \frac{1}{\sqrt{2\pi\sigma^2}} e^{-(y-\lambda(k_1, \cdots, k_m))^2/2\sigma^2}. \tag{9.1}$$

This is a conditional distribution model defined on the parameter space

$$\Lambda = \{(\sigma^2, \lambda(1, \cdots, 1), \cdots, \lambda(c_1, \cdots, c_m)) \mid \sigma^2 > 0\} \tag{9.2}$$

and the number of free parameters is equal to $(\prod_{j=1}^{m} c_j + 1)$.

Reparametrizing (9.1) by

$$\lambda(k_1, \cdots, k_m) = \theta_0 + \theta_1(k_1) \cdots + \theta_m(k_m), \quad k_j \in [c_j] \quad j=1, \cdots, m$$

we get a simpler model

MODEL$(1, 2, \cdots, m)$:

$$f(y \mid \theta) = \frac{1}{\sqrt{2\pi\sigma^2}} \exp\{-(y - \theta_0 - \sum_{j=1}^{m} \theta_j(k_j))^2/2\sigma^2\}, \tag{9.3}$$

which we call an analysis of variance model. To correctly indentify the grand mean level θ_0, the parameter space for this model has been chosen to be

$$\Theta = \left\{ \theta = (\sigma^2, \theta_0, \theta_1(1), \cdots, \theta_1(c_1), \cdots, \theta_m(1), \cdots, \theta_m(c_m)) \right.$$
$$\left. \mid \sigma^2 > 0, \sum_{k_j=1}^{c_j} \theta_j(k_j) = 0, j = 1, \cdots, m \right\} \tag{9.4}$$

and the number of free parameters is $(\sum_{j=1}^{m}(c_j-1)+2)$. Constraints are necessary to make the correspondance between Λ and Θ one to one.

9.1.1 *Notation*

Suppose that n observations $(y_i, x_{1i}, \ldots, x_{mi})$, $i=1, \ldots, n$ of $(Y, X_1, \ldots, X_m)$ are given. We call y_i the value of sample i. When $(x_{1i}, \ldots, x_{mi})$ belongs to category $(k_1, \ldots, k_m)$, we say that sample i belongs to category $(k_1, \ldots, k_m)$. Re-numbering the samples in each category, we define the notation

$$y(i, k_1, \cdots, k_m) = (\text{ the value of the } i\text{-th sample of the category } (k_1, \cdots, k_m)). \quad (9.5)$$

Let $n(k_1, \ldots, k_m)$ denote the number of samples which belongs to category $(k_1, \ldots, k_m)$. Then, since each sample belongs to one and only one category, we have

$$\sum_{k_1=1}^{c_1} \cdots \sum_{k_m=1}^{c_m} n(k_1, k_2, \cdots, k_m) = n \,. \quad (9.6)$$

Using the notation (9.5), the total sum y_0 of the sample values is given by

$$y_0 = \sum_{k_1=1}^{c_1} \cdots \sum_{k_m=1}^{c_m} \sum_{i=1}^{n(k_1, \cdots, k_m)} y(i, k_1, \cdots, k_m). \quad (9.7)$$

In the sequel, for brevity, we denote $\sum_{k_j=1}^{c_j}$ and $\sum_{i=1}^{n(k_1, \cdots, k_m)}$ by $\sum_{k_j}$ and $\sum_{i}$, respectively. Let $n_j(k_j)$ denote the number of samples whose j-th characteristic X_j belongs to category k_j, then $n_j(k_j)$ is given by

$$n_j(k_j) = \sum_{k_1}\cdots\sum_{k_{j-1}}\sum_{k_{j+1}}\cdots\sum_{k_m} n(k_1, \cdots, k_{j-1}, k_j, k_{j+1}, \cdots, k_m). \tag{9.8}$$

When we exclude the j-th category from the summation like in (9.8), we use the notation:

$$n_j(k_j) = \sum_{k_1}\cdots\sum_{k_j}^{k_j}\cdots\sum_{k_m} n(k_1, \cdots, k_j, \cdots, k_m).$$

With this notation the sum $y_j(k_j)$ of the values of samples whose j-th characteristic X_j belongs to k_j is given by

$$y_j(k_j) = \sum_{k_1}\cdots\sum_{k_j}^{k_j}\cdots\sum_{k_m}\sum_{i} y(i, k_1, \cdots, k_j, \cdots, k_m) \; .$$

For [Example 9.1], we have $m=2$, $c_1=c_2=4$, $[c_1]=[c_2] =\{1, 2, 3, 4\}$ and $n(k_1, k_2)=1$, $k_j \in [c_j]$, $j=1, 2$. According to the notation (9.5), we have $y(1, 2, 3)=231$, $y(1, 4, 3)=248$, etc.. The values of samples whose X_1 belongs to category '2' are listed on the second row {240, 249, 231, 314} of Table 9.1. The values of samples whose X_2 belongs to '3' is listed on the third column {254, 231, 280, 248} of the same table. Thus we have the following row sums:

$$\begin{aligned} y_1(1) &= 268+233+254+281 = 1036 \\ y_1(2) &= 240+249+231+314 = 1034 \\ y_1(3) &= 256+250+280+291 = 1077 \\ y_1(4) &= 265+250+248+271 = 1034 \end{aligned} \tag{9.9}$$

Summation over the columns yields

$$\begin{aligned} y_2(1) &= 268+240+256+265 = 1029 \\ y_2(2) &= 233+249+250+250 = 982 \end{aligned}$$

$$y_2(3) = 254+231+280+248 = 1013$$

$$y_2(4) = 281+314+291+271 = 1157 \tag{9.10}$$

The total sum is given by

$$y_0 = 1029+982+1013+1157 = 4181. \tag{9.11}$$

We also have $n_1(1)=n_1(2)=n_1(3)=n_1(4)=4$ and $n_2(1)=n_2(2)=n_2(3)=n_2(4)=4$ in this case.

9.1.2 Maximum Likelihood Estimation of Parameters

For the set of data

$$S = \{y(i, k_1, \cdots, k_m) \mid i = 1, \cdots, n(k_1, \cdots, k_m), k_j \in [c_j], j = 1, \cdots, m\}, \tag{9.12}$$

the conditional log likelihood of *MODEL(1, 2, ⋯, m)* defined by (9.3) is

$$\begin{aligned} l(\theta) &= \sum_{k_1}\cdots\sum_{k_m}\sum_{i} \log f(y(i, k_1, \cdots, k_m) \mid \theta) \\ &= \sum_{k_1}\cdots\sum_{k_m}\sum_{i} \Big[-\frac{1}{2}\log 2\pi - \frac{1}{2}\log \sigma^2 \\ &\qquad -\frac{1}{2\sigma^2}\{y(i, k_1 \cdots, k_m) - (\theta_0+\theta_1(k_1)+\cdots+\theta_m(k_m))\}^2\Big] \\ &= -\frac{n}{2}\log 2\pi - \frac{n}{2}\log \sigma^2 \\ &\qquad -\frac{1}{2\sigma^2}\sum_{k_1}\cdots\sum_{k_m}\sum_{i}\{y(i, k_1, \cdots, k_m) - \theta_0 - \theta_1(k_1) - \cdots - \theta_m(k_m)\}^2. \end{aligned} \tag{9.13}$$

The maximum likelihood estimator $\hat{\theta}$ is defined as the value of θ which maximizes (9.13) under the constraints

$$\sum_{k_j}\theta_j(k_j) = 0 \quad j = 1, \cdots, m \ . \tag{9.14}$$

Since the calculation of $\hat{\theta}$ is rather complicated for the general case, we assume here the following assumption:

[Assumption of balanced experiment]

(a1) The value of

$$n_j(k_j) = \sum_{k_1}\cdots\overset{k_j}{\sum_{k_j}}\cdots\sum_{k_m} n(k_1, \cdots, k_j, \cdots, k_m)$$

is a constant, n_j, independent of k_j.

(a2) Let $1 \leq j < l \leq m$ then $n_{jl}(k_j, k_l)$ defined by

$$n_{jl}(k_j, k_l) = \sum_{k_1}\cdots\overset{k_j}{\sum_{k_j}}\cdots\overset{k_l}{\sum_{k_l}}\cdots\sum_{k_m} n(k_1, \cdots, k_j, \cdots, k_l, \cdots, k_m)$$

is a constant, n_{jl}, independent of (k_j, k_l).

(*) It should be noted that Program VARMOD given in Part III is free from these assumptions.

We have already verified that the data of [Example 9.1] satisfy the Assumption (a1) with $n_1=n_2=4$. This set of data also satisfies (a2) with $n_{12}=1$.

Under these two assumptions, and with v^2 defined by

$$v^2 = \sum_{k_1} \ldots \sum_{k_m}\sum_{i} y(i, k_1, \ldots, k_m)^2 \tag{9.15}$$

the maximum likelihood estimator $\hat{\theta}=(\hat{\sigma}^2, \hat{\theta}_0, \hat{\theta}_1(1), \cdots, \hat{\theta}_m(c_m))$ is given by

$$\hat{\theta}_0 = \frac{y_0}{n}, \tag{9.16}$$

$$\hat{\theta}_j(k_j) = \frac{y_j(k_j)}{n_j} - \hat{\theta}_0 \qquad (k_j \in [c_j],\ j=1, 2, \cdots, m), \tag{9.17}$$

$$w_0^2 = v^2 - y_0\hat{\theta}_0, \tag{9.18}$$

$$w_j^2 = \sum_{k_j} y_j(k_j)\hat{\theta}_j(k_j) \qquad (j=1, \cdots, m), \tag{9.19}$$

and

$$\hat{\sigma}^2 = \frac{1}{n}(w_0^2 - w_1^2 - \cdots - w_m^2). \tag{9.20}$$

Hence the maximum log likelihood is obtained as

$$l(\hat{\theta}) = -\frac{n}{2}\log 2\pi - \frac{n}{2}\log \hat{\sigma}^2 - \frac{n}{2}. \tag{9.21}$$

The derivation is given in the following. (Those who are not interested in the details may skip the rest of this section.)

[Proof]

Under the assumptions of balanced experiment, we have for $1 \leq j \leq m$,

$$\sum_{k_1}\cdots\sum_{k_m}\sum_{i}\theta_0\theta_j(k_j) = n_j\theta_0\sum_{k_j}\theta_j(k_j) = 0, \tag{9.22}$$

$$\sum_{k_1}\cdots\sum_{k_m}\sum_{i}\theta_j(k_j)\theta_i(k_i) = n_{ji}\sum_{k_j}\theta_j(k_j)\sum_{k_l}\theta_l(k_l) = 0, \tag{9.23}$$

$$\sum_{k_1}\cdots\sum_{k_m}\sum_{i}y(i, k_1, \cdots, k_m)\theta_j(k_j) = \sum_{k_j}y_j(k_j)\theta_j(k_j) \tag{9.24}$$

and

$$\sum_{k_1}\cdots\sum_{k_m}\sum_{i}y(i, k_1, \cdots, k_m)\theta_0 = y_0\theta_0. \tag{9.25}$$

Thus $l(\theta)$ reduces to

$$l(\theta) = -\frac{n}{2}\log 2\pi - \frac{n}{2}\log\sigma^2$$
$$-\frac{1}{2\sigma^2}[v^2 + n\theta_0^2 + n_1\sum_{k_1}\theta_1^2(k_1) + \cdots + n_m\sum_{k_m}\sigma_m^2(k_m)$$
$$-2\{y_0\theta_0 + \sum_{k_1}y_1(k_1)\theta_1(k_1) + \cdots + \sum_{k_m}y_m(k_m)\theta_m(k_m)\}]. \quad (9.26)$$

To maximize this log likelihood under the constraints (9.14), we introduce Lagrangean multipliers α_j, $j=1, \ldots, m$. We first maximize the augmented log likelihood

$$l(\theta) + \alpha_1\sum_{k_1}\theta_1(k_1) + \cdots + \alpha_m\sum_{k_m}\theta_m(k_m)$$
$$= -\frac{n}{2}\log 2\pi - \frac{n}{2}\log\sigma^2$$
$$-\frac{1}{2\sigma^2}[v^2 + (n\theta_0^2 - 2y_0\theta_0)$$
$$+\sum_{k_1}\{n_1\theta_1^2(k_1) - 2(\sigma^2\alpha_1 + y_1(k_1))\theta_1(k_1)\}$$
$$+\cdots$$
$$+\sum_{k_m}\{n_m\theta_m^2(k_m) - 2(\sigma^2\alpha_m + y_m(k_m))\theta_m(k_m)\}] \quad (9.27)$$

to find $\hat{\theta}$ as a function of $(\alpha_1, \ldots, \alpha_m)$, and then determine $(\alpha_1, \ldots, \alpha_m)$ so as to satisfy the constraints (9.14).

Setting the partial derivative of (9.27) with respect to θ_0 equal to zero, we have

$$n\theta_0 = y_0$$

and hence

$$\hat{\theta}_0 = \frac{y_0}{n}. \quad (9.28)$$

By equating the partial derivative of (9.27) with respect to

$\theta_j(k_j)$ to zero, we get

$$n_j\theta_j(k_j) = \sigma^2\alpha_j + y_j(k_j) \ . \tag{9.29}$$

Summing both sides of this equation over k_j, we have

$$0 = \sum_{k_j=1}^{c_j}(\sigma^2\alpha_j + y_j(k_j)) = c_j\sigma^2\alpha_j + y_0 \ ,$$

and then

$$\sigma^2\alpha_j = -\frac{y_0}{c_j} \ .$$

From this and (9. 29) we have

$$\hat{\theta}_j(k_j) = \frac{y_j(k_j)}{n_j} - \frac{y_0}{n_jc_j} = \frac{y_j(k_j)}{n_j} - \hat{\theta}_0 \ , \tag{9.30}$$

where we have used the relation

$$n_jc_j = \sum_{k_j=1}^{c_j} n_j = \sum_{k_j=1}^{c_j} n_j(k_j) = n \ .$$

By defferentiating (9. 26) with respect to σ^2 and equating the resultant to zero, we have

$$\hat{\sigma}^2 = \frac{1}{n}[v^2 + n\hat{\theta}_0^2 + n_1\sum_{k_1}\hat{\theta}_1^2(k_1) + \cdots + n_m\sum_{k_m}\hat{\theta}_m^2(k_m)$$

$$-2\{y_0\hat{\theta}_0 + \sum_{k_1}y_1(k_1)\hat{\theta}_1(k_1) + \cdots + \sum_{k_m}y_m(k_m)\hat{\theta}_m(k_m)]. \tag{9.31}$$

Substituting $y_0 = n\hat{\theta}_0$, $y_j(k_j) = n_j\hat{\theta}_j(k_j) + n_j\hat{\theta}_0$ into (9. 31), and using the equality $\sum_{k_j}\hat{\theta}_j(k_j) = 0$, we obtain

$$\hat{\sigma}^2 = \frac{v^2}{n} - \hat{\theta}_0^2 - \frac{n_1}{n}\sum_{k_1}\hat{\theta}_1^2(k_1) - \cdots - \frac{n_m}{n}\sum_{k_m}\hat{\theta}_m^2(k_m), \quad (9.32)$$

or equivalently,

$$\hat{\sigma}^2 = \frac{1}{n}\{v^2 - y_0\hat{\theta}_0 - \sum_{k_1} y_1(k_1)\hat{\theta}_1(k_1) - \cdots - \sum_{k_m} y_m(k_m)\hat{\theta}_m(k_m)\}. \quad (9.33)$$

Defining w_0^2 and w_j^2 by

$$w_0^2 = v^2 - y_0\hat{\theta}_0 \quad (9.34)$$

$$w_j^2 = \sum_{k_j} y_j(k_j)\hat{\theta}_j(k_j) \quad (j=1, 2, \cdots, m), \quad (9.35)$$

we have

$$\hat{\sigma}^2 = \frac{1}{n}\{w_0^2 - w_1^2 - \cdots - w_m^2\}. \quad (9.36)$$

9.2 *AIC for Restricted Models*

Let $\{j_1, j_2, \ldots, j_r\}$ be any subset of $\{1, 2, \ldots, m\}$. Then we can define restricted models whose explanatory variables are $X_{j_1}, \ldots, X_{j_r}$:

MODEL$(j_1, j_2, \cdots, j_r)$:

$$f(y\,|\,\theta) = \frac{1}{\sqrt{2\pi\sigma^2}} e^{-(y-\theta_0-\theta_1(k_1)-\cdots-\theta_m(k_m))^2/2\sigma^2}, \quad (9.37)$$

where

$$\theta_j(k_j) = 0 \quad k_j=1, \cdots, c_j \quad j \notin \{j_1, j_2, \cdots, j_r\}. \quad (9.38)$$

The model whose number of explanatory variables is zero is denoted by *MODEL*(ϕ). From the derivation given in Section

9.1, it is shown, under the assumption of balanced experiment, that the maximum likelihood estimators of the free parameters of $MODEL(j_1, \ldots, j_r)$ are calculated as follows:

$$\tilde{\theta}_0 = \frac{y_0}{n} \tag{9.39}$$

$$\tilde{\theta}_j(k_j) = \frac{y_j(k_j)}{n_j} - \tilde{\theta}_0 \tag{9.40}$$

$$w_0 = v^2 - y_0\tilde{\theta}_0 \tag{9.41}$$

$$w_j^2 = \sum_{k_j} (k_j)\tilde{\theta}_j(k_j) \tag{9.42}$$

$$\hat{\theta}_0 = \tilde{\theta}_0 \tag{9.43}$$

$$\hat{\theta}_j(k_j) = \begin{cases} \tilde{\theta}_j(k_j) & j \in \{j_1, \cdots, j_r\} \\ 0 & j \notin \{j_1, \cdots, j_r\} \end{cases} \tag{9.44}$$

The maximum likelihood estimator of σ^2 is given by

$$\hat{\sigma}^2(j_1, \cdots, j_r) = \frac{1}{n}\{w_0^2 - \sum_{j \in J} w_j^2\}, \tag{9.45}$$

where $J = \{j_1, \ldots, j_r\}$.

Since the maximum log likelihood of $MODEL(j_1, \ldots, j_r)$ is

$$l(\hat{\theta}) = -\frac{n}{2}\log 2\pi - \frac{n}{2}\log \hat{\sigma}^2(j_1, \cdots, j_r) - \frac{n}{2} \tag{9.46}$$

and the number of free parameters of this model is

$$\sum_{j \in J}(c_j - 1) + 2,$$

the AIC for $MODEL(j_1, \ldots, j_r)$ is given by

$AIC(j_1, \cdots, j_r)$

$$=-2l(\hat{\theta})+2(\text{the number of free parameters})$$
$$=n\log 2\pi+n\log\hat{\sigma}^2(j_1, \cdots, j_r)+n+2\{\sum_{j\in J}(c_j-1)+2\}. \quad (9.47)$$

9.3 *Numerical Examples*

9.3.1 *A Case of Balanced Experiment*

We will solve the problem of [Example 9.1]. The maximum likelihood estimates of the parameters of *MODEL(1, 2)* are obtained from (9.39 ~ 45) and (9.9 ~ 11) as follows:

$$\hat{\theta}_0=\frac{y_0}{n}=\frac{4181}{16}=261.3125$$

$$\hat{\theta}_1(1)=\frac{y_1(1)}{n_1}-\hat{\theta}_0=\frac{1036}{4}-261.3125=-2.3125$$

$$\hat{\theta}_1(2)=\frac{y_1(2)}{n_2}-\hat{\theta}_0=\frac{1034}{4}-261.3125=-2.8125$$

$$\hat{\theta}_1(3)=\frac{y_1(3)}{n_3}-\hat{\theta}_0=\frac{1077}{4}-261.3125=\ 7.9375$$

$$\hat{\theta}_1(4)=\frac{y_1(4)}{n_1}-\hat{\theta}_0=\frac{1034}{4}-261.3125=-2.8125$$

$$\hat{\theta}_2(1)=\frac{y_2(1)}{n_2}-\hat{\theta}_0=\frac{1029}{4}-261.3125=-4.0625$$

$$\hat{\theta}_2(2)=\frac{y_2(2)}{n_2}-\hat{\theta}_0=\frac{982}{4}-261.3125=-15.8125$$

$$\hat{\theta}_2(3)=\frac{y_2(3)}{n_2}-\hat{\theta}_0=\frac{1013}{4}-261.3125=-8.0625$$

$$\hat{\theta}_2(4)=\frac{y_2(4)}{n_2}-\hat{\theta}_0=\frac{1157}{4}-261.3125=27.9375$$

$$v^2=268^2+233^2+\cdots+271^2=1099935$$

$$w_0^2=v^2-y_0\hat{\theta}_0=1099935-4181\times261.3125=7387.4370$$

$$w_1^2=\sum_{k_1=1}^{4}y_1(k_1)\hat{\theta}_1(k_1)=336.6875$$

$$w_2^2=\sum_{k_2=1}^{4}y_2(k_2)\hat{\theta}_2(k_2)=4448.1875$$

$$\hat{\sigma}^2(1,2)=\frac{1}{n}\{w_0^2-w_1^2-w_2^2\}$$

$$=\frac{7387.4370-336.6875-4448.1875}{16}=162.6601\ .$$

Therefore AIC for this model is

$$AIC(1,2) = 16\log 2\pi+16\log 162.6601+16\times(3+3+2)$$
$$= 16\times1.83788+16\times5.09166+16+16 = 142.9\ .$$

MODEL(2) represents the hypothesis that there is no difference between the four materials. By using

$$\hat{\sigma}^2(2) = \frac{1}{n}(w_0^2-w_2^2) = \frac{7387.4370-4448.1875}{16} = 183.7031\ ,$$

AIC for this model is given by

$$AIC(2) = 16\log 2\pi+16\log 183.7031+16+2\times(3+2)$$
$$= 16\times1.83788+16\times5.21332+16+10 = 138.8\ .$$

The comparison of *AIC(1, 2)* and *AIC(2)* leads us to the conclusion that *MODEL(2)* is the better one. The difference of AIC values, 4.1, is significant.

This result should not be interpreted that there is no difference between materials. Common sense tells us that wear resistance should vary with materials. The above results mean that we could not detect the difference between materials because of the too small sample size or too much noise. It is not reasonable to judge the qualities of these four materials based on the estimated values, $\hat{\theta}_1(1)$, $\hat{\theta}_1(2)$, $\hat{\theta}_1(3)$ and $\hat{\theta}_1(4)$, which were obtaind when we assumed *MODEL(1, 2)*.

AIC's for *MODEL(1)* and *MODEL(ø)* are also calculated easily. The results are summarized as follows.

MODEL	Number of free parameters	Estimate of σ^2	*AIC*	Difference in *AIC*
(2)	5	183.70	138.8	
(1, 2)	8	162.66	142.9	4.1
(ø)	2	461.71	147.6	4.7
(1)	5	440.67	152.8	5.2

The best model among the four tested is *MODEL(2)*. Note that $AIC(\phi)>AIC(2)$. This means that the contribution of X_2 to the variation of the data is significant. From the values of $\theta_2(1), \ldots, \theta_2(4)$ we conclude that the mass reduction is most severe when material is attached to the position '4' .

9.3.2 A Case of Un-balanced Experiment

We cannot apply (9.39)~(9.47) for this case. It can be shown, however, that the expression for $AIC(j_1, j_2, \ldots, j_r)$ is still given by

$$AIC(j_1, \cdots, j_r) = n \log 2\pi + n \log \hat{\sigma}^2(j_1, \cdots, j_r) + n + 2\{\sum_{j \in J}(c_j - 1) + 2\},$$

where the only difference is in the procedure to calculate the estimate $\hat{\sigma}^2(j_1, \ldots, j_r)$ of σ^2.

[*Example 9.2*] A random sample of housewives in Tokyo was

drawn and each housewife was asked to answer the following questions.

1) How much do you spend on preparing a supper for your family? Answer in hundred yen units.

( ) hundred yen.

2) How many persons are in your family? Choose the appropriate category.

1. less than three
2. three to four
3. more than four

3) Is the income level of your family

1. less than 2, 500 thousand yen a year ?
2. 2, 500 to 3, 500 thousand yen a year ?
3. more than 3, 500 thousand yen a year ?

4) Are you interested in local politics ?

1. Yes
2. No

A total of 284 responses were obtained (Tables 9.2 and 9.3). We would expect that the expenses of a supper increases as the size of the family increases. Is there any relation between supper expenses and interest in politics ?

Table 9.3 indicates that this set of data does not satisfy conditions (a1) and (a2) of the balanced experiment. Table 9.4 shows the results obtained by applying the program VARMOD to the data.

We see from this table that if we stick to single variable models the best one is *MODEL(1)*, but the *AIC* decreases significantly when one more explanatory variable X_2 is added. It is also apparent that interest in the politics has little to do with supper expenses. Note that the estimate of θ_0 varies from

Table 9.2 Supper expenses (Y)

		The size of family (X_1)								
		1			2			3		
		X_2			Income level (X_2)			X_2		
		1	2	3	1	2	3	1	2	3
X_3	1	10, 10, 10 7, 8, 3 10, 10, 8 8, 15, 8	15	8	8, 15, 16, 20, 15 13, 15, 25, 15, 15 10, 10, 13, 15	12, 50, 18, 20, 10 10, 18, 15, 15, 20 1, 15, 20, 25, 20 15, 15, 20	20, 20, 13, 8, 20 15, 15, 10, 25, 20 20, 16, 20, 15, 10 30, 20, 20	15, 30	20, 30 20, 8	18, 20 30, 15 30, 20 30
	2	8, 20, 10 20, 10, 7 12, 20, 10 20, 2	20 10 5 15 13 15	35	10, 20, 12, 10, 20 20, 15, 20, 20, 12 15, 10, 15, 20, 12 25, 17, 15, 12, 10 8, 15, 15, 15, 30 7, 15, 20, 20, 20 15, 20, 15, 5, 15 10, 10, 6, 8, 20 15, 13, 5, 15, 20 15, 20, 15, 15, 15	20, 15, 15, 10, 20 20, 15, 15, 18, 10 15, 15, 25, 20, 25 15, 15, 15, 10, 6 15, 12, 10, 10, 10 16, 13, 10, 7, 3 20, 13, 15, 8, 25 20, 15, 15, 8, 10 8, 13, 10, 20, 20 20, 18, 13, 20, 20 10, 16, 15	15, 8, 30, 10, 15 20, 30, 30, 30, 20 30, 30, 30, 20, 10 20, 25, 30, 20, 15 22, 20, 20, 15, 20 15, 25, 20, 15, 30 2, 10, 20, 15, 20 30, 12, 20, 20, 30 10, 10, 10, 15, 15 15, 15, 20, 15, 15	50, 14 15, 50 10, 20 10, 20 15	25, 20 12, 16 20, 13 15, 25 25, 20 8, 20 22, 15 12, 9	30, 20 20, 20 10, 10 30, 15 30, 30 15

Table 9.3 $n(k_1, k_2, k_3)$

		k_1									Total $n_3(k_3)$
		1			2			3			
		k_2			k_2			k_2			
		1	2	3	1	2	3	1	2	3	
k_3	1	12	1	1	14	18	18	2	4	7	77
	2	11	6	1	50	53	50	9	16	11	207
Total $n_1(k_1)$		32			203			49			284

k_2	1	2	3
$n_2(k_2)$	98	98	88

model to model. This never occurs when the experiment is balanced.

Table 9.4 Analysis of the supper expenses data

MODEL	Number of free parameters	$\hat{\sigma}^2$	*AIC*	Difference in *AIC*	$\hat{\theta}_0$
(1, 2)	6	6.72^2	1899.8		16.5
(1, 2, 3)	7	6.72^2	1901.8	2.0	16.5
(1)	4	6.91^2	1911.8	10.0	16.2
(1, 3)	5	6.91^2	1913.8	2.0	16.2
(2)	4	6.98^2	1917.6	3.8	16.7
(2, 3)	5	6.98^2	1919.4	1.8	16.6
(ϕ)	2	7.26^2	1935.6	16.2	16.6
(3)	3	7.25^2	1937.5	1.9	16.5

The following is the table of estimates of $\theta_1(k_1)$, $k_1=1, 2, 3$ and $\theta_2(k_2)$, $k_2=1, 2, 3$ when *MODEL(1, 2)* is assumed.

Table 9.5 Effects of X_1 and X_2 on supper expenses

k	1	2	3
$\hat{\theta}_1(k)$	-3.58	-0.14	3.72
$\hat{\theta}_2(k)$	-1.23	-1.15	2.38

This table implies that supper expenses increases as the size of the family increases and that richer families enjoy more expensive suppers. The estimate of σ^2 for this model is 6.72^2.

9.4* *Latin Square Design*

An investigation into the wear resistance of vulcanised rubber was carried out. The main interest had been in the effects of qualities of the raw rubber(X_1), the quality of the filler(X_2) and the method of pretreatment of the rubber(X_3). The results are summarized in Table 9.6 (from Davies, id.). The gaps in the table indicate missing observations.

Table 9.6 Latin Square Design

		The raw rubber (X_1)								
		1			2			3		
		X_2			X_2			X_2		
		1	2	3	1	2	3	1	2	3
X_3	1	104				-61				-10
	2		118				72	76		
	3			160	228				-106	

Since

$$n_1(1) = n_1(2) = n_1(3) = 3 = n_1$$
$$n_2(1) = n_2(2) = n_2(3) = 3 = n_2$$
$$n_3(1) = n_3(2) = n_3(3) = 3 = n_3$$

$$n_{12}(1, 1) = \cdots = n_{12}(3, 3) = 1 = n_{12}$$
$$n_{13}(1, 1) = \cdots = n_{13}(3, 3) = 1 = n_{13}$$
$$n_{23}(1, 1) = \cdots = n_{23}(3, 3) = 1 = n_{23} ,$$

this set of data satisfy the assumption of balanced experiment.

As shown in Table 9.6, we sometimes omit some of experiments on purpose. An abbreviated form for Table 9.6 is given in Table 9.7, where the values of X_1 are placed in parenthesis after each value of the objective variable.

Table 9.7 $Y(k_3)$

	$k_1 \backslash k_2$	X_2 1	2	3
X_3	1	104(1)	-61(2)	-10(3)
	2	76(3)	118(1)	72(2)
	3	228(2)	-106(3)	160(1)

Extracting the numbers in parentheses, we have the matrix

$$\begin{bmatrix} 1 & 2 & 3 \\ 3 & 1 & 2 \\ 2 & 3 & 1 \end{bmatrix} . \tag{9.53}$$

A characteristic of this matrix is that a number never repeats itself in a row or column. In general, if a $c \times c$ matrix L is constructed so that each row and column is a permutation of $(1, 2, \ldots, c)$, then L is called a Latin square of order c. (9.53) is an example of Latin square of order 3. The (i, j) element of L is denoted by l_{ij}.

Let X_j, $j=1, 2, 3$ be variables which take their values in $\{1, 2, \ldots, c\}$. If we want to investigate the relation between Y and (X_1, X_2, X_3), and measure the value of Y for each combination of (X_1, X_2, X_3), we would need c^3 experiments. However, if we measure the value of Y only for the combinations

$$(X_1, X_2, X_3) = (l_{ij}, i, j) \quad 1 \leqq i \leqq c, \quad 1 \leqq j \leqq c \ ,$$

then the number of experiments reduces to c^2, resulting in a simpler data analysis with lower computaional cost. Note that data so obtained always satisfies the conditions for the balanced experiment. This type of design of experiment is called the Latin square design.

PROBLEM

9.1 Find the best fitting model for the data given in Table 9.6 among *MODEL*(∅), . . . , *MODEL*(*1*, *2*, *3*). (Restrict your attention to those models with less than 5 free parameters since the number of data points is only 9.)

PART III

FORTRAN PROGRAMS

III-1 SEARCH FOR AN OPTIMAL HISTOGRAM (CATDAP-11)

This program CATDAP-11 automatically constructs an optimal histogram for a given set of data. For the input required, see the relevant comments in the program. The possible output from this program is as follows:

1) an optimal histogram (exemplified by Figures 5.3 and 5.4)
2) a list of better models (exemplified by Table 5.7).

As an example, for the data given in [Example 5.7] this program produced the output given in Figure 5.4 and Table 5.7.

```
      PROGRAM HIST                                                      00000010
C      CATEGORICAL DATA ANALYSIS PROGRAM PACKAGE 11  (CATDAP11)         00000020
C                  -----  SEARCH FOR AN OPTIMAL HISTOGRAM               00000030
C                                                                       00000040
C                                                       OCT. 1,1981     00000050
C                                     DESIGNED BY YOSIYUKI SAKAMOTO     00000060
C                                     PROGRAMMED BY KOUICHI KATSURA     00000070
C                                                                       00000080
C                          THE INSTITUTE OF STATISTICAL MATHEMATICS 000000090
C                   1-4-6, MINAMI-AZABU, MINATO-KU, TOKYO, JAPAN 106    00000100
C                                                                       00000110
C                                                                       00000120
C      THIS PROGRAM AUTOMATICALLY CONSTRUCTS AN OPTIMAL HISTOGRAM FOR   00000130
C      A GIVEN DATA SET.                                                00000140
C     -----------------------------------------------------------------00000150
C                                                                       00000160
C     INPUT REQUIRED:                                                   00000170
C                                                                       00000180
C      EVERY INPUT SHOULD BE PUNCHED WITH FORMAT IN EACH PARENTHESIS.   00000190
C                                                                       00000200
C     DATA NO. 1 :    FORMAT (3I5,F10.1,I5,F10.1)                       00000210
C             N   :   SAMPLE SIZE                                       00000220
C             ITY :   =0 FOR A SMAPLE OF UNGROUPED MEASUREMENTS         00000230
C                     =1 FOR THAT OF GROUPED MEASUREMENTS               00000240
C             IN  :   INPUT DATA DEVICE SPECIFICATION  (IN=5:CARD READER)00000250
C             X1  :   ACCURACY OF MEASUREMENT OF MAIN DATA              00000260
C                     FOR EXAMPLE, PUT X1 AS 0.1 IF THE DATA ARE 18.3,  00000270
C                     19.2,15.1, ... .                                  00000280
C             K   :   TOTAL NUMBER OF INTERVALS IN CASE 'ITY=1.'   NO   00000290
C                     INPUT REQUIRED IF 'ITY=0.'                        00000300
C             AM1 :   MINIMUM VALUE OF CELL MID-POINTS IN CASE 'ITY=1.' 00000310
C                     NO INPUT REQUIRED IF 'ITY=0.'                     00000320
C     DATA NO. 2:    FORMAT (20A4)                                      00000330
C              TITLE(I), I=1,20 :  TITLE OF THE DATA                    00000340
C     DATA NO. 3:    FORMAT (20A4)                                      00000350
C               FMT(I), I=1,20 : INPUT DATA FORMAT SPECIFICATION        00000360
C                     STATEMENT  --- FOR EXAMPLE, '(8F3.1)'             00000370
C     DATA NO. 4                                                        00000380
C               MAIN DATA                                               00000390
C                                                                       00000400
C     -----------------------------------------------------------------00000410
C                                                                       00000420
C      OUTPUTS:                                                         00000430
C       1) OPTIMAL HISTGRAM.                                            00000440
C          THIS OUTPUT CONTAINS                                         00000450
C          TITLE;                                                       00000460
C          MINIMUM OF MEASUREMENTS; INTERVAL LENGTH IN INITIAL          00000470
C          CATEGORIZATION; MAXIMUM OF MEASUREMENTS; TOTAL NUMBER OF     00000480
C          CLASSES  IN INITIAL CATEGORIZATION;                          00000490
C          OPTIMAL HISTOGRAM ( WITH CLASS BOUNDARY, CELL FREQUENCY      00000500
C          AND ESTIMATED PROBABILITY OF EACH CLASS );                   00000510
C          SAMPLE SIZE;                                                 00000520
C          MAICE-VALUE.                                                 00000530
C       2) LIST OF BETTER MODELS ARRANGED IN ASCENDING ORDER OF AIC.    00000540
C          THIS OUTPUT CONTAINS THE RANKING OF MODEL, MODEL, NUMBER OF  00000550
C          PARAMETERS, AIC VALUE AND THE DIFFERENCE OF AIC VALUE.       00000560
C     -----------------------------------------------------------------00000570
C                                                                       00000580
      IMPLICIT REAL*8 (A-H,O-Z)                                         00000590
      DIMENSION X(10000),NN(1000),NX(1000),NC(1000),NXX(1000),NCC(1000),00000600
     1          NX1(1000),NC1(1000),FMT(20),TITLE(20),Z(1000),AXX(1000),00000610
     2          NAA(4,1000)                                             00000620
C                                                                       00000630
C      DATA INPUT                                                       00000640
C                                                                       00000650
      READ(5,1001) N,ITY,IN,X1,K,AM1                                    00000660
      READ(5,1002) (TITLE(I),I=1,20)                                    00000670
      READ(5,1002) FMT                                                  00000680
 1001 FORMAT(3I5,F10.1,I5,F10.1)                                        00000690
 1002 FORMAT(20A4)                                                      00000700
      KM=0                                                              00000710
      XN=N                                                              00000720
      AICMI=10.**10                                                     00000730
      IF(KM.EQ.0) KM=1000                                               00000740
      IF(X1.EQ.0.) X1=1.                                                00000750
      IF(IN.EQ.0) IN=5                                                  00000760
      IF(ITY.EQ.1) GO TO 10                                             00000770
C                                                                       00000780
C      MAIN DATA INPUT, CHOICE OF THE INITIAL CLASS INTERVAL AND        00000790
C      CONSTRUCTION OF THE INITIAL FREQUENCY TABLE (IN CASE 'ITY=0')    00000800
C                                                                       00000810
      READ(IN,FMT) (X(I),I=1,N)                                         00000820
      AMIN=10.**10                                                      00000830
      AMAX=-10.**10                                                     00000840
      SUMX=0.0                                                          00000850
      DO 25 I=1,N                                                       00000860
      IF(AMIN.GE.X(I)) AMIN=X(I)                                        00000870
      IF(AMAX.LE.X(I)) AMAX=X(I)                                        00000880
      SUMX=SUMX+X(I)                                                    00000890
   25 CONTINUE                                                          00000900
      AMAX=AMAX+X1/2.                                                   00000910
      AMIN=AMIN-X1/2.                                                   00000920
      DO 20 I=1,1000                                                    00000930
      NN(I)=0                                                           00000940
      NC(I)=0                                                           00000950
   20 CONTINUE                                                          00000960
```

```
      LLL=2*SQRT(XN)-1                                                  00000970
      AI=X1                                                             00000980
      AM1=AMIN                                                          00000990
      MM=(AMAX-AMIN)/X1+0.5                                             00001000
      MM=MIN(MM,200,LLL)                                                00001010
      XX=(AMAX-AMIN)/MM                                                 00001020
      AI=XX                                                             00001030
      K=MM                                                              00001040
      AM2=AMAX                                                          00001050
      KK=K                                                              00001060
      DO 70 J=1,N                                                       00001070
      DO 30 I=1,K                                                       00001080
      AM=AM1+AI*I                                                       00001090
      IF(AM.LE.X(J)) GO TO 30                                           00001100
      NN(I)=NN(I)+1                                                     00001110
      GO TO 70                                                          00001120
   30 CONTINUE                                                          00001130
   70 CONTINUE                                                          00001140
      GO TO 75                                                          00001150
C      MAIN DATA INPUT IN CASE 'ITY=1'                                  00001160
C                                                                       00001170
   10 CONTINUE                                                          00001180
      READ(IN,FMT) (NN(I),I=1,K)                                        00001190
      XX=X1                                                             00001200
      AM2=AM1+K*XX                                                      00001210
   75 CONTINUE                                                          00001220
C                                                                       00001230
C     PRINTING THE MAX. AND MIN. VALUES OF CELL MIDPOINTS AND THE CLASS 00001240
C     INTERVALS IN THE ABOVE INITIAL FREQUENCY TABLE                    00001250
C                                                                       00001260
      N1=0                                                              00001270
      NCT=0                                                             00001280
      NJ=K-1                                                            00001290
      WRITE(6,1004) TITLE                                               00001300
 1004 FORMAT(1H1/1H ,10X,'TITLE :',10X,20A4)                            00001310
C     WRITE(6,1005)                                                     00001320
 1005 FORMAT(1H0,'      INITIAL FREQUENCY TABLE                         00001330
C     WRITE(6,1006) (NN(I),I=1,K)                                       00001340
 1006 FORMAT((1H ,4(5I5,3X)))                                           00001350
      WRITE(6,1007)                                                     00001360
 1007 FORMAT(1H0,10X,'INITIAL CATEGORIZATION :'                         00001370
     1        /1H0,5X,' *** MIN OF DATA        INTERVAL LENGTH      MAX OF 00001380
     2DATA        NO. OF INTIAL CATEGORIES ***'/)                       00001390
      WRITE(6,1008) AM1,XX,AM2,K                                        00001400
 1008 FORMAT(1H ,3E20.5,10X,I10)                                        00001410
C                                                                       00001420
C      POOLING OF CALSSES (CASE OF EQUAL INTERVAL LENGTH)               00001430
C                                                                       00001440
   40 CONTINUE                                                          00001450
      N1=N1+1                                                           00001460
      II=0                                                              00001470
      DO 80 J=1,K                                                       00001480
      NX(J)=0                                                           00001490
      NC(J)=N1                                                          00001500
      DO 80 I=1,N1                                                      00001510
      II=II+1                                                           00001520
      NX(J)=NX(J)+NN(II)                                                00001530
      IF(II.EQ.K) GO TO 90                                              00001540
   80 CONTINUE                                                          00001550
   90 KK=J                                                              00001560
      NC(KK)=I                                                          00001570
      DO 50 I=1,KK                                                      00001580
      IF(NX(I).EQ.0) GO TO 61                                           00001590
   50 CONTINUE                                                          00001600
C                                                                       00001610
C      COMPUTATION OF THE VALUE OF AIC                                  00001620
C                                                                       00001630
      IM=KK-1                                                           00001640
      AIC=0.0                                                           00001650
      DO 95 I=1,KK                                                      00001660
      N2=NC(I)                                                          00001670
      AIC=-2*(NX(I)*LOG(NX(I)/(N2*XN))    )+AIC                         00001680
   95 CONTINUE                                                          00001690
      AIC=AIC+2*IM                                                      00001700
      IF(AIC.GT.AICMI+20.0) GO TO 300                                   00001710
      NCT=NCT+1                                                         00001720
      AXX(NCT)=AIC                                                      00001730
      NAA(1,NCT)=NC(1)                                                  00001740
      IF(KK.GT.2) NAA(2,NCT)=NC(2)                                      00001750
      NAA(3,NCT)=NC(KK)                                                 00001760
      NAA(4,NCT)=KK                                                     00001770
  300 CONTINUE                                                          00001780
      IF(AICMI.LE.AIC) GO TO 60                                         00001790
C                                                                       00001800
C       RETAINING OF MINIMUM AIC                                        00001810
C                                                                       00001820
      AICMI=AIC                                                         00001830
      KKK=KK                                                            00001840
      DO 250 I=1,KK                                                     00001850
      NXX(I)=NX(I)                                                      00001860
      NCC(I)=NC(I)                                                      00001870
  250 CONTINUE                                                          00001880
   60 CONTINUE                                                          00001890
      IF(KK.LE.3) GO TO 61                                              00001900
      AICMK=AIC                                                         00001910
      CALL TANTEN(NX,NC,KK,XN,AICMK, AIC,KKX,NX1,NC1,NCT,NAA,AXX)       00001920
```

```
      IF(KK.GT.KM)GO TO 61                                              00001930
      IF(AICMI.LE.AIC) GO TO 61                                         00001940
      AICMI=AIC                                                         00001950
      KKK=KKX                                                           00001960
      DO 251 I=1,KKX                                                    00001970
      NXX(I)=NX1(I)                                                     00001980
      NCC(I)=NC1(I)                                                     00001990
  251 CONTINUE                                                          00002000
   61 CONTINUE                                                          00002010
      IF(N1.EQ.1) GO TO 40                                              00002020
C                                                                       00002030
C      SHIFT OF CELL BOUNDARIES                                         00002040
C                                                                       00002050
      N11=N1-1                                                          00002060
  200 CONTINUE                                                          00002070
      II=0                                                              00002080
      IF(N11.EQ.0) GO TO 140                                            00002090
      NX(1)=0                                                           00002100
      NC(1)=N11                                                         00002110
      DO 170 I=1,N11                                                    00002120
      II=II+1                                                           00002130
      NX(1)=NN(II)+NX(1)                                                00002140
  170 CONTINUE                                                          00002150
      DO 180 J=2,K                                                      00002160
      NX(J)=0                                                           00002170
      NC(J)=N1                                                          00002180
      DO 180 I=1,N1                                                     00002190
      II=II+1                                                           00002200
      NX(J)=NX(J)+NN(II)                                                00002210
      IF(II.EQ.K) GO TO 190                                             00002220
  180 CONTINUE                                                          00002230
  190 KK=J                                                              00002240
      NC(KK)=I                                                          00002250
      DO 150 I=1,KK                                                     00002260
      IF(NX(I).EQ.0) GO TO 211                                          00002270
  150 CONTINUE                                                          00002280
C                                                                       00002290
C      COMPUTATION OF THE VALUE OF AIC                                  00002300
C                                                                       00002310
      IM=KK-1                                                           00002320
      AIC=0.0                                                           00002330
      DO 195 I=1,KK                                                     00002340
      N2=NC(I)                                                          00002350
      AIC=-2*(NX(I)*LOG(NX(I)/(N2*XN))    )+AIC                         00002360
  195 CONTINUE                                                          00002370
      AIC=AIC+2*IM                                                      00002380
      IF(AIC.GT.AICMI+20.0) GO TO 310                                   00002390
      NCT=NCT+1                                                         00002400
      AXX(NCT)=AIC                                                      00002410
      NAA(1,NCT)=NC(1)                                                  00002420
      IF(KK.GT.2) NAA(2,NCT)=NC(2)                                      00002430
      NAA(3,NCT)=NC(KK)                                                 00002440
      NAA(4,NCT)=KK                                                     00002450
  310 CONTINUE                                                          00002460
      IF(AICMI.LE.AIC) GO TO 210                                        00002470
      AICMI=AIC                                                         00002480
      KKK=KK                                                            00002490
      DO 260 I=1,KK                                                     00002500
      NXX(I)=NX(I)                                                      00002510
      NCC(I)=NC(I)                                                      00002520
  260 CONTINUE                                                          00002530
  210 CONTINUE                                                          00002540
      IF(KK.LE.3) GO TO 211                                             00002550
      AICMK=AIC                                                         00002560
      CALL TANTEN(NX,NC,KK,XN,AICMK,AIC,KKX,NX1,NC1,NCT,NAA,AXX)        00002570
      IF(KK.GT.KM)GO TO 211                                             00002580
      IF(AICMI.LE.AIC) GO TO 211                                        00002590
      AICMI=AIC                                                         00002600
      KKK=KKX                                                           00002610
      DO 252 I=1,KKX                                                    00002620
      NXX(I)=NX1(I)                                                     00002630
      NCC(I)=NC1(I)                                                     00002640
  252 CONTINUE                                                          00002650
  211 CONTINUE                                                          00002660
      N11=N11-1                                                         00002670
      GO TO 200                                                         00002680
  140 CONTINUE                                                          00002690
      IF(N1.GE.NJ) GO TO 100                                            00002700
      GO TO 40                                                          00002710
  100 CONTINUE                                                          00002720
      CALL HISTG(AM1,XX,X1,NXX,NCC,KKK,N,AICMI,Z,NN)                    00002730
      CALL AICPR(NCT,NAA,AXX,AICMI)                                     00002740
      STOP                                                              00002750
      END                                                               00002760
      SUBROUTINE TANTEN(N1,N2,KK,XN,AICC,ACC,KKK,NX,NCC,NCT,NAA,AXX)    00002770
C                                                                       00002780
C      THIS SUBROUTINE DOES WITH THE POOLING OF THE TWO EXTREME CLASSES.00002790
C                                                                       00002800
      IMPLICIT REAL*8 (A-H,O-Z)                                         00002810
      DIMENSION N1(1000),N2(1000),NN(1000),NC(1000),NX(1000),NCC(1000)  00002820
      DIMENSION NY(1000),NCY(1000),AXX(1000),NAA(4,1000)                00002830
      AICMM=AICC                                                        00002840
      ACC=AICC                                                          00002850
      II=1                                                              00002860
      KN=KK/2                                                           00002870
  100 CONTINUE                                                          00002880
```

```
      AICM=10.**10                                                      00002890
      II1=II+1                                                          00002900
      NN(1)=0                                                           00002910
      NC(1)=0                                                           00002920
      DO 20 I=1,II1                                                     00002930
      NN(1)=NN(1)+N1(I)                                                 00002940
      NC(1)=NC(1)+N2(I)                                                 00002950
   20 CONTINUE                                                          00002960
      DO 30 I=1,II1                                                     00002970
      KK1=KK-II-I                                                       00002980
      IF(KK1.LT.2) KK1=1                                                00002990
      IF(KK1.LT.2) GO TO 45                                             00003000
      DO 40 J=2,KK1                                                     00003010
      NN(J)=N1(II+J)                                                    00003020
      NC(J)=N2(II+J)                                                    00003030
   40 CONTINUE                                                          00003040
   45 CONTINUE                                                          00003050
      KK2=KK1+1                                                         00003060
      NN(KK2)=0                                                         00003070
      NC(KK2)=0                                                         00003080
      DO 50 J=1,I                                                       00003090
      IF(II+KK1+J.GT.KK) GO TO 50                                       00003100
      NN(KK2)=NN(KK2)+N1(II+KK1+J)                                      00003110
      NC(KK2)=NC(KK2)+N2(II+KK1+J)                                      00003120
   50 CONTINUE                                                          00003130
      IF(NC(KK2).EQ.0) GO TO 30                                         00003140
      AIC=0.                                                            00003150
      DO 60 J=1,KK2                                                     00003160
      N4=NC(J)                                                          00003170
      AIC=-2.*(NN(J)*LOG(NN(J)/(N4    *XN)))+AIC                        00003180
   60 CONTINUE                                                          00003190
      AIC=AIC+2*KK1                                                     00003200
      IF(AIC.GT.AICMM+20.0) GO TO 300                                   00003210
      NCT=NCT+1                                                         00003220
      AXX(NCT)=AIC                                                      00003230
      NAA(1,NCT)=NC(1)                                                  00003240
      IF(KK2.GT.2) NAA(2,NCT)=NC(2)                                     00003250
      NAA(3,NCT)=NC(KK2)                                                00003260
      NAA(4,NCT)=KK2                                                    00003270
  300 CONTINUE                                                          00003280
      IF(AICM.LT.AIC) GO TO 70                                          00003290
      AICM=AIC                                                          00003300
      KKY=KK2                                                           00003310
      DO 80 J=1,KK2                                                     00003320
      NY(J)=NN(J)                                                       00003330
      NCY(J)=NC(J)                                                      00003340
   80 CONTINUE                                                          00003350
   70 CONTINUE                                                          00003360
   30 CONTINUE                                                          00003370
      DO 110 I=1,II                                                     00003380
      II1=II+1-I                                                        00003390
      IF(II1.LT.1) II1=1                                                00003400
      NN(1)=0                                                           00003410
      NC(1)=0                                                           00003420
      DO 120 J=1,II1                                                    00003430
      NN(1)=NN(1)+N1(J)                                                 00003440
      NC(1)=NC(1)+N2(J)                                                 00003450
  120 CONTINUE                                                          00003460
      KK1=KK-II-II1                                                     00003470
      IF(KK1.LT.2) KK1=1                                                00003480
      IF(KK1.LT.2) GO TO 123                                            00003490
      DO 125 J=2,KK1                                                    00003500
      NN(J)=N1(II1+J-1)                                                 00003510
      NC(J)=N2(II1+J-1)                                                 00003520
  125 CONTINUE                                                          00003530
  123 CONTINUE                                                          00003540
      KK2=KK1+1                                                         00003550
      NN(KK2)=0                                                         00003560
      NC(KK2)=0                                                         00003570
      III=II+1                                                          00003580
      DO 130 J=1,III                                                    00003590
      IF(II1+KK1-1+J.GT.KK) GO TO 130                                   00003600
      NN(KK2)=NN(KK2)+N1(II1+KK1-1+J )                                  00003610
      NC(KK2)=NC(KK2)+N2(II1+KK1-1+J )                                  00003620
  130 CONTINUE                                                          00003630
      AIC=0.                                                            00003640
      DO 140 J=1,KK2                                                    00003650
      N4=NC(J)                                                          00003660
      AIC=-2.*(NN(J)*LOG(NN(J)/(N4    *XN)))+AIC                        00003670
  140 CONTINUE                                                          00003680
      AIC=AIC+2*KK1                                                     00003690
      IF(AIC.GT.AICMM+20.0) GO TO 310                                   00003700
      NCT=NCT+1                                                         00003710
      AXX(NCT)=AIC                                                      00003720
      NAA(1,NCT)=NC(1)                                                  00003730
      IF(KK2.GT.2) NAA(2,NCT)=NC(2)                                     00003740
      NAA(3,NCT)=NC(KK2)                                                00003750
      NAA(4,NCT)=KK2                                                    00003760
  310 CONTINUE                                                          00003770
      IF(AICM.LT.AIC) GO TO 150                                         00003780
      AICM=AIC                                                          00003790
      KKY=KK2                                                           00003800
      DO 160 J=1,KK2                                                    00003810
      NY(J)=NN(J)                                                       00003820
      NCY(J)=NC(J)                                                      00003830
  160 CONTINUE                                                          00003840
```

```
  150 CONTINUE                                                          00003850
  110 CONTINUE                                                          00003860
      IF(II.GE.KN) GO TO 200                                            00003870
      IF(AICM.GT.AICMM) GO TO 200                                       00003880
      AICMM=AICM                                                        00003890
      ACC=AICM                                                          00003900
      KKK=KKY                                                           00003910
      DO 170 I=1,KKK                                                    00003920
      NX(I)=NY(I)                                                       00003930
      NCC(I)=NCY(I)                                                     00003940
  170 CONTINUE                                                          00003950
      II=II+1                                                           00003960
      GO TO 100                                                         00003970
  200 CONTINUE                                                          00003980
      RETURN                                                            00003990
      END                                                               00004000
      SUBROUTINE HISTG(AM1,XX,X1,NX,NC,KK,N,AICMI,Z,NN)                 00004010
C                                                                       00004020
C      THIS SUBROUTINE PRODUCES A HISTOGRAM ON LINE PRINTER.            00004030
C                                                                       00004040
      IMPLICIT REAL*8 (A-H,O-Z)                                         00004050
      INTEGER*2 ST                                                      00004060
      REAL FMT1,FMT2                                                    00004070
      DIMENSION NX(1000),Z(100),NC(1000),NN(1000),FMT1(20),FMT2(20)     00004080
      DIMENSION Y(11),RANGE(20)                                         00004090
      DATA ST/'*'/                                                      00004100
      DATA FMT1/'(1H+','3X, ','2(  ','F8. ','    ',',2X)',')   '/       00004110
      DATA FMT2/'   0','   1','   2','   3','   4','   5'/             00004120
      DATA KR,RANGE/11,0.01,0.02,0.03,0.04,0.05,0.1,0.2,0.3,0.4,0.5,1./ 00004130
      WRITE(6,1001)                                                     00004140
      WRITE(6,1002)                                                     00004150
      X2=X1/2.0D0                                                       00004160
      DO 20 I=1,5                                                       00004170
      IJ=X2*10.0D0**(I-1)                                               00004180
      IF(IJ.NE.0) GO TO 30                                              00004190
   20 CONTINUE                                                          00004200
      I=5                                                               00004210
   30 IX=I                                                              00004220
      FMT1(5)=FMT2(IX)                                                  00004230
      AMAX=0.0                                                          00004240
      DO 10 I=1,KK                                                      00004250
      IF(AMAX.LT.FLOAT(NX(I))/NC(I)/N) AMAX=FLOAT(NX(I))/NC(I)/N        00004260
   10 CONTINUE                                                          00004270
      IF(AMAX.LE.RANGE(1)) AMAX1=RANGE(1)                               00004280
      DO 70 I=2,KR                                                      00004290
      IF(AMAX.GT.RANGE(I-1).AND.AMAX.LE.RANGE(I)) AMAX1=RANGE(I)        00004300
   70 CONTINUE                                                          00004310
      IF(AMAX.GT.RANGE(KR)) AMAX1=RANGE(KR)                             00004320
      AB=AM1                                                            00004330
      IL=0                                                              00004340
      DO 60 I=1,11                                                      00004350
      Y(I)=AMAX1*(I-1)/10                                               00004360
   60 CONTINUE                                                          00004370
      WRITE(6,1008) (Y(I),I=1,11)                                       00004380
      WRITE(6,1007)                                                     00004390
      DO 40 I=1,KK                                                      00004400
      AI1=AMAX1/80                                                      00004410
      NIJ=(NX(I))/AI1/NC(I)/N+0.5                                       00004420
      P=NX(I)/(FLOAT(N)*NC(I))                                          00004430
      NCX=NC(I)                                                         00004440
      Z(I)=AB+NCX*XX                                                    00004450
      DO 45 L=1,NCX                                                     00004460
      IL=IL+1                                                           00004470
      AB1=AB                                                            00004480
      AB=AB+XX                                                          00004490
      IF(NIJ.NE.0) WRITE(6,1005) IL,NN(IL),P,(ST,J=1,NIJ)               00004500
      IF(NIJ.EQ.0) WRITE(6,1005) IL,NN(IL),P                            00004510
      WRITE(6,1006)                                                     00004520
      WRITE(6,FMT1) AB1,AB                                              00004530
   45 CONTINUE                                                          00004540
   40 CONTINUE                                                          00004550
      WRITE(6,1007)                                                     00004560
      WRITE(6,1003) N                                                   00004570
      WRITE(6,1004) AICMI                                               00004580
   50 CONTINUE                                                          00004590
 1001 FORMAT(///1H0,50X,   'OPTIMAL FREQUENCY HISTOGRAM  ')             00004600
 1002 FORMAT(1H0,'NO.',6X,'CLASS   ',6X,'FREQUENCIES',3X,'ESTIMATES'/   00004610
     1       1H ,9X,'BOUNDARIES')                                       00004620
 1003 FORMAT(1H ,8X,' TOTAL',8X,I7)                                     00004630
 1004 FORMAT(/1H0, 8X,' MAICE ',F13.2)                                  00004640
 1005 FORMAT(1H ,I3,8X,1X,'~',8X,I8,5X,F10.5,' |',80A1)                 00004650
 1006 FORMAT(1H+,45X,10('|',7X),'|')                                    00004660
 1007 FORMAT(1H ,45X,10('+-------'),'+')                                00004670
 1008 FORMAT(1H ,42X,11(F6.4,2X))                                       00004680
      RETURN                                                            00004690
      END                                                               00004700
      SUBROUTINE AICPR(NCT,NAA,AXX,AICMI)                               00004710
C                                                                       00004720
C     THIS SUBROUTINE PRODUCES THE LIST OF GOOD MODELS IN ASCENDING     00004730
C     ORDER OF AIC.                                                     00004740
C                                                                       00004750
      REAL *8 AXX,AMIN,AICMI,DF,AI                                      00004760
      DIMENSION NAA(4,1000),AXX(1000)                                   00004770
      WRITE(6,1001)                                                     00004780
      NO=0                                                              00004790
      DO 10 I=1,NCT                                                     00004800
```

```
      II=I                                                             00004810
      AMIN=AXX(I)                                                      00004820
      DO 20 J=I+1,NCT                                                  00004830
      IF(AMIN.LT.AXX(J)) GO TO 20                                      00004840
      II=J                                                             00004850
      AMIN=AXX(J)                                                      00004860
   20 CONTINUE                                                         00004870
      IF(AMIN.GT.AICMI+20.0) GO TO 40                                  00004880
      J=II                                                             00004890
      AXX(J)=AXX(I)                                                    00004900
      AXX(I)=AMIN                                                      00004910
      DO 30 K=1,4                                                      00004920
      NA=NAA(K,J)                                                      00004930
      NAA(K,J)=NAA(K,I)                                                00004940
      NAA(K,I)=NA                                                      00004950
   30 CONTINUE                                                         00004960
      IF(I.EQ.1) DF=0.0                                                00004970
      IF(I.NE.1) DF=AXX(I)-AI                                          00004980
      AI=AXX(I)                                                        00004990
      ND=NAA(4,I)-1                                                    00005000
      IF(I.EQ.1) GO TO 60                                              00005010
      DO 50 K=1,4                                                      00005020
      IF(NAA(K,I-1).NE.NAA(K,I)) GO TO 60                              00005030
   50 CONTINUE                                                         00005040
      GO TO 10                                                         00005050
   60 CONTINUE                                                         00005060
      NO=NO+1                                                          00005070
      IF(ND.GT.1) WRITE(6,1002) NO,NAA(1,I),NAA(2,I),NAA(3,I),ND,AXX(I),00005080
     1                          DF                                     00005090
      IF(ND.EQ.1) WRITE(6,1003) NO,NAA(1,I),NAA(3,I),ND,AXX(I),DF      00005100
   10 CONTINUE                                                         00005110
   40 CONTINUE                                                         00005120
      RETURN                                                           00005130
 1001 FORMAT(1H1//1H ,2X,'NO.',10X,'MODEL',5X,'NO. OF FREE PARAMETERS', 00005140
     1             8X,'A I C',8X,'DIFFERRENCE OF AIC'/)                00005150
 1002 FORMAT(1H ,I5,5X,'(',2(I3,','),I3,')',5X,I5,13X,F15.2,5X,F15.2)  00005160
 1003 FORMAT(1H ,I5,5X,'(',(I3,','),4X,I3,')',5X,I5,13X,F15.2,5X,F15.2) 00005170
      END                                                              00005180
```

III-2 SEARCH FOR AN OPTIMAL TWO-WAY TABLE (CATDAP-01 PART 1)

The program CATDAP-01 PART 1 is a part of the FORTRAN program CATDAP-01 which was developed for the analysis of contingency tables described in Chapter 6. This program searches for the best single explanatory variable which has the most effective information on a specific response variable in all possible two-way tables. Unlike CATDAP-02, this program cannot perform the search required for an optimal combination and categorization of the variables. For the input required and the possible output, see the comments in the program. Input and output examples for a sample size of 100 are shown after the source list. In this example we discarded the sample in which the variable "AGE" takes the value 4 and considered the sixth variable "BORN AGAIN" and the subsequent variables as response variables one after another. (For reasons of space, we have given only the two-way tables in which the sixth variable is considered as the response variable.)

```
      PROGRAM CATDAP                                                    00000010
C     CATEGORICAL DATA ANALYSIS PROGRAM PACKAGE 01  PART 1              00000020
C                                         (CATDAP 01 PART 1)            00000030
C                                            (LAST VERSION)  APR.22,1982 00000040
C                                                                       00000050
C                                  DESIGNED BY Y.SAKAMOTO               00000060
C                                  PROGRAMMED BY K.KATSURA AND Y.SAKAMOTO 00000070
C                                THE INSTITUTE OF STATISTICAL MATHEMATICS 00000080
C                        4-6-7, MINAMI-AZABU, MINATO-KU, TOKYO, JAPAN 106 00000090
C                                                                       00000100
C       THIS IS A BASIC PROGRAM SEARCHING FOR THE BEST SINGLE           00000110
C       PREDICTOR (EXPLANATORY VARIABLE) ON WHICH A SPECIFIC VARIABLE   00000120
C       (RESPONSE VARIABLE) HAS THE STRONGEST DEPENDENCE.   THIS        00000130
C       PROGRAM CAN SIMULTANEOUSLY HANDLE MANY VARIABLES AS RESPONSE    00000140
C       VARIABLES. UNLIKE CATDAP-02, CATDAP-01 DOES NOT PERFORM AN      00000150
C       AUTOMATIC POOLING OF CATEGORIES OF EACH VARIABLE.   FOR THE     00000160
C       COMPLETE SOURCE LIST OF CATDAP-01 AND 02, SEE "K. KATSURA       00000170
C       AND Y.SAKAMOTO (1980)  CATDAP, A CATEGORICAL DATA ANALYSIS      00000180
C       PROGRAM PACKAGE,  COMPUTER SCIENCE MONOGRAPHS, NO.14, INST.     00000190
C       STATIST. MATH.                                                  00000200
C                                                                       00000210
C-----------------------------------------------------------------------00000220
C                                                                       00000230
C       INPUTS REQUIRED:                                                00000240
C                                                                       00000250
C       EVERY INPUT SHOULD BE PUNCHED WITH FORMAT IN PARENTHESIS.       00000260
C                                                                       00000270
C       DATA NO. 1:     FORMAT (20I4)                                   00000280
C         NSAMP:  SAMPLE SIZE                                           00000290
C         N:  TOTAL NUMBER OF VARIABLES                                 00000300
C         L:  TOTAL NUMBER OF RESPONSE VARIABLES. (L CANNOT EXCEED N.)  00000310
C         RECODE:  TOTAL NUMBER OF VARIABLES CONTAINING CATEGORIES TO BE00000320
C                 RECODED, IF ANY.                                      00000330
C         ICROSS:  NUMBER OF TWO-WAY TABLES REQUESTED AS OUTPUTS.   SET 00000340
C                  -1 IF ALL TABLES DESIRED.   REGARDED IT AS 10 UNLESS 00000350
C                  OTHERWISE SPECIFIED.                                 00000360
C         IEXP: =1  TO REGARD AS EXPLANATORY VARIABLES ONLY THE         00000370
C                   VARIABLES SPECIFIED BY 'FACE2(I)'                   00000380
C               =0  OTHERWISE                                           00000390
C         N1:  TOTAL NUMBER OF EXPLANATORY VARIABLES IN CASE 'IEXP=1'.  00000400
C              PUT 0 IF 'IEXP=0'.                                       00000410
C         JSAMP:  NUMBER OF DATA TO BE SKIPPED TO REQUIRED DATA         00000420
C         IN:  INPUT DATA DEVICE SPECIFICATION  (IN=5: CARD READER)     00000430
C         ISKIP1:  NUMBER OF VARIABLES SPECIFIED BY 'ISKIP(1,I)' TO     00000440
C                  DELETE A PART OF DATA.   PUT IT AS 0 IF ALL THE      00000450
C                  DATA ARE UTILIZED.                                   00000460
C                                                                       00000470
C       DATA NO. 2:     FORMAT (20I4)                                   00000480
C         ITEM1(I),ITEM2(I), I=1,N:   MINIMUM AND MAXIMUM VALUES OF CODES00000490
C                                     OF EACH VARIABLE                  00000500
C                                                                       00000510
C       DATA NO. 3:     FORMAT (20I4)                                   00000520
C         ICONV(I,J), I=1,RECODE, J=1,20:  VARIABLE NUMBER AND RECODED  00000530
C                                          NUMBERS IN CASE 'RECODE>0'.  00000540
C              ICONV(I,1):  VARIABLE NUMBER                             00000550
C              ICONV(I,J), J=2,20:  RECODED NUMBERS CORRESPONDING TO    00000560
C                                   THE ORIGINAL CODES 1,2,...  ,19.    00000570
C           BY THIS INPUT, THE ORIGINAL CODE 'J-1' IS TO BE             00000580
C           CONVERTED INTO THE CODE 'ICONV(I,J)'.                       00000590
C                                                                       00000600
C       DATA NO. 4:     FORMAT (20I4)                                   00000610
C         FACE(I),I=1,L:  VARIABLE NUMBERS OF RESPONSE VARIABLES        00000620
C                                                                       00000630
C       DATA NO. 5:     FORMAT (20I4)                                   00000640
C         FACE2(I),I=1,N1:  VARIABLE NUMBERS OF EXPLANATORY VARIABLES IN00000650
C                           CASE 'IEXP1'.   NO INPUT REQUIRED IF        00000660
C                           'IEXP=0'.                                   00000670
C                                                                       00000680
C       DATA NO. 6:     FORMAT (20A4)                                   00000690
C         FMT(I),I=1,20:  INPUT DATA FORMAT SPECIFICATION STATEMENT     00000700
C                         ----FOR EXAMPLE----  (10I1)                   00000710
C                                                                       00000720
C       DATA NO. 7:     FORMAT (10A1)                                   00000730
C         TITLE(I,K),I=1,10, K=1,N:  TITLES OF VARIABLES                00000740
C                                                                       00000750
C       DATA NO. 8:     FORMAT (20I4)                                   00000760
C         ISKIP(J,I),J=1,20, I=1,ISKIP1:                                00000770
C              ISKIP(1,I):  VARIABLE NUMBER                             00000780
C              ISKIP(2,I):  NUMBER OF CODES                             00000790
C              ISKIP(K,I), K=3,20: CODES                                00000800
C           CATDAP-01 DISCARDS THE RECORDS IN WHICH THE VARIABLE        00000810
C           SPECIFIED BY 'ISKIP(1,I)' HAS ONE OF THE VALUES             00000820
C           SPECIFIED BY 'ISKIP(K,I)'  (K=3,4,...,'ISKIP(2,I)'+2).      00000830
C                                                                       00000840
C       NO. 9:  ORIGINAL DATA                                           00000850
C-----------------------------------------------------------------------00000860
C     POSSIBLE OUTPUTS:                                                 00000870
C                                                                       00000880
C        <PART 1>                                                       00000890
C         1) LIST OF EXPLANATORY VARIABLES ARRANGED IN ORDER OF THE     00000900
C            STRENGTH OF DEPENDENCE OF A RESPONSE VARIABLE; EXPLANATORY 00000910
C            VARIABLE, NUMBER OF CATEGORIES OF EACH EXPLANATORY VARIABLE00000920
C            , VALUE OF AIC, DIFFERENCE OF AIC                          00000930
C         2) THE CORRESPONDING TWO-WAY TABLES                           00000940
C            THE ABOVE OUTPUTS 1) AND 2) ARE PRINTED OUT EVERY RESPONSE 00000950
C            VARIABLE.                                                  00000960
```

```
C         3) SUMMARY OF THE AIC'S                                           00000970
C         4) GRAY SHADING DISPLAY OF ALL THE AIC'S                          00000980
C       -----------------------------------------------------------------00000990
C                                                                           00001000
C                                                                           00001010
C     ***  PART  1 ***                                                      00001020
C                                                                           00001030
C     DATA INPUT,RECODE AND CONSTRUCTION OF TWO-WAY TABLES                  00001040
      IMPLICIT REAL*8(A-H,O-Z)                                              00001050
      INTEGER *2 IA                                                         00001060
      INTEGER ALIM,RECODE                                                   00001070
      REAL *4 FMT                                                           00001080
      DIMENSION IA(350000),A(20000),FMT(100)                                00001090
      IALIM=350000                                                          00001100
      ALIM=20000                                                            00001110
      READ(5,1001) NSAMP,N,L,RECODE,ICROSS,IEXP,N1,JSAMP,IN,ISKIP1          00001120
      NFM=1                                                                 00001130
      N22=2*N+1                                                             00001140
      N24=4*N                                                               00001150
      READ(5,1001) (IA(I),I=N22,N24)                                        00001160
      N2=0                                                                  00001170
      N4=0                                                                  00001180
      DO 10 I=1,N                                                           00001190
      N23=2*I+N22-1                                                         00001200
      INN=I+N                                                               00001210
      IA(I)=IA(N23-1)                                                       00001220
      IA(INN)=IA(N23)                                                       00001230
      N2=N2+(IA(INN)-IA(I)+1)                                               00001240
      N5=IA(INN)-IA(I)+1                                                    00001250
      N4=MAX(N4,N5)                                                         00001260
   10 CONTINUE                                                              00001270
      I1=N+1                                                                00001280
      I2=I1+N                                                               00001290
      I0=2*N                                                                00001300
      IF(RECODE.EQ.0) GO TO 15                                              00001310
      DO 20 K=1,RECODE                                                      00001320
      I01=I0+1                                                              00001330
      I02=I0+20                                                             00001340
      READ(5,1001) (IA(I),I=I01,I02)                                        00001350
      I0=I0+20                                                              00001360
   20 CONTINUE                                                              00001370
   15 CONTINUE                                                              00001380
      I3=I2+20*RECODE+1                                                     00001390
      I3L=I3+L-1                                                            00001400
      READ(5,1001) (IA(I),I=I3,I3L)                                         00001410
      I4=I3+L                                                               00001420
      IF(IEXP.NE.1) GO TO 30                                                00001430
      I4N=I4+N1-1                                                           00001440
      READ(5,1001) (IA(I),I=I4,I4N)                                         00001450
   30 CONTINUE                                                              00001460
      I5=I4+N1                                                              00001470
      NN2=NFM*20                                                            00001480
      READ(5,1002) (FMT(I),I=1,NN2)                                         00001490
      I0=I5-1                                                               00001500
      DO 40 K=1,N                                                           00001510
      I01=I0+1                                                              00001520
      I010=I0+10                                                            00001530
      READ(5,1003) (IA(I),I=I01,I010)                                       00001540
      I0=I0+10                                                              00001550
   40 CONTINUE                                                              00001560
      I6=I5+10*N                                                            00001570
      I0=I6-1                                                               00001580
      IF(ISKIP1.EQ.0) GO TO 50                                              00001590
      DO 45 K=1,ISKIP1                                                      00001600
      I01=I0+1                                                              00001610
      I02=I0+20                                                             00001620
      READ(5,1001) (IA(I),I=I01,I02)                                        00001630
      I0=I0+20                                                              00001640
   45 CONTINUE                                                              00001650
   50 CONTINUE                                                              00001660
      I7=I6+20*ISKIP1                                                       00001670
      I8=I7+N                                                               00001680
      I9=I8+N                                                               00001690
      I10=I9+N2                                                             00001700
      I11=I10+N2                                                            00001710
      I12=I11+N                                                             00001720
      I13=I12+10*N                                                          00001730
      I14=I13+N                                                             00001740
      I15=I14+N                                                             00001750
      I16=I15+N*N                                                           00001760
      I17=I16+N                                                             00001770
      I18=I17+N                                                             00001780
      I19=I18+N                                                             00001790
      I20=I19+N2*N2                                                         00001800
      I21=I20+N                                                             00001810
      I22=I21+N                                                             00001820
      IMAX=I22+N*3                                                          00001830
      J1=N+1                                                                00001840
      J2=J1+N4                                                              00001850
      J3=J2+N4*N4                                                           00001860
      J4=J3+N4                                                              00001870
      J5=J4+N                                                               00001880
      J6=J5+N                                                               00001890
      J7=J6+N                                                               00001900
      J8=J7+N*N                                                             00001910
      J9=J8+N2                                                              00001920
```

```
      JMAX=J9                                                           00001930
      NNN=IALIM-I22                                                     00001940
      N1=N1+1                                                           00001950
      ISKIP1=ISKIP1+1                                                   00001960
      RECODE=RECODE+1                                                   00001970
      IF(IMAX.GT.IALIM.OR.JMAX.GT.ALIM) GO TO 60                        00001980
      CALL CAT1(NSAMP,N,L,RECODE,ICROSS,IEXP,N1,JSAMP,IN,ISKIP1,FMT,N2, 00001990
     1          N4,IA(1),IA(I1),IA(I2),IA(I3),IA(I4),IA(I5),IA(I6),     00002000
     2          IA(I7),IA(I8),IA(I9),IA(I10),IA(I11),IA(I12),IA(I13),   00002010
     3          IA(I14),IA(I15),IA(I16),IA(I17),IA(I18),IA(I19),IA(I20),00002020
     4          IA(I21),A(1),A(J1),A(J2),A(J3),A(J4),A(J5),A(J6),A(J7), 00002030
     5          A(J8),IA(I22),NNN)                                      00002040
      WRITE(6,1004) IMAX,JMAX                                           00002050
      GO TO 70                                                          00002060
   60 WRITE(6,1005) IALIM,IMAX,ALIM,JMAX                                00002070
      STOP 10                                                           00002080
   70 CONTINUE                                                          00002090
 1001 FORMAT(20I4)                                                      00002100
 1002 FORMAT(20A4)                                                      00002110
 1003 FORMAT(10A1)                                                      00002120
 1004 FORMAT(1H ,5X,'USED MEMORIES'                                     00002130
     1      //1H ,I10,2X,'WORDS  (FOR ''IA'' )'/1H0,I10,2X,             00002140
     2       'WORDS  (FOR ''A'' )')                                     00002150
 1005 FORMAT(1H ,' IA OR A DIMENSION OVER ',4I10)                       00002160
 1006 FORMAT(8F10.4)                                                    00002170
      STOP                                                              00002180
      END                                                               00002190
      SUBROUTINE CAT1(NSAMP,N,L,RECODE,ICROSS,IEXP,N3,JSAMP,IN,ISKIP1,  00002200
     1                FMT,N2,N4,ITEM1,ITEM2,ICONV,FACE,FACE2,TITLE,     00002210
     2                ISKIP,ITEM,IDATA,TOTALR,TOTALC,HYO,HY,IZ,NNK,IMX, 00002220
     3                IDF,IDF1,NO1,IA,INDE,MIN,AMIN,TPAR,P,PT,AIC1,AIC2,00002230
     4                AIC3,AIMIN,TCC,IW,NNN)                            00002240
      IMPLICIT REAL*8(A-H,O-Z)                                          00002250
      INTEGER * 2 ITEM,ITEM1,ITEM2,IDATA,TOTALR,TOTALC,FACE,FACE2,IMX,  00002260
     1            TITLE,ICONV,HYO,HY,INDE,IZ,NNK,IDF,IDF1,NO1,IA,MIN,   00002270
     2            ISKIP,IW                                              00002280
      INTEGER     RECODE                                                00002290
      REAL * 4    FMT,FM1,FM2,FM3,FM4,FM5,FAA,FAC,FAD,BL                00002300
      DIMENSION ITEM(N),ITEM1(N),ITEM2(N),IDATA(N),TOTALR(N2),HYO(N),   00002310
     1          TOTALC(N2),FACE(L),FACE2(N3),TITLE(10,N),HY(10,N),IZ(N),00002320
     2          NNK(N),IMX(N,N),NO1(N),IA(N2,N2),ICONV(20,RECODE),      00002330
     3          INDE(N),MIN(N),FAA(5),FAC(10),IDF(N),AIMIN(N,N),        00002340
     4          FMT(100),FM1(7),FM2(11),FM3(7),FM4(11),FM5(7),IDF1(N),  00002350
     5          AMIN(N),TPAR(N4),P(N4,N4),PT(N4),FAD(3),TCC(N2),        00002360
     6          AIC1(N),AIC2(N),AIC3(N),ISKIP(20,ISKIP1),IW(NNN)        00002370
      DATA FM1/'(1H+',',60X',',12I','6,6X',4H,'TO,4HTAL',')'/,          00002380
     1     FM2/', 1I',', 2I',', 3I',', 4I',', 5I',', 6I',', 7I',', 8I', 00002390
     2         ', 9I',',10I',',11I'/                                    00002400
      DATA FM3/'(1H+',',57X',',I2,','2X  ',',12F','6.1,','I5)'/,        00002410
     1     FM4/', 2F',', 3F',', 4F',', 5F',', 6F',', 7F',', 8F',        00002420
     2         ', 9F',',10F',',11F',',12F'/                             00002430
      DATA FM5/'(1H+',',56X',4H,'TO,4HTAL',',12F','6.1,','I5)'/         00002440
      DATA FAA/'(1H+','    ','    ',',F10', '.2) '/                     00002450
      DATA FAC/',13X',',25X',',37X',',49X',',61X',',73X',',85X',        00002460
     1         ',97X',',57X',',69X'/                                    00002470
      DATA FAD/',30X',',64X',',50X'/                                    00002480
      DATA BL/'    '/                                                   00002490
      NWW=0                                                             00002500
      N1=N3-1                                                           00002510
      RECODE=RECODE-1                                                   00002520
      ISKIP1=ISKIP1-1                                                   00002530
      NL=N                                                              00002540
      SAMP=NSAMP                                                        00002550
      IF(IN.EQ.0) IN=5                                                  00002560
      DO 60 I=1,N                                                       00002570
      ITEM(I)=ITEM2(I)-ITEM1(I)+1                                       00002580
   60 CONTINUE                                                          00002590
C                                                                       00002600
C     INITIAL CLEARING                                                  00002610
C                                                                       00002620
      DO 70 J=1,N2                                                      00002630
      TOTALR(J)=0                                                       00002640
      TOTALC(J)=0                                                       00002650
      DO 70 I=1,N2                                                      00002660
   70 IA(I,J)=0                                                         00002670
      DO 80 I=1,N                                                       00002680
      DO 80 J=1,N                                                       00002690
      AIMIN(I,J)=0.                                                     00002700
   80 CONTINUE                                                          00002710
C                                                                       00002720
C     SKIP TO DATA REQUIRED                                             00002730
C                                                                       00002740
      IF(JSAMP.EQ.0) GO TO 90                                           00002750
      DO 100 I=1,JSAMP                                                  00002760
      READ(IN,1001)                                                     00002770
  100 CONTINUE                                                          00002780
   90 CONTINUE                                                          00002790
C                                                                       00002800
C     DATA INPUT, RECODE AND CONSTRUCTION OF TWO-WAY TABLES             00002810
C                                                                       00002820
      NNS=0                                                             00002830
      DO 110 K=1,NSAMP                                                  00002840
      READ(IN,FMT) (IDATA(I),I=1,N)                                     00002850
      IF(ISKIP1.EQ.0) GO TO 150                                         00002860
      DO 160 IS1=1,ISKIP1                                               00002870
      ISKIP2=ISKIP(1,IS1)                                               00002880
```

```
      ISKIP3=ISKIP(2,IS1)                                               00002890
      DO 170 IS2=1,ISKIP3                                               00002900
      IF(IDATA(ISKIP2).EQ.ISKIP(IS2+2,IS1)) GO TO 110                   00002910
  170 CONTINUE                                                          00002920
  160 CONTINUE                                                          00002930
  150 NNS=NNS+1                                                         00002940
      IF(RECODE.EQ.0) GO TO 180                                         00002950
      DO 190 I=1,N                                                      00002960
      DO 200 IK=1,RECODE                                                00002970
      IF(ICONV(1,IK).NE.I) GO TO 200                                    00002980
      IK1=IDATA(I)+1                                                    00002990
      IF(IDATA(I).EQ.0) IK1=11                                          00003000
      IDATA(I)=ICONV(IK1,IK)                                            00003010
      GO TO 190                                                         00003020
  200 CONTINUE                                                          00003030
  190 CONTINUE                                                          00003040
  180 CONTINUE                                                          00003050
      DO 210 I=1,N                                                      00003060
      IF(IDATA(I).EQ.0) IDATA(I)=ITEM(I)                                00003070
      IF(IDATA(I).GT.ITEM(I)) IDATA(I)=ITEM(I)                          00003080
  210 CONTINUE                                                          00003090
      IF(IEXP.NE.1) N1=N                                                00003100
      II=0                                                              00003110
      DO 240 I2=1,L                                                     00003120
      I3=FACE(I2)                                                       00003130
      I=IDATA(I3)+II                                                    00003140
      JJ=0                                                              00003150
      DO 250 J3=1,N1                                                    00003160
      J2=J3                                                             00003170
      IF(IEXP.EQ.1) J2=FACE2(J3)                                        00003180
      J=IDATA(J2)+JJ                                                    00003190
      IF(I2.EQ.1)TOTALR(J)=TOTALR(J)+1                                  00003200
      IF(J3.EQ.1)TOTALC(I)=TOTALC(I)+1                                  00003210
      IA(I,J)=IA(I,J)+1                                                 00003220
  250 JJ=JJ+ITEM(J2)                                                    00003230
  240 II=II+ITEM(I3)                                                    00003240
  110 CONTINUE                                                          00003250
      NSAMP=NNS                                                         00003260
      SAMP=NNS                                                          00003270
      DO 360 I=1,N                                                      00003280
      IDF(I)=ITEM2(I)-ITEM1(I)+1                                        00003290
  360 CONTINUE                                                          00003300
      IF(IEXP.EQ.1) N=N1                                                00003310
      I1=1                                                              00003320
      I3=FACE(1)                                                        00003330
      I2=ITEM(I3)                                                       00003340
      DO 370 IK=1,L                                                     00003350
      J1=1                                                              00003360
      J2=ITEM(1)                                                        00003370
      IF(IEXP.EQ.1) J3=FACE2(1)                                         00003380
      IF(IEXP.EQ.1) J2=ITEM(J3)                                         00003390
C                                                                       00003400
C     INITIAL CLEARING                                                  00003410
C                                                                       00003420
      DO 380 K=1,NL                                                     00003430
      IDF1(K)=0                                                         00003440
      MIN(K)=0                                                          00003450
      AIC1(K)=0.0                                                       00003460
      AIC3(K)=0.0                                                       00003470
  380 AIC2(K)=0.0                                                       00003480
C                                                                       00003490
C      COMPUTATION OF AIC'S                                             00003500
C                                                                       00003510
      K10=0                                                             00003520
      DO 390 K21=1,N                                                    00003530
      K=K21                                                             00003540
      IF(IEXP.EQ.1) K=FACE2(K21)                                        00003550
      NO1(K21)=J1                                                       00003560
      IF(K.EQ.I3)GO TO 400                                              00003570
      K=K21                                                             00003580
      TSMP=0.                                                           00003590
      DO 410 J=J1,J2                                                    00003600
      TCC(J)=0.                                                         00003610
      DO 410 I=I1,I2                                                    00003620
      IF(IA(I,J).EQ.0) TSMP=TSMP+0.5                                    00003630
      IF(IA(I,J).EQ.0) TCC(J)=TCC(J)+0.5                                00003640
  410 CONTINUE                                                          00003650
      DO 420 I=I1,I2                                                    00003660
      TR=0.                                                             00003670
      DO 430 J=J1,J2                                                    00003680
      IF(IA(I,J).EQ.0) TR=TR+0.5                                        00003690
  430 CONTINUE                                                          00003700
      TT=TOTALC(I)+TR                                                   00003710
      DO 420 J=J1,J2                                                    00003720
      IF(IA(J,I).EQ.0) IZ(K)=IZ(K)+1                                    00003730
      AAA=IA(I,J)                                                       00003740
      IF(IA(I,J).EQ.0) AAA=0.5                                          00003750
      AIC1(K)=AIC1(K)+AAA*LOG(AAA/(TT*(TOTALR(J)+TCC(J)))*(SAMP+TSMP))  00003760
  420 CONTINUE                                                          00003770
      DO 440 J=J1,J2                                                    00003780
      IF(TOTALR(J).EQ.0) GO TO 440                                      00003790
      AIC2(K)=AIC2(K)+(TOTALR(J)+TCC(J))*(LOG((TOTALR(J)+TCC(J))/       00003800
     1         (SAMP+TSMP)))                                            00003810
  440 CONTINUE                                                          00003820
C                                                                       00003830
C     INDEPENDENCE TEST BY AIC                                          00003840
```

```
C                                                                       00003850
      IDE=I2-I1+J2-J1                                                   00003860
      IDF1(K)=(I2-I1+1)*(J2-J1+1)-1                                     00003870
      SA=-2.*(-AIC1(K)-IDE+IDF1(K))                                     00003880
      IF(SA.GT.0.0)K10=K10+1                                            00003890
  400 CONTINUE                                                          00003900
      AIC3(K)=-2.0*(AIC1(K)-IDF1(K)+IDE)                                00003910
      J1=J2+1                                                           00003920
      IF(K21.EQ.N) GO TO 390                                            00003930
      IF(IEXP.EQ.1) J3=FACE2(K21+1)                                     00003940
      IF(IEXP.EQ.1) J2=J2+ITEM(J3)                                      00003950
      IF(IEXP.NE.1) J2=J2+ITEM(K21+1)                                   00003960
  390 CONTINUE                                                          00003970
      INDE(IK)=K10                                                      00003980
      K10=N-1                                                           00003990
      IF(IEXP.EQ.1) K10=N                                               00004000
C                                                                       00004010
C     ARRANGING IN ASCENDING ORDER OF AIC'S                             00004020
C                                                                       00004030
      K11=K10                                                           00004040
      DO 450 K2=1,N                                                     00004050
      K3=K2-1                                                           00004060
      AMINN=1.D10                                                       00004070
      DO 510 K4=1,N                                                     00004080
      K=K4                                                              00004090
      IF(IEXP.EQ.1) K=FACE2(K)                                          00004100
      IF(K.EQ.I3) K7=K4                                                 00004110
      IF(K.EQ.I3) GO TO 510                                             00004120
      IF(K3.EQ.0) GO TO 500                                             00004130
      DO 490 K5=1,K3                                                    00004140
      IF(K4.EQ.MIN(K5)) GO TO 510                                       00004150
  490 CONTINUE                                                          00004160
  500 CONTINUE                                                          00004170
      IF(AMINN.LT.AIC3(K4)) GO TO 510                                   00004180
      AMINN=AIC3(K4)                                                    00004190
      K6=K4                                                             00004200
  510 CONTINUE                                                          00004210
      IF(AMINN.EQ.1.D10) K6=K7                                          00004220
      AMIN(K2)=AMINN                                                    00004230
      MIN(K2)=K6                                                        00004240
  450 CONTINUE                                                          00004250
      IF(IEXP.NE.1) MIN(N)=I3                                           00004260
      NN=K11+1                                                          00004270
      NNK(IK)=NN                                                        00004280
      IMX(1,IK)=I3                                                      00004290
      DO 530 I=2,NN                                                     00004300
      M1=MIN(I-1)                                                       00004310
      IF(IEXP.EQ.1) M1=FACE2(M1)                                        00004320
      IF(I3.EQ.M1) NNK(IK)=NN-1                                         00004330
      IF(I3.EQ.M1) GO TO 530                                            00004340
      IMX(I,IK)=M1                                                      00004350
      IF(AMIN(I-1).LT.0.) MMM=I                                         00004360
  530 CONTINUE                                                          00004370
      IF(NWW.LT.0) NNK(IK)=MMM                                          00004380
C                                                                       00004390
C     PRINTING ORDERED AIC'S                                            00004400
C                                                                       00004410
      WRITE(6,1002) (TITLE(I,I3),I=1,10)                                00004420
      WRITE(6,1014)                                                     00004430
      AS=0.                                                             00004440
      DO 540 K2=1,N                                                     00004450
      I4=MIN(K2)                                                        00004460
      I=I4                                                              00004470
  560 CONTINUE                                                          00004480
      AMMIN=AMIN(K2)                                                    00004490
      IF(AMMIN.EQ.1.D10) AMIN(K2)=0.0                                   00004500
      IF(IEXP.EQ.1) AIMIN(I,IK)=AMIN(K2)                                00004510
      IF(IEXP.NE.1) AIMIN(IK,I)=AMIN(K2)                                00004520
      IF(K10.LT.K2) GO TO 540                                           00004530
      IF(AMMIN.EQ.1.D10) GO TO 540                                      00004540
      IF(K2.NE.1) AS=AMIN(K2)-AS                                        00004550
      IF(IEXP.EQ.1) I4=FACE2(I4)                                        00004560
      WRITE(6,1003) K2,(TITLE(I,I4),I=1,10),IDF(I4),AMIN(K2),AS         00004570
      AS=AMIN(K2)                                                       00004580
  540 CONTINUE                                                          00004590
C                                                                       00004600
C     'ICROSS' TWO-WAY TABLES PRINT OUT                                 00004610
C                                                                       00004620
      WRITE(6,1016)                                                     00004630
      IF(ICROSS.EQ.0.AND.K11.GT.10) K11=10                              00004640
      IF(ICROSS.GT.0.AND.K11.GT.ICROSS) K11=ICROSS                      00004650
      IF(IEXP.EQ.1) WRITE(6,1004) ((TITLE(I,I3),I=1,10),J=1,2)          00004660
      IF(IEXP.NE.1) WRITE(6,1005) ((TITLE(I,I3),I=1,10),J=1,2)          00004670
      DO 570 K2=1,K11                                                   00004680
      I4=MIN(K2)                                                        00004690
      J1=NO1(I4)                                                        00004700
      IF(IEXP.EQ.1) I4=FACE2(I4)                                        00004710
      IF(I3.EQ.I4) GO TO 570                                            00004720
      J2=J1+ITEM(I4)-1                                                  00004730
      JJ=0                                                              00004740
      DO 580 J=J1,J2                                                    00004750
      TP=0                                                              00004760
      JJ=JJ+1                                                           00004770
      IF(TOTALR(J).NE.0) TP=100.0/TOTALR(J)                             00004780
      TC=0                                                              00004790
      II=0                                                              00004800
```

```
      DO 590 I=I1,I2                                                    00004810
      II=II+1                                                           00004820
      P(II,JJ)=IA(I,J)*TP                                               00004830
  590 TC=TC+P(II,JJ)                                                    00004840
  580 TPAR(JJ)=TC                                                       00004850
      PTT=0                                                             00004860
      II=0                                                              00004870
      DO 600 I=I1,I2                                                    00004880
      II=II+1                                                           00004890
      PT(II)=TOTALC(I)/SAMP*100.0                                       00004900
  600 PTT=PTT+PT(II)                                                    00004910
      IR=ITEM(I3)                                                       00004920
      IR1=1                                                             00004930
      IR2=IR                                                            00004940
      IF(IR2.GT.10) IR2=10                                              00004950
      IR3=IR2                                                           00004960
      I11=I1                                                            00004970
      I21=I1+IR2-1                                                      00004980
  610 WRITE(6,1006)(IPR,IPR=IR1,IR2)                                    00004990
      FM1(3)=FM2(IR3)                                                   00005000
      WRITE(6,FM1)(IPR,IPR=IR1,IR2)                                     00005010
      WRITE(6,1004) ((TITLE(I,I4),I=1,10),I8=1,2)                       00005020
      JJ=0                                                              00005030
      DO 620 J=J1,J2                                                    00005040
      JJ=JJ+1                                                           00005050
      WRITE(6,1007) JJ,(IA(I,J),I=I11,I21),TOTALR(J)                    00005060
      FM3(5)=FM4(IR3)                                                   00005070
  620 WRITE(6,FM3)JJ,(P(I5,JJ),I5=IR1,IR2),TPAR(JJ),TOTALR(J)           00005080
      WRITE(6,1008)(TOTALC(I5),I5=I11,I21),NSAMP                        00005090
      FM5(5)=FM4(IR3)                                                   00005100
      WRITE(6,FM5)(PT(I5),I5=IR1,IR2),PTT,NSAMP                         00005110
      IF(IR2.EQ.IR) GO TO 570                                           00005120
      IR1=IR2+1                                                         00005130
      IR2=IR2+10                                                        00005140
      IF(IR2.GT.IR) IR2=IR                                              00005150
      IR3=MOD(IR2,10)                                                   00005160
      IF(IR3.EQ.0)IR3=10                                                00005170
      I11=I21+1                                                         00005180
      I21=I21+IR3                                                       00005190
      WRITE(6,1009)                                                     00005200
      GO TO 610                                                         00005210
  570 CONTINUE                                                          00005220
      I1=I2+1                                                           00005230
      IF(IK.EQ.L) GO TO 370                                             00005240
      I3=FACE(IK+1)                                                     00005250
      I2=I2+ITEM(I3)                                                    00005260
  370 CONTINUE                                                          00005270
      LA=L                                                              00005280
      NA=N                                                              00005290
      IF(IEXP.NE.1) N1=L                                                00005300
      IF(IEXP.NE.1) L=N                                                 00005310
      IF(IEXP.NE.1) N=N1                                                00005320
C                                                                       00005330
C     'SUMMARY OF AIC'S OF THE TWO-WAY TABLES' PRINT OUT                00005340
C                                                                       00005350
      IS=1                                                              00005360
      IE=10                                                             00005370
  890 IF(IE.GE.N1) IE=N1                                                00005380
      DO 900 I=1,10                                                     00005390
      IH=0                                                              00005400
      DO 910 IJ=IS,IE                                                   00005410
      IH=IH+1                                                           00005420
      IF(IEXP.EQ.1) II=FACE2(IJ)                                        00005430
      IF(IEXP.NE.1) II=FACE(IJ)                                         00005440
      HY(I,IH)=TITLE(I,II)                                              00005450
  910 CONTINUE                                                          00005460
  900 CONTINUE                                                          00005470
      WRITE(6,1010)                                                     00005480
      WRITE(6,1015)                                                     00005490
      WRITE(6,1012) ((HY(I,II),I=1,10),II=1,IH)                         00005500
      WRITE(6,1011)                                                     00005510
      DO 920   IK=1,L                                                   00005520
      IF(IEXP.EQ.1) II=FACE(IK)                                         00005530
      IF(IEXP.NE.1) II=IK                                               00005540
      WRITE(6,1013) (TITLE(I,II),I=1,10)                                00005550
      FAA(3)=BL                                                         00005560
      IH=0                                                              00005570
      DO 930 I=IS,IE                                                    00005580
      IF(IEXP.EQ.1) III=FACE2(I)                                        00005590
      IF(IEXP.NE.1) III=FACE(I)                                         00005600
      IH=IH+1                                                           00005610
      IF(IH.GT.8) FAA(3)=FAD(3)                                         00005620
      FAA(2)=FAC(IH)                                                    00005630
      IF(III.NE.II) WRITE(6,FAA) AIMIN(I,IK)                            00005640
  930 CONTINUE                                                          00005650
  920 CONTINUE                                                          00005660
      IF(IE.EQ.N1) GO TO 880                                            00005670
      IS=IE+1                                                           00005680
      IE=IS+9                                                           00005690
      GO TO 890                                                         00005700
  880 CONTINUE                                                          00005710
C                                                                       00005720
C     'GRAY SHADING DISPLAY OF ALL THE AIC'S  ' PRINT OUT               00005730
C                                                                       00005740
      WRITE(6,1010)                                                     00005750
      CALL PROUT(AIMIN,IW(1),N1,L,NL,TITLE,HYO,HY,FACE,FACE2,IEXP,      00005760
```

```
     1              NSAMP,LA,N3)                                        00005770
      L=LA                                                              00005780
      N=NA                                                              00005790
      WRITE(6,1010)                                                     00005800
  320 CONTINUE                                                          00005810
 1001 FORMAT(20I4)                                                      00005820
 1002 FORMAT(1H1/1H ,'  LIST OF EXPLANATORY VARIABLES ARRANGED IN',     00005830
     1       ' ASCENDING ORDER OF AIC'//1H ,'RESPONSE VARIABLE  : '     00005840
     2       ,'(',10A1,')'/)                                            00005850
 1003 FORMAT(1H ,I5,10X,10A1,21X,I5,10X,F10.2,5X,F10.2)                 00005860
 1004 FORMAT(1H0,'(',10A1,')',42X,'(',10A1,')')                         00005870
 1005 FORMAT(//1H ,5X,'(',10A1,')',42X,'(',10A1,')')                    00005880
 1006 FORMAT(1H0,9X,12I4)                                               00005890
 1007 FORMAT(1H ,I6,3X,12I4)                                            00005900
 1008 FORMAT(1H ,3X,'TOTAL',1X,12I4)                                    00005910
 1009 FORMAT(///)                                                       00005920
 1010 FORMAT(1H1)                                                       00005930
 1011 FORMAT(1H+,'EXPLANATORY '/1H ,'VARIABLES'/)                       00005940
 1012 FORMAT(1H ,14X, 9(1X,10A1,1X),10A1)                               00005950
 1013 FORMAT(1H ,'(',10A1,')')                                          00005960
 1014 FORMAT(1H ,'  NO.',2X,'EXPLANATORY VARIABLE     ',4X,'NUMBER OF CA00005970
     1TEGORIES',8X,'  A I C  ',4X,'DIFFERENCE OF AIC'/1H ,36X,'OF EXPLAN00005980
     2ATORY VARIABLE'/)                                                 00005990
 1015 FORMAT(1H0,10X,' SUMMARY OF AIC''S FOR THE TWO-WAY TABLES '//)    00006000
 1016 FORMAT(1H1/1H ,'  TWO-WAY TABLES ARRANGED IN ASCENDING ORDER OF AI00006010
     1C')                                                               00006020
      RETURN                                                            00006030
      END                                                               00006040
      SUBROUTINE PROUT(A,KA,N,M,ND,TITLE,HYO,HY,FACE,FACE2,IEXP,NSAMP,  00006050
     1                 LL,N3)                                           00006060
C                                                                       00006070
C     THIS SUBROUTINE PRINTS OUT 'GRAY SHADING OF ALL THE AIC'S'.       00006080
C                                                                       00006090
      IMPLICIT REAL*8(A-H,O-Z)                                          00006100
      INTEGER *2 KA,KX,KBL,TITLE,HYO,HY,FACE,FACE2                      00006110
      DIMENSION A(ND,ND),KA(ND,3),KX(5,3),B(4),TITLE(10,ND),HYO(ND),    00006120
     1          HY(10,ND),FACE(LL),FACE2(N3)                            00006130
      DATA KX/'X','X','X','-',' ',                                      00006140
     1        'H','H',' ',' ',' ',                                      00006150
     2        'I',' ',' ',' ',' '/                                      00006160
      DATA KBL/' '/                                                     00006170
      DATA B/-0.1D0,-0.05D0,-0.01D0,0.D0/                               00006180
      N11=1                                                             00006190
  100 CONTINUE                                                          00006200
      WRITE(6,1002)                                                     00006210
      N12=N                                                             00006220
      IF(N12.GE.N11+99) N12=N11+99                                      00006230
      DO 110 I=1,10                                                     00006240
      DO 120 IJ=N11,N12                                                 00006250
      IF(IEXP.EQ.1) II=FACE2(IJ)                                        00006260
      IF(IEXP.NE.1) II=FACE(IJ)                                         00006270
      HYO(IJ)=TITLE(I,II)                                               00006280
      HY(I,IJ)=TITLE(I,II)                                              00006290
  120 CONTINUE                                                          00006300
      WRITE(6,1011) (HYO(J),J=N11,N12)                                  00006310
      IF(I.EQ.9) WRITE(6,1008)                                          00006320
      IF(I.EQ.10) WRITE(6,1009)                                         00006330
  110 CONTINUE                                                          00006340
      WRITE(6,1011)                                                     00006350
      DO 130 IK=1,M                                                     00006360
      IF(IEXP.EQ.1) II=FACE(IK)                                         00006370
      IF(IEXP.NE.1) II=IK                                               00006380
      DO 20 J=N11,N12                                                   00006390
      IF(IEXP.EQ.1) JJ=FACE2(J)                                         00006400
      IF(IEXP.NE.1) JJ=FACE(J)                                          00006410
      IF(II.EQ.JJ) GO TO 35                                             00006420
      DO 30 K=1,4                                                       00006430
      IF(A(J,IK)/NSAMP.GT.B(K)) GO TO 30                                00006440
      DO 60 L=1,3                                                       00006450
      KA(J,L)=KX(K,L)                                                   00006460
   60 CONTINUE                                                          00006470
      GO TO 20                                                          00006480
   30 CONTINUE                                                          00006490
   35 CONTINUE                                                          00006500
      DO 70 L=1,3                                                       00006510
      KA(J,L)=KBL                                                       00006520
   70 CONTINUE                                                          00006530
   20 CONTINUE                                                          00006540
      WRITE(6,1012) (TITLE(I,II),I=1,10)                                00006550
      DO 40 L=1,3                                                       00006560
      WRITE(6,1001) (KA(J,L),J=N11,N12)                                 00006570
   40 CONTINUE                                                          00006580
  130 CONTINUE                                                          00006590
      IF(N12.EQ.N) GO TO 140                                            00006600
      N11=N12+1                                                         00006610
      WRITE(6,1010)                                                     00006620
      GO TO 100                                                         00006630
  140 CONTINUE                                                          00006640
      WRITE(6,1003)                                                     00006650
      DO 80 I=1,5                                                       00006660
      IF(I.EQ.1) WRITE(6,1006) B(I)                                     00006670
      IF(I.GT.1.AND.I.LT.5) WRITE(6,1004) B(I-1),B(I)                   00006680
      IF(I.EQ.5) WRITE(6,1007) B(I-1)                                   00006690
      DO 90 J=1,3                                                       00006700
      WRITE(6,1005) KX(I,J)                                             00006710
   90 CONTINUE                                                          00006720
```

```
   80 CONTINUE                                                            00006730
 1001 FORMAT(1H+,14X,100A1)                                               00006740
 1002 FORMAT(1H ,'   GRAY SHADING DISPLAY OF ALL THE AIC''S '//           00006750
     1      1H ,'                    RESPONSE VARIABLES'/)                00006760
 1003 FORMAT(///1H ,'< N O T E >'/)                                       00006770
 1004 FORMAT(1H ,4X,':',F10.3,' < AIC/NSAMP <',F10.3)                     00006780
 1005 FORMAT(1H+,2X,A1)                                                   00006790
 1006 FORMAT(1H ,4X,':',10X,'   AIC/NSAMP <',F10.3)                       00006800
 1007 FORMAT(1H ,4X,':',F10.3,' < AIC/NSAMP  ')                           00006810
 1008 FORMAT(1H+,1X,'EXPLANATORY')                                        00006820
 1009 FORMAT(1H+,1X,'VARIABLES')                                          00006830
 1010 FORMAT(1H1)                                                         00006840
 1011 FORMAT(1H ,14X,100(A1))                                             00006850
 1012 FORMAT(1H ,'(',10A1,')')                                            00006860
      RETURN                                                              00006870
      END                                                                 00006880
```

```
               +------------------------------------------------------------------------------------+
DATA NO. 1 :   | 100  10   5   9  -1   0   0   0   5   1                                         |
DATA NO. 2 :   |   1   2   1   4   1   4   1   3   1   5   1   2   1   2   1   2   1   2   1   3|
DATA NO. 3 :   |   2   1   1   2   2   3   3   4   4   4                                         |
               |   3   1   2   3   3   2   4   4   4                                             |
               |   4   1   1   2   2   3                                                         |
               |   5   1   1   2   3   4   4   5   5   5                                         |
               |   6   1   2   2   2                                                             |
               |   7   1   2   2   2                                                             |
               |   8   1   2   2   2                                                             |
               |   9   1   2   2   2                                                             |
               |  10   1   2   3   3   3                                                         |
DATA NO. 4 :   |   6   7   8   9  10                                                             |
DATA NO. 6 :   |(10I4)                                                                           |
DATA NO. 7 :   |SEX                                                                              |
               |AGE                                                                              |
               |POL. PARTY                                                                       |
               |EDUCATION                                                                        |
               |OCCUPATION                                                                       |
               |BORN AGAIN                                                                       |
               |DIFFICULT                                                                        |
               |PLEASURE                                                                         |
               |WOMEN JOB                                                                        |
               |MONEY                                                                            |
DATA NO. 8 :   |   2   1   4                                                                     |
DATA NO. 9 :   |   2   7   1   2   4   1   1   1   1   1                                         |
               |   2   7   1   1   1   1   2   2   1   5                                         |
               |   2   2   1   5   5   2   2   1   1   1                                         |
               |   1   6   3   3   2   1   1   2   1   1                                         |
               |   1   6   1   3   1   1   1   1   1   1                                         |
               |   2   3   1   2   7   2   1   1   1   3                                         |
               |   2   2   6   2   4   2   1   2   1   1                                         |
               |   2   6   6   2   7   2   1   1   2   2                                         |
               |   2   1   1   3   4   1   1   2   1   1                                         |
               |   2   9   1   2   2   2   2   2   2   1                                         |
               |   2   8   1   2   7   1   4   1   4   1                                         |
               |   1   5   3   2   1   2   2   1   1   1                                         |
               |   2   9   6   2   8   2   2   3   1   1                                         |
               |   1   5   6   2   5   4   1   3   1   1                                         |
               |   2   5   3   1   1   1   2   1   2   2                                         |
               |   1   4   1   3   1   1   1   1   1   2                                         |
               |   2   8   1   5   7   2   1   1   1   1                                         |
               |   2   6   1   2   1   1   3   4   1   3                                         |
               |   1   8   1   2   4   1   3   1   1   1                                         |
               |   1   9   6   1   4   1   1   2   4   5                                         |
               |   2   3   1   3   2   2   1   1   1   2                                         |
               |   1   3   1   4   2   1   1   1   1   3                                         |
               |   1   5   7   3   4   1   1   1   2   3                                         |
               |   1   1   2   3   4   1   1   2   2   2                                         |
               |   1   4   1   2   2   1   1   1   1   3                                         |
               |   2   4   4   3   7   1   1   1   1   2                                         |
               |   1   7   2   2   4   1   2   1   2   2                                         |
               |   2   5   1   3   7   1   2   1   1   2                                         |
               |   1   2   2   4   3   1   2   1   1   2                                         |
               |   1   1   6   4   5   1   3   3   3   2                                         |
               |   1   9   1   2   2   1   1   1   1   2                                         |
               |   1   6   1   2   3   1   1   1   1   1                                         |
               |   1   6   1   3   2   1   3   1   3   1                                         |
               |   2   1   6   3   7   2   1   2   1   2                                         |
               |   1   9   1   1   8   1   1   1   1   1                                         |
               |   2   1   1   3   3   1   2   1   1   2                                         |
               |   1   1   8   3   1   2   2   1   1   3                                         |
               |   2   4   7   3   7   2   1   1   1   2                                         |
               |   1   6   1   3   2   1   1   1   1   2                                         |
               |   1   1   3   2   4   1   4   4   1   1                                         |
               |   2   3   5   3   7   1   2   1   1   2                                         |
               |   1   4   1   4   3   1   1   1   1   2                                         |
               |   2   2   6   3   7   2   1   1   1   2                                         |
               |   1   5   8   1   4   1   4   1   4   1                                         |
               |   1   5   3   2   4   1   1   3   3   1                                         |
               |   2   1   6   3   7   2   1   1   1   3                                         |
               |   2   1   6   3   7   2   3   3   1   2                                         |
               |   2   3   1   3   7   1   1   1   1   3                                         |
               |   1   9   8   1   8   1   2   1   2   2                                         |
               |   1   5   6   3   2   1   1   1   2   1                                         |
               |   2   7   1   3   3   1   4   1   1   2                                         |
               |   1   5   6   3   3   1   3   3   1   2                                         |
               |   2   5   3   1   7   2   2   1   2   1                                         |
               |   2   7   3   1   4   1   2   1   1   1                                         |
               |   1   5   3   3   3   1   2   1   1   3                                         |
               |   2   1   1   3   8   1   2   1   1   2                                         |
               |   2   9   6   1   1   1   1   1   1   5                                         |
               |   2   5   1   3   7   2   1   1   1   1                                         |
               |   1   3   1   2   2   1   3   1   2   1                                         |
               |   1   1   6   3   3   3   1   1   1   2                                         |
               |   2   5   8   3   2   1   1   1   1   1                                         |
               |   2   4   6   2   7   2   1   2   1   3                                         |
               |   2   6   1   2   1   2   2   2   1   1                                         |
               |   2   4   6   2   7   2   4   4   4   5                                         |
```

1	6	7	4	3	1	4	4	4	2
2	4	1	3	7	3	2	1	1	1
1	4	6	3	2	1	1	1	1	2
1	1	1	3	3	1	1	3	1	3
2	1	6	3	8	1	1	1	1	2
2	4	6	2	4	1	2	2	2	1
2	7	1	4	7	1	1	2	1	2
1	3	6	2	5	1	1	4	1	1
2	6	6	2	4	1	3	1	1	1
1	4	1	2	2	1	1	1	2	1
1	3	6	3	5	1	4	1	1	2
1	2	3	3	4	1	1	1	1	2
2	5	6	3	2	2	3	1	1	1
1	3	6	2	1	1	1	1	1	1
1	9	3	1	1	3	3	1	3	1
2	2	6	3	7	2	4	3	1	2
2	9	1	1	8	4	2	1	4	3
1	5	3	3	3	1	4	1	1	1
2	1	6	3	7	2	1	1	1	3
1	7	1	2	2	1	1	1	2	1
2	1	8	3	3	1	2	1	1	3
2	3	3	3	7	1	1	1	1	1
1	4	6	3	4	1	1	1	1	1
2	3	3	3	7	1	2	1	3	2
1	2	6	3	3	1	1	1	1	2
2	7	6	2	1	2	2	1	1	3
2	2	3	2	4	1	4	1	1	3
2	4	3	2	1	2	1	1	1	1
2	8	6	2	1	1	2	1	1	1
2	2	6	3	2	1	1	1	2	1
1	9	5	5	1	3	4	4	1	5
2	3	8	2	4	1	2	1	4	1
2	4	1	3	7	2	2	1	1	3
1	4	8	3	2	1	1	1	1	3
1	7	7	4	5	1	1	1	1	2
2	3	5	4	7	1	2	1	1	3

LIST OF EXPLANATORY VARIABLES ARRANGED IN ASCENDING ORDER OF AIC

RESPONSE VARIABLE : (BORN AGAIN)

NO.	EXPLANATORY VARIABLE	NUMBER OF CATEGORIES OF EXPLANATORY VARIABLE	A I C	DIFFERENCE OF AIC
1	OCCUPATION	5	-6.52	0.0
2	SEX	2	-6.21	0.31
3	EDUCATION	3	-1.49	4.71
4	WOMEN JOB	2	0.79	2.28
5	PLEASURE	2	0.94	0.16
6	DIFFICULT	2	1.99	1.04
7	AGE	4	3.00	1.01
8	MONEY	3	3.01	0.01
9	POL. PARTY	4	3.18	0.17

TWO-WAY TABLES ARRANGED IN ASCENDING ORDER OF AIC

(BORN AGAIN)

(OCCUPATION)	1	2	
1	17	9	26
2	11	1	12
3	15	1	16
4	4	2	6
5	12	13	25
TOTAL	59	26	85

(SEX)	1	2	
1	33	6	39
2	26	20	46
TOTAL	59	26	85

(EDUCATION)	1	2	
1	25	12	37
2	34	11	45
3	0	3	3
TOTAL	59	26	85

(WOMEN JOB)	1	2	
1	41	21	62
2	18	5	23
TOTAL	59	26	85

(PLEASURE)	1	2	
1	45	17	62
2	14	9	23
TOTAL	59	26	85

(DIFFICULT)	1	2	
1	28	12	40
2	31	14	45
TOTAL	59	26	85

(AGE)	1	2	
1	14	10	24
2	11	2	13
3	18	7	25
4	16	7	23
TOTAL	59	26	85

(MONEY)	1	2	
1	25	12	37
2	22	7	29
3	12	7	19
TOTAL	59	26	85

(POL. PARTY)	1	2	
1	22	8	30
2	5	1	6
3	11	3	14
4	21	14	35
TOTAL	59	26	85

(BORN AGAIN)

(OCCUPATION)	1	2	TOTAL	
1	65.4	34.6	100.0	26
2	91.7	8.3	100.0	12
3	93.8	6.3	100.0	16
4	66.7	33.3	100.0	6
5	48.0	52.0	100.0	25
TOTAL	69.4	30.6	100.0	85

(SEX)	1	2	TOTAL	
1	84.6	15.4	100.0	39
2	56.5	43.5	100.0	46
TOTAL	69.4	30.6	100.0	85

(EDUCATION)	1	2	TOTAL	
1	67.6	32.4	100.0	37
2	75.6	24.4	100.0	45
3	0.0	100.0	100.0	3
TOTAL	69.4	30.6	100.0	85

(WOMEN JOB)	1	2	TOTAL	
1	66.1	33.9	100.0	62
2	78.3	21.7	100.0	23
TOTAL	69.4	30.6	100.0	85

(PLEASURE)	1	2	TOTAL	
1	72.6	27.4	100.0	62
2	60.9	39.1	100.0	23
TOTAL	69.4	30.6	100.0	85

(DIFFICULT)	1	2	TOTAL	
1	70.0	30.0	100.0	40
2	68.9	31.1	100.0	45
TOTAL	69.4	30.6	100.0	85

(AGE)	1	2	TOTAL	
1	58.3	41.7	100.0	24
2	84.6	15.4	100.0	13
3	72.0	28.0	100.0	25
4	69.6	30.4	100.0	23
TOTAL	69.4	30.6	100.0	85

(MONEY)	1	2	TOTAL	
1	67.6	32.4	100.0	37
2	75.9	24.1	100.0	29
3	63.2	36.8	100.0	19
TOTAL	69.4	30.6	100.0	85

(POL. PARTY)	1	2	TOTAL	
1	73.3	26.7	100.0	30
2	83.3	16.7	100.0	6
3	78.6	21.4	100.0	14
4	60.0	40.0	100.0	35
TOTAL	69.4	30.6	100.0	85

SUMMARY OF AIC'S FOR THE TWO-WAY TABLES

EXPLANATORY VARIABLES	BORN AGAIN	DIFFICULT	PLEASURE	WOMEN JOB	MONEY
(SEX)	-6.21	0.66	1.50	-0.86	3.18
(AGE)	3.00	4.59	1.60	1.27	0.43
(POL. PARTY)	3.18	0.34	5.10	4.54	3.80
(EDUCATION)	-1.49	1.19	3.65	-1.60	-10.69
(OCCUPATION)	-6.52	6.74	4.84	2.62	0.43
(BORN AGAIN)		1.99	0.94	0.79	3.01
(DIFFICULT)	1.99		1.99	0.07	3.90
(PLEASURE)	0.94	1.99		1.98	3.99
(WOMEN JOB)	0.79	0.07	1.98		2.11
(MONEY)	3.01	3.90	3.99	2.11	

GRAY SHADING DISPLAY OF ALL THE AIC'S

```
               RESPONSE VARIABLES

               BDPWM
               OILOO
               RFEMN
               NFAEE
                ISNY
               ACU
               GURJ
               ALEO
 EXPLANATORY   IT B
 VARIABLES     N

(SEX        )  M  X
(AGE        )
(POL. PARTY)
(EDUCATION )   X  X■
(OCCUPATION)   M
(BORN AGAIN)
(DIFFICULT )
(PLEASURE   )
(WOMEN JOB )
(MONEY      )
```

```
< N O T E >

  ■ :             AIC/NSAMP <    -0.100
  M :    -0.100 < AIC/NSAMP <    -0.050
  X :    -0.050 < AIC/NSAMP <    -0.010
  - :    -0.010 < AIC/NSAMP <     0.0
    :     0.0   < AIC/NSAMP
```

III-3 Regression Analysis Program (REGRES)

The program for the estimation of the parameters and the computation of AIC is composed of the following parts.

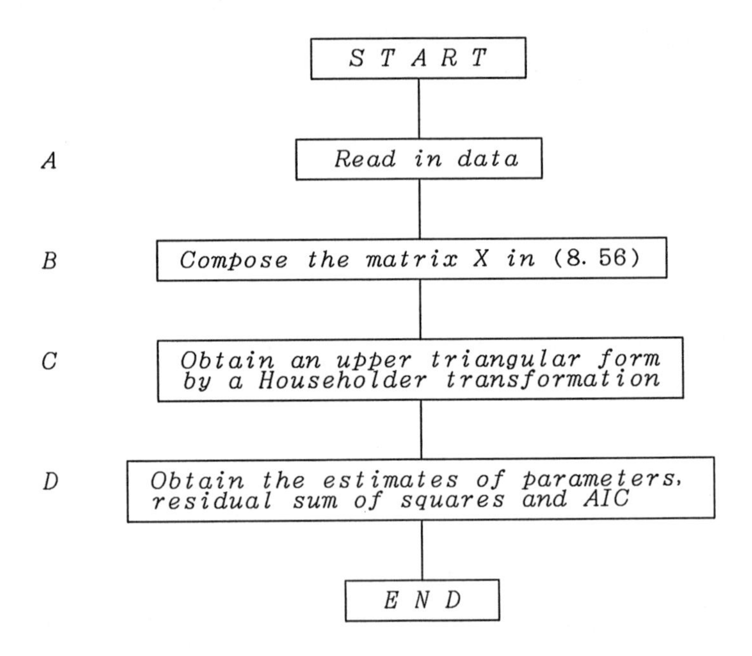

The subroutines for C and D are common for all types of regression model. We can apply these to the fitting of various models by properly modifying the subroutines A and B.

The program REGRES is obtained by simplifying the subroutines of the program package TIMSAC-78. The necesasary inputs to the program are as follows;

Card	Explanation	Format
1	Number of regressors, K	(I5)
2	Title of data	(20A4)
3	Data length, N	(I5)
4	Reading format of y_i and x_{ij}	(20A4)
5	Objective variable (y_i) and regressors ($x_{i1}, \ldots, x_{ik}$)	Specified by the fourth card

The outputs of the program are (k+2) models;

$$
\begin{aligned}
&MODEL(-1): && y = \varepsilon \\
&MODEL(\phi): && y = a_0 + + \varepsilon \\
&MODEL(x_1): && y = a_0 + a_1 x_1 + \varepsilon \\
&MODEL(x_1, x_2): && y = a_0 + a_1 x_1 + a_2 x_2 + \varepsilon \\
&\quad\vdots && \\
&MODEL(x_1, x_2, \cdots, x_K): && y = a_0 + a_1 x_1 + \cdots + a_K x_K + \varepsilon
\end{aligned}
$$

Here ε is the normal random variable with mean 0 and variance σ^2. The parameter estimates and AIC are printed out for each model. We note that, with this input, the program does not fit a regression model such as

$$MODEL(x_1, x_3): \quad y = a_0 + a_1 x_1 + a_3 x_3 + \varepsilon$$

To obtain such models the user should reorder the explanatory variables in data cards or use the subset regression program SUBREG which is shown after the program REGRESS. (This program was not given in the original Japanease edition.)

The sample inputs are the data used in [Example 8.2]. Compare the sample outputs and Table 8.3.

```
      PROGRAM REGRES                                                    00000010
C                                                                       00000020
C     THIS PROGRAM FITS REGRESSION MODEL BY THE MINIMUM AIC PROCEDURE   00000030
C                                                                       00000040
C     INPUTS REQUIRED:                                                  00000050
C         K:        NUMBER OF REGRESSORS                                00000060
C         TITLE:    TITLE OF DATA                                       00000070
C         N:        NUMBER OF OBSERVATIONS                              00000080
C         FORMAT:   READING FORMAT                                      00000090
C         Z,Y:      OBSERVATIONS OF REGRESANDS AND REGRESSORS           00000100
C                   (Z(I),Y(I,J),J=1,...,K) (I=1,...,N)                 00000110
C                                                                       00000120
C                                                                       00000130
      IMPLICIT REAL*8(A-H,O-Z)                                          00000140
      DIMENSION X(500,22), A(21), Z(500), D(500), Y(500,20)             00000150
      MJ=500                                                            00000160
C                                                                       00000170
      WRITE(6,2)                                                        00000180
      READ(5,1) K                                                       00000190
C                                                                       00000200
C          ORIGINAL DATA LOADING                                        00000210
C                                                                       00000220
      CALL DATARG(Z,Y,MJ,N,K)                                           00000230
C                                                                       00000240
C          DATA MATRIC SET UP                                           00000250
C                                                                       00000260
      CALL MATREG(Z,Y,N,K,MJ,X)                                         00000270
C                                                                       00000280
C          HOUSEHOLDER TRANSFORMATION                                   00000290
C                                                                       00000300
      CALL HUSHLD(X,D,MJ,N,K+2)                                         00000310
C                                                                       00000320
C                                                                       00000330
C          MAICE PROCEDURE                                              00000340
      CALL MODEL(X,N,K+1,MJ,KMIN,A,SIG2,AICM)                           00000350
C                                                                       00000360
C                                                                       00000370
      STOP                                                              00000380
    1 FORMAT(16I5)                                                      00000390
    2 FORMAT(1H0,'PROGRAM REGRES  ...  REGRESSION MODEL FITTING.')      00000400
      END                                                               00000410


      SUBROUTINE DATARG(Z,Y,MJ,N,K)                                     00000420
C                                                                       00000430
C     THIS SUBROUTINE READS IN THE ORIGINAL DATA WHICH IS COMPOSED      00000440
C     OF TITLE, DATA LENGTH, READING FORMAT AND OBSERVATIONS.           00000450
C                                                                       00000460
C     INPUT:                                                            00000470
C        K:       NUMBER OF REGRESSORS                                  00000480
C                                                                       00000490
C     INPUTS FROM CARD READER:                                          00000500
C        TITLE:   TITLE OF DATA                                         00000510
C        N:       DATA LENGTH                                           00000520
C        FORMAT:  READING FORMAT                                        00000530
C        X(I):    REGRESSAND (I=1,...,N)                                00000540
C        Y(I,J):  REGRESSORS (I=1,...,N;J=1,...,K)                      00000550
C                                                                       00000560
C     OUTPUTS:                                                          00000570
C        X, Y, N, TITLE                                                 00000580
C                                                                       00000590
      IMPLICIT REAL*8(A-H,O-Z)                                          00000600
      DIMENSION Z(1), Y(MJ,1), FORMAT(10), TITLE(10)                    00000610
      DATA A/1H /                                                       00000620
C                                                                       00000630
      READ(5,1) TITLE                                                   00000640
      READ(5,2) N                                                       00000650
      READ(5,1) FORMAT                                                  00000660
      DO 10 I=1,N                                                       00000670
   10 READ(5,FORMAT) Z(I), (Y(I,J),J=1,K)                               00000680
C                                                                       00000690
      WRITE(6,3) TITLE                                                  00000700
      WRITE(6,4) N                                                      00000710
      WRITE(6,5) FORMAT                                                 00000720
      WRITE(6,7) ((A,I),I=1,K)                                          00000730
      DO 20 I=1,N                                                       00000740
   20 WRITE(6,6) I, Z(I), (Y(I,J),J=1,K)                                00000750
C                                                                       00000760
      RETURN                                                            00000770
    1 FORMAT(10A8)                                                      00000780
    2 FORMAT(16I5)                                                      00000790
    3 FORMAT(1H0,'ORIGINAL DATA',5X,10A8)                               00000800
    4 FORMAT(1H ,'N (DATA LENGTH) =',I4)                                00000810
    5 FORMAT(1H ,'READING FORMAT =',10A8)                               00000820
    6 FORMAT(1H ,I3,10F13.5)                                            00000830
    7 FORMAT(1H0,6X,'REGRESSAND',5X,'REGRESSORS',/,'   I',5X,'Z(I)',    00000840
     *       2X,8(A7,'Y(I,',I1,1H)))                                    00000850
      END                                                               00000860


      SUBROUTINE MATREG(Z,Y,N,K,MJ,X)                                   00000870
C                                                                       00000880
C     THIS SUBROUTINE BUILDS DATA MATRIX X FOR FITTING REGRESSION MODEL.00000890
C                                                                       00000900
C      INPUTS:                                                          00000910
C        Z:       VECTOR OF REGRESSAND                                  00000920
C        Y:       MATRIX OF REGRESSORS                                  00000930
C        N:       DATA LENGTH                                           00000940
C        K:       NUMBER OF REGRESSORS                                  00000950
C        MJ:      DIMENSION OF THE FIRST ARRAY OF X                     00000960
```

```
C                                                                       00000970
C     OUTPUTS:                                                          00000980
C         X:      N*(K+2) DATA MATRIX                                   00000990
C                                                                       00001000
      IMPLICIT REAL*8(A-H,O-Z)                                          00001010
      DIMENSION X(MJ,1), Y(MJ,1), Z(1)                                  00001020
C                                                                       00001030
      DO 20 I=1,N                                                       00001040
      DO 10 J=1,K                                                       00001050
   10 X(I,J+1)=Y(I,J)                                                   00001060
      X(I,1)=1.0D0                                                      00001070
   20 X(I,K+2)=Z(I)                                                     00001080
C                                                                       00001090
      RETURN                                                            00001100
      END                                                               00001110


      SUBROUTINE HUSHLD(X,D,MJ,N,K)                                     00001120
C                                                                       00001130
C     THIS SUBROUTINE TRANSFORMS MATRIX X TO AN UPPER TRIANGULAR        00001140
C     FORM BY HOUSEHOLDER TRANSFORMATION.                               00001150
C                                                                       00001160
C     INPUTS:                                                           00001170
C         X:      N*K DATA MATRIX                                       00001180
C         D:      WORKING AREA                                          00001190
C         MJ:     DIMENSION OF THE FIRST ARRAY                          00001200
C         N:      NUMBER OF RAWS OF X, NOT GREATER THAN MJ.             00001210
C         K:      NUBER OF COLUMNS OF X                                 00001220
C     OUTPUTS:                                                          00001230
C         X:      SQUARE ROOT OF DATA COVARIANCE MATRIX (UPPER TRIANGULAR 00001240
C                 MATRIX)                                               00001250
C                                                                       00001260
      IMPLICIT REAL*8(A-H,O-Z)                                          00001270
      DIMENSION X(MJ,1), D(1)                                           00001280
C                                                                       00001290
          TOL = 1.0D-30                                                 00001300
C                                                                       00001310
      DO 100 II=1,K                                                     00001320
      H=0.0D0                                                           00001330
      DO 10 I=II,N                                                      00001340
      D(I)=X(I,II)                                                      00001350
   10 H=H+D(I)**2                                                       00001360
      IF(H.GT.TOL) GO TO 20                                             00001370
      G=0.0D0                                                           00001380
      GO TO 100                                                         00001390
   20 G=DSQRT(H)                                                        00001400
      F=X(II,II)                                                        00001410
      IF(F.GE.0.0D0) G=-G                                               00001420
      D(II)=F-G                                                         00001430
      H=H-F*G                                                           00001440
C                                                                       00001450
      IF(II.EQ.K) GO TO 100                                             00001460
      DO 60 J=II+1,K                                                    00001470
      S=0.0D0                                                           00001480
      DO 40 I=II,N                                                      00001490
   40 S=S+D(I)*X(I,J)                                                   00001500
      S=S/H                                                             00001510
      DO 50 I=II,N                                                      00001520
   50 X(I,J)=X(I,J)-D(I)*S                                              00001530
   60 CONTINUE                                                          00001540
  100 X(II,II)=G                                                        00001550
C                                                                       00001560
      RETURN                                                            00001570
      END                                                               00001580


      SUBROUTINE MODEL(X,N,K,MJ,KMIN,AMAICE,VARM,AICM)                  00001590
C                                                                       00001600
C     THIS SUBROUTINE FITS REGRESSION MODEL BY THE MINIMUM AIC PROCEDURE00001610
C     USING THE OUTPUT OF SUBROUTINE HUSHLD.                            00001620
C                                                                       00001630
C     INPUTS:                                                           00001640
C         X:      UPPER TRIANGULAR MATRIX                               00001650
C         N:      DATA LENGTH                                           00001660
C         K:      NUMBER OF REGRESSORS                                  00001670
C         MJ:     DIMENSION OF THE FIRST ARRAY OF X                     00001680
C                                                                       00001690
C     OUTPUS:                                                           00001700
C         KMIN:   MAICE ORDER                                           00001710
C         A(I):   REGRESSION COEFFICIENTS (I=1,...,KMIN)                00001720
C         VARM:   RESIDUAL VARIANCE                                     00001730
C         AICM:   MINIMUN AIC                                           00001740
C                                                                       00001750
      IMPLICIT REAL*8(A-H,O-Z)                                          00001760
      DIMENSION AMAICE(1), X(MJ,1), A(21), AIC(21), VAR(21)             00001770
      DATA PAI2/6.2831853D0/                                            00001780
C                                                                       00001790
C          COMPUTE RESIDUAL VARIANCE AND AIC                            00001800
C                                                                       00001810
      SUM=0.0D0                                                         00001820
      DO 10 I=1,K+1                                                     00001830
      J=K+2-I                                                           00001840
      SUM=SUM+X(J,K+1)**2                                               00001850
      VAR(J)=SUM/N                                                      00001860
   10 AIC(J) = N*(DLOG(PAI2) + 1.0D0) + N*DLOG( VAR(J) ) + 2.D0*(J)     00001870
C                                                                       00001880
C          FIND MINIMUM OF AIC                                          00001890
C                                                                       00001900
      KMIN = 0                                                          00001910
      VARM = VAR(1)                                                     00001920
```

```
      AICM = AIC(1)                                                     00001930
      DO 20 I=1,K                                                       00001940
      IF(AIC(I+1).GE.AICM) GO TO 20                                     00001950
      KMIN = I                                                          00001960
      VARM = VAR(I+1)                                                   00001970
      AICM = AIC(I+1)                                                   00001980
   20 CONTINUE                                                          00001990
C                                                                       00002000
C           PRINT OUT AIC AND RESIDUAL VARIANCE                         00002010
C                                                                       00002020
      WRITE(6,1)                                                        00002030
      DO 30 I=1,K+1                                                     00002040
      J = I-1                                                           00002050
   30 WRITE(6,2) J, AIC(I), VAR(I)                                      00002060
      WRITE(6,3) KMIN, AICM, VARM                                       00002070
      WRITE(6,6)                                                        00002080
C                                                                       00002090
C           COMPUTE REGRESSION COEFFICEINTS                             00002100
C                                                                       00002110
      DO 100 M=1,K                                                      00002120
C                                                                       00002130
        A(M) = X(M,K+1)/X(M,M)                                          00002140
        IF(M.EQ.1) GO TO 60                                             00002150
C                                                                       00002160
        DO 50 I=1,M-1                                                   00002170
        MI =M-I                                                         00002180
        SUM=X(MI,K+1)                                                   00002190
        DO 40 J=MI+1,M                                                  00002200
   40   SUM=SUM-A(J)*X(MI,J)                                            00002210
   50   A(MI)=SUM/X(MI,MI)                                              00002220
C                                                                       00002230
   60   WRITE(6,4) M                                                    00002240
        WRITE(6,5) (A(I),I=1,M)                                         00002250
C                                                                       00002260
      IF(M .NE. KMIN) GO TO 100                                         00002270
      DO 70 I=1,M                                                       00002280
   70 AMAICE(I)=A(I)                                                    00002290
  100 CONTINUE                                                          00002300
C                                                                       00002310
      RETURN                                                            00002320
    1 FORMAT(1H0,'  ORDER',5X,'AIC',8X,'VARIANCE',/,1H ,34(1H_))        00002330
    2 FORMAT(1H ,I5,F12.2,D17.6)                                        00002340
    3 FORMAT(1H0,'MAICE ORDER =',I3,5X,'MINIMUM AIC =',F10.2,5X,        00002350
     *       'RESIDUAL VARIANCE =',D20.10)                              00002360
    4 FORMAT(1H0,'ORDER (M) =',I3)                                      00002370
    5 FORMAT(1H ,10F13.8)                                               00002380
    6 FORMAT(1H0,'REGRESSION COEFFICIENTS')                             00002390
      END                                                               00002400
```

```
PROGRAM REGRES  ...  REGRESSION MODEL FITTING.

ORIGINAL DATA      MEAN LOWEST TEMPERETURE
N (DATA LENGTH) =  20
READING FORMAT =(4F10.0)

        REGRESSAND     REGRESSORS
  I       Z(I)         Y(I,1)        Y(I,2)        Y(I,3)
  1     -8.00000      45.42000     141.68000       2.80000
  2    -13.60000      43.77000     142.37000     111.90000
  3     -9.50000      43.05000     141.33000      17.20000
  4     -5.40000      40.82000     140.78000       3.00000
  5     -6.70000      39.70000     141.17000     155.20000
  6     -3.20000      38.27000     140.90000      38.90000
  7     -0.10000      36.55000     136.65000      26.10000
  8     -5.50000      36.67000     138.20000     418.20000
  9     -7.60000      36.15000     137.25000     560.20000
 10    -10.00000      36.33000     138.55000     999.10000
 11     -0.90000      35.17000     136.97000      51.10000
 12     -4.70000      35.52000     137.83000     481.80000
 13     -0.40000      35.68000     139.77000       5.30000
 14      0.50000      35.48000     134.23000       7.10000
 15     -0.60000      35.02000     135.73000      41.40000
 16      0.20000      34.37000     132.43000      29.30000
 17      1.50000      33.58000     130.38000       2.50000
 18      2.00000      31.57000     130.55000       4.30000
 19      0.10000      33.55000     133.53000       1.90000
 20     13.50000      26.23000     127.68000      34.90000

  ORDER      AIC          VARIANCE
 __________________________________
    0       133.12     0.411870D+02
    1       130.48     0.326606D+02
    2       106.10     0.873395D+01
    3       105.68     0.773650D+01
    4        84.78     0.246201D+01

MAICE ORDER =  4     MINIMUM AIC =      84.78     RESIDUAL VARIANCE =     0.2462007451D+01

REGRESSION COEFFICIENTS

ORDER (M) =  1
  -2.92000000

ORDER (M) =  2
  39.02193799  -1.14454736

ORDER (M) =  3
  94.25810957  -0.69346246  -0.52422719

ORDER (M) =  4
  38.32469458  -1.17154060   0.02306095  -0.00982962
```

The following two programs show driver programs and necessary subroutines for the fitting of polynomial regression model and autoregressive model, respectively.

```
C     DRIVER PROGRAM FOR POLYNOMIAL REGRESSION MODEL FITTING            00000010
C                                                                       00000020
C     INPUTS REQUIRED:                                                  00000030
C         K:        MAXIMUM ORDER OF POLYNOMINAL                        00000040
C         TITLE:    TITLE                                               00000050
C         N:        NUMBER OF OBSERVATIONS                              00000060
C         FORMAT:   READING FORMAT                                      00000070
C         Z,Y:      OBSERVATIONS                                        00000080
C                   (Z(I),Y(I)) (I=1,...,N)                             00000090
      IMPLICIT REAL*8(A-H,O-Z)                                          00000100
      PARAMETER( MJ=500 )                                               00000110
      DIMENSION X(MJ,22), A(21), Z(MJ), D(MJ), Y(MJ)                    00000120
C                                                                       00000130
      WRITE(6,2)                                                        00000140
      READ(5,1) K                                                       00000150
      CALL DATAPL(Z,Y,N)                                                00000160
      CALL MATPOL(Z,Y,N,K,MJ,X)                                         00000170
      CALL HUSHLD(X,D,MJ,N,K+2)                                         00000180
      CALL MODEL(X,N,K+1,MJ,KMIN,A,SIG2,AICM)                           00000190
      STOP                                                              00000200
    1 FORMAT(16I5)                                                      00000210
    2 FORMAT(1H0,'PROGRAM POLNOM  ...  POLYNOMINAL REGRESSION MODEL',   00000220
     *       ' FITTING.')                                               00000230
  600 FORMAT( 1H0,'SD =',D15.7,5X,'AIC =',F10.3 )                       00000240
      END                                                               00000250


      SUBROUTINE DATAPL(Z,Y,N)                                          00000260
C                                                                       00000270
C     THIS SUBROUTINE READS IN THE ORIGINAL DATA WHICH IS COMPOSED      00000280
C     OF TITLE, DATA LENGTH, READING FORMAT AND OBSERVATIONS.           00000290
C                                                                       00000300
C     INPUTS FROM CARD READER:                                          00000310
C         TITLE:  TITLE OF DATA                                         00000320
C         N:      NUMBER OF OBSERVATIONS                                00000330
C         FORMAT: READING FORMAT                                        00000340
C         Z,Y:    OBSERVATIONS OF REGRESSAND AND INDEPENDENT VARIABLE   00000350
C                 (Z(I),Y(I)) (I=1,...,N)                               00000360
C                                                                       00000370
      IMPLICIT REAL*8(A-H,O-Z)                                          00000380
      DIMENSION Z(1), Y(1), FORMAT(10), TITLE(10)                       00000390
C                                                                       00000400
      READ(5,1) TITLE                                                   00000410
      READ(5,2) N                                                       00000420
      READ(5,1) FORMAT                                                  00000430
      DO 10 I=1,N                                                       00000440
   10 READ(5,FORMAT) Y(I), Z(I)                                         00000450
C                                                                       00000460
      WRITE(6,3) TITLE                                                  00000470
      WRITE(6,4) N                                                      00000480
      WRITE(6,5) FORMAT                                                 00000490
      WRITE(6,7)                                                        00000500
      DO 20 I=1,N                                                       00000510
   20 WRITE(6,6) I, Y(I), Z(I)                                          00000520
C                                                                       00000530
      RETURN                                                            00000540
    1 FORMAT(10A8)                                                      00000550
    2 FORMAT(16I5)                                                      00000560
    3 FORMAT(1H0,'ORIGINAL DATA',5X,10A8)                               00000570
    4 FORMAT(1H ,'N (DATA LENGTH) =',I4)                                00000580
    5 FORMAT(1H ,'READING FORMAT =',10A8)                               00000590
    6 FORMAT(1H ,I5,10F13.5)                                            00000600
    7 FORMAT(1H0,10X,'INDEPENDENT',3X,'REGRESSAND',/,13X,'VARIABLE')    00000610
      END                                                               00000620


      SUBROUTINE MATPOL(Z,Y,N,K,MJ,X)                                   00000630
C                                                                       00000640
C     THIS SUBROUTINE SETS DATA MATRIX UP FOR THE FITTING               00000650
C     OF POLYNOMINAL REGRESSION MODEL.                                  00000660
C                                                                       00000670
C     INPUTS:                                                           00000680
C         Z:      VECTOR OF REGRESSAND                                  00000690
C         Y:      VECTOR OF INDEPENDENT VARIABLE                        00000700
C         N:      NUMBER OF OBSERVATIONS                                00000710
C         K:      MAXIMUM ORDER OF POLYNOMINAL                          00000720
C         MJ:     DIMENSION OF THE FIRST ARRAY OF X                     00000730
C     OUTPUTS:                                                          00000740
C         X:      N*(K+2) DATA MATRIX                                   00000750
C                                                                       00000760
      IMPLICIT REAL*8(A-H,O-Z)                                          00000770
      DIMENSION X(MJ,1), Y(MJ,1), Z(1)                                  00000780
C                                                                       00000790
      DO 20 I=1,N                                                       00000800
      XX=1.0D0                                                          00000810
      X(I,1)=XX                                                         00000820
      DO 10 J=1,K                                                       00000830
      XX=XX*Y(I,1)                                                      00000840
   10 X(I,J+1)=XX                                                       00000850
   20 X(I,K+2)=Z(I)                                                     00000860
C                                                                       00000870
      RETURN                                                            00000880
      END                                                               00000890
```

```
C     DRIVER PROGRAM FOR AUTOREGRESSIVE MODEL FITTING                   00000010
C                                                                       00000020
C     INPUTS REQUIRED                                                   00000030
C        K:          MAXIMUM TIME LAG OF AUTOREGRESSION                 00000040
C        TITLE:      TITLE OF DATA                                      00000050
C        N:          NUMBER OF OBSERVATIONS                             00000060
C        FORMAT:     READING FORMAT                                     00000070
C        Z(I):       TIME SERIES (I=1,...,N)                            00000080
C                                                                       00000090
      IMPLICIT REAL*8(A-H,O-Z)                                          00000100
      PARAMETER(MJ=500)                                                 00000110
      DIMENSION  Z(MJ), X(MJ,21), A(20), D(MJ)                          00000120
C                                                                       00000130
      WRITE(6,2)                                                        00000140
      READ(5,1) K                                                       00000150
      CALL DATAAR(Z,N)                                                  00000160
      CALL MATARM(Z,N,K,MJ,X)                                           00000170
      CALL HUSHLD(X,D,MJ,N-K,K+1)                                       00000180
      CALL MODEL(X,N-K,K,MJ,KMIN,A,SIG2,AICM)                           00000190
C                                                                       00000200
      STOP                                                              00000210
    1 FORMAT(16I5)                                                      00000220
    2 FORMAT(1H0,'PROGRAM ARFIT  ...  AUTOREGESSIVE MODEL ')            00000230
      END                                                               00000240


      SUBROUTINE DATAAR(Z,N)                                            00000250
C                                                                       00000260
C     THIS SUBROUTINE READS IN THE ORIGINAL DATA WHICH IS COMPOSED      00000270
C     OF TITLE, DATA LENGTH, READING FORMAT AND OBSERVATIONS.           00000280
C                                                                       00000290
C     INPUTS FROM CARD READER:                                          00000300
C          TITLE:  TITLE OF DATA                                        00000310
C          N:      NUMBER OF OBSERVATIONS                               00000320
C          FORMAT: READING FORMAT                                       00000330
C          Z(I):   TIME SERIES (I=1,...,N)                              00000340
C                                                                       00000350
C     OUTPUTS:                                                          00000360
C          X, N                                                         00000370
C                                                                       00000380
      IMPLICIT REAL*8(A-H,O-Z)                                          00000390
      DIMENSION Z(1), FORMAT(10), TITLE(10)                             00000400
C                                                                       00000410
      READ(1,1) TITLE                                                   00000420
      READ(1,2) N                                                       00000430
      READ(1,1) FORMAT                                                  00000440
      READ(1,FORMAT) (Z(I),I=1,N)                                       00000450
C                                                                       00000460
      WRITE(6,3) TITLE                                                  00000470
      WRITE(6,4) N                                                      00000480
      WRITE(6,5) FORMAT                                                 00000490
      WRITE(6,7)                                                        00000500
      WRITE(6,6) (Z(I),I=1,N)                                           00000510
      SUM=0.0D0                                                         00000520
      DO 10 I=1,N                                                       00000530
   10 SUM=SUM+Z(I)                                                      00000540
      ZMEAN=SUM/N                                                       00000550
      DO 20 I=1,N                                                       00000560
   20 Z(I)=Z(I)-ZMEAN                                                   00000570
      WRITE(6,8)                                                        00000580
      WRITE(6,6) (Z(I),I=1,N)                                           00000590
C                                                                       00000600
      RETURN                                                            00000610
    1 FORMAT(10A8)                                                      00000620
    2 FORMAT(16I5)                                                      00000630
    3 FORMAT(1H0,'ORIGINAL DATA',5X,10A8)                               00000640
    4 FORMAT(1H ,'N (DATA LENGTH) =',I4)                                00000650
    5 FORMAT(1H ,'READING FORMAT =',10A8)                               00000660
    6 FORMAT(1H ,10F13.5)                                               00000670
    7 FORMAT(1H0,'ORIGINAL DATA')                                       00000680
    8 FORMAT(1H0,'MEAN DELETED')                                        00000690
      END                                                               00000700


      SUBROUTINE MATARM(Z,N,K,MJ,X)                                     00000710
C                                                                       00000720
C     THIS SUBROUTINE MAKES DATA MATRIX X FOR FITTING AUTOREGRESSIVE    00000730
C     MODEL.                                                            00000740
C                                                                       00000750
C     INPUTS:                                                           00000760
C          Z:      VECTOR OF TIME SERIES                                00000770
C          N:      NUMBER OF OBSERVATIONS                               00000780
C          K:      MAXIMUM TIME LAG                                     00000790
C          MJ:     DIMENSION OF THE FIRST ARRAY OF X                    00000800
C                                                                       00000810
C     OUTPUTS:                                                          00000820
C          X:      N*(K+1) DATA MATRIX                                  00000830
C                                                                       00000840
      IMPLICIT REAL*8(A-H,O-Z)                                          00000850
      DIMENSION X(MJ,1), Z(1)                                           00000860
C                                                                       00000870
      DO 20 I=K+1,N                                                     00000880
      DO 10 J=1,K                                                       00000890
   10 X(I-K,J)=Z(I-J)                                                   00000900
   20 X(I-K,K+1)=Z(I)                                                   00000910
C                                                                       00000920
      RETURN                                                            00000930
      END                                                               00000940
```

The following subset regression program (SUBREG) finds the best model among all possible combination of explanatory variables. The inputs required are the same as the ones for program REGRES. SUBREG is a modified version of the subset autoregression program listed in TIMSAC-78 (Akaike et al., 1979).

```
      PROGRAM SUBREG                                                    00000010
C                                                                       00000020
C     THIS PROGRAM FITS REGRESSION MODEL BY THE MINIMUM AIC PROCEDURE   00000030
C                                                                       00000040
C     INPUTS REQUIRED:                                                  00000050
C         K:        NUMBER OF REGRESSORS                                00000060
C         TITLE:    TITLE OF DATA                                       00000070
C         N:        NUMBER OF OBSERVATIONS                              00000080
C         FORMAT:   READING FORMAT                                      00000090
C         Z,Y:      OBSERVATIONS OF REGRESANDS AND REGRESSORS           00000100
C                   (Z(I),Y(I,J),J=1,...,K) (I=1,...,N)                 00000110
C                                                                       00000120
C                                                                       00000130
      IMPLICIT REAL*8(A-H,O-Z)                                          00000140
      DIMENSION X(500,22), Z(500), D(500), Y(500,20)                    00000150
      MJ=500                                                            00000160
C                                                                       00000170
      WRITE(6,2)                                                        00000180
      READ(5,1) K                                                       00000190
C                                                                       00000200
C          ORIGINAL DATA LOADING                                        00000210
C                                                                       00000220
      CALL DATARG(Z,Y,MJ,N,K)                                           00000230
C                                                                       00000240
C          DATA MATRIC SET UP                                           00000250
C                                                                       00000260
      CALL MATREG(Z,Y,N,K,MJ,X)                                         00000270
C                                                                       00000280
C          HOUSEHOLDER TRANSFORMATION                                   00000290
C                                                                       00000300
      CALL HUSHLD(X,D,MJ,N,K+2)                                         00000310
C                                                                       00000320
C          MAICE PROCEDURE                                              00000330
C                                                                       00000340
      CALL SUBSET(X,N,K+1,MJ)                                           00000350
C                                                                       00000360
C                                                                       00000370
      STOP                                                              00000380
    1 FORMAT(16I5)                                                      00000390
    2 FORMAT(1H0,'PROGRAM SUBREG  ...  SUBSET REGRESSION MODEL.')       00000400
      END                                                               00000410


      SUBROUTINE  DATARG(Z,Y,MJ,N,K)                                    00000420
C                                                                       00000430
C     THIS SUBROUTINE READS IN THE ORIGINAL DATA WHICH IS COMPOSED      00000440
C     OF TITLE, DATA LENGTH, READING FORMAT AND OBSERVATIONS.           00000450
C                                                                       00000460
C     INPUT:                                                            00000470
C        K:      NUMBER OF REGRESSORS                                   00000480
```

```
C                                                                       00000490
C     INPUTS FROM CARD READER:                                          00000500
C        TITLE:   TITLE OF DATA                                         00000510
C        N:       DATA LENGTH                                           00000520
C        FORMAT:  READING FORMAT                                        00000530
C        X(I):    REGRESSAND (I=1,...,N)                                00000540
C        Y(I,J):  REGRESSORS (I=1,...,N;J=1,...,K)                      00000550
C                                                                       00000560
C     OUTPUTS:                                                          00000570
C        X, Y, N, TITLE                                                 00000580
C                                                                       00000590
      IMPLICIT REAL*8(A-H,O-Z)                                          00000600
      DIMENSION Z(1), Y(MJ,1), FORMAT(10), TITLE(10)                    00000610
      DATA A/1H /                                                       00000620
C                                                                       00000630
      READ(5,1) TITLE                                                   00000640
      READ(5,2) N                                                       00000650
      READ(5,1) FORMAT                                                  00000660
      DO 10 I=1,N                                                       00000670
   10 READ(5,FORMAT) Z(I), (Y(I,J),J=1,K)                               00000680
C                                                                       00000690
      WRITE(6,3) TITLE                                                  00000700
      WRITE(6,4) N                                                      00000710
      WRITE(6,5) FORMAT                                                 00000720
      WRITE(6,7) ((A,I),I=1,K)                                          00000730
      DO 20 I=1,N                                                       00000740
   20 WRITE(6,6) I, Z(I), (Y(I,J),J=1,K)                                00000750
C                                                                       00000760
      RETURN                                                            00000770
    1 FORMAT(10A8)                                                      00000780
    2 FORMAT(16I5)                                                      00000790
    3 FORMAT(1H0,'ORIGINAL DATA',5X,10A8)                               00000800
    4 FORMAT(1H ,'N (DATA LENGTH) =',I4)                                00000810
    5 FORMAT(1H ,'READING FORMAT =',10A8)                               00000820
    6 FORMAT(1H ,I3,10F13.5)                                            00000830
    7 FORMAT(1H0,6X,'REGRESSAND',5X,'REGRESSORS',/,'   I',5X,'Z(I)',    00000840
     *       2X,8(A7,'Y(I,',I1,1H)))                                    00000850
      END                                                               00000860


      SUBROUTINE  MATREG(Z,Y,N,K,MJ,X)                                  00000870
C                                                                       00000880
C     THIS SUBROUTINE BUILDS DATA MATRIX X FOR FITTING REGRESSION MODEL.00000890
C                                                                       00000900
C      INPUTS:                                                          00000910
C        Z:     VECTOR OF REGRESSAND                                    00000920
C        Y:     MATRIX OF REGRESSORS                                    00000930
C        N:     DATA LENGTH                                             00000940
C        K:     NUMBER OF REGRESSORS                                    00000950
C        MJ:    DIMENSION OF THE FIRST ARRAY OF X                       00000960
C                                                                       00000970
C     OUTPUTS:                                                          00000980
C        X:     N*(K+2) DATA MATRIX                                     00000990
C                                                                       00001000
      IMPLICIT REAL*8(A-H,O-Z)                                          00001010
      DIMENSION X(MJ,1), Y(MJ,1), Z(1)                                  00001020
C                                                                       00001030
      DO 20 I=1,N                                                       00001040
      DO 10 J=1,K                                                       00001050
   10 X(I,J+1)=Y(I,J)                                                   00001060
      X(I,1)=1.0D0                                                      00001070
   20 X(I,K+2)=Z(I)                                                     00001080
C                                                                       00001090
      RETURN                                                            00001100
      END                                                               00001110


      SUBROUTINE  HUSHLD(X,D,MJ,N,K)                                    00001120
C                                                                       00001130
C     THIS SUBROUTINE TRANSFORMS MATRIX X TO AN UPPER TRIANGULAR        00001140
C     FORM BY HOUSEHOLDER TRANSFORMATION.                               00001150
C                                                                       00001160
C     INPUTS:                                                           00001170
C        X:     N*K DATA MATRIX                                         00001180
C        D:     WORKING AREA                                            00001190
C        MJ:    DIMENSION OF THE FIRST ARRAY                            00001200
C        N:     NUMBER OF RAWS OF X, NOT GREATER THAN MJ.               00001210
C        K:     NUBER OF COLUMNS OF X                                   00001220
C     OUTPUTS:                                                          00001230
C        X:     SQUARE ROOT OF DATA COVARIANCE MATRIX (UPPER TRIANGULAR 00001240
C               MATRIX)                                                 00001250
C                                                                       00001260
      IMPLICIT REAL*8(A-H,O-Z)                                          00001270
      DIMENSION X(MJ,1), D(1)                                           00001280
C                                                                       00001290
           TOL = 1.0D-30                                                00001300
C                                                                       00001310
      DO 100 II=1,K                                                     00001320
      H=0.0D0                                                           00001330
      DO 10 I=II,N                                                      00001340
      D(I)=X(I,II)                                                      00001350
   10 H=H+D(I)**2                                                       00001360
      IF(H.GT.TOL) GO TO 20                                             00001370
      G=0.0D0                                                           00001380
      GO TO 100                                                         00001390
   20 G=DSQRT(H)                                                        00001400
      F=X(II,II)                                                        00001410
      IF(F.GE.0.0D0) G=-G                                               00001420
      D(II)=F-G                                                         00001430
      H=H-F*G                                                           00001440
```

```
C                                                                       00001450
      DO 30 I=II+1,N                                                    00001460
   30 X(I,II) = 0.0D0                                                   00001470
      IF(II.EQ.K) GO TO 100                                             00001480
      DO 60 J=II+1,K                                                    00001490
      S=0.0D0                                                           00001500
      DO 40 I=II,N                                                      00001510
   40 S=S+D(I)*X(I,J)                                                   00001520
      S=S/H                                                             00001530
      DO 50 I=II,N                                                      00001540
   50 X(I,J)=X(I,J)-D(I)*S                                              00001550
   60 CONTINUE                                                          00001560
  100 X(II,II)=G                                                        00001570
C                                                                       00001580
      RETURN                                                            00001590
      END                                                               00001600



      SUBROUTINE  SUBSET( X,N,K,MJ )                                    00001610
C                                                                       00001620
C     THIS SUBROUTINE FINDS BEST KMAX SUBSET REGRESSION MODELS FOR EACH 00001630
C     NUMBER M OF REGRESSORS ( M=1,2,...,K ), BY USING THE OUTPUT OF    00001640
C     SUBROUTINE HUSHLD.                                                00001650
C                                                                       00001660
C       INPUTS:                                                         00001670
C          X:      UPPER TRIANGULAR MATRIX , OUTPUT OF SUBROUTINE HUSHLD00001680
C          N:      DATA LENGTH                                          00001690
C          K:      UPPER LIMIT OF THE NUMBER OF REGRESSORS              00001700
C          MJ:     ABSOLUTE DIMENSION OF X                              00001710
C                                                                       00001720
      IMPLICIT  REAL*8(A-H,O-Z)                                         00001730
      DIMENSION  X(MJ,1), A(20), D(20), SD(20), AIC1(20), ISD(20)       00001740
      DIMENSION  IND(20), JND(20), KND(20), MND(20), LND(20), B(20,20)  00001750
      DIMENSION  SDM(200)                                               00001760
      DATA  PAI2/6.2831853D0/                                           00001770
C                                                                       00001780
      KMAX = 5                                                          00001790
      IPR = 0                                                           00001800
      K1 = K + 1                                                        00001810
      KM1 = K - 1                                                       00001820
      DN = N                                                            00001830
      DO 5  I=1,K1                                                      00001840
      IND(I) = I                                                        00001850
      JND(I) = I                                                        00001860
      KND(I) = 1                                                        00001870
      MND(I) = I                                                        00001880
    5 CONTINUE                                                          00001890
C                                                                       00001900
      WRITE( 6,606 )                                                    00001910
C           .....  FULL ORDER MODEL  .....                              00001920
C                                                                       00001930
      OSD = X(K1,K1)*X(K1,K1)/DN                                        00001940
      AIC = N*(DLOG(PAI2) + 1.0D0 +  DLOG( OSD )) + 2.0D0*K1            00001950
      AIC1(K1) = AIC                                                    00001960
      WRITE( 6,601 )      K                                             00001970
      WRITE( 6,605 )                                                    00001980
      WRITE( 6,604 )      OSD , AIC                                     00001990
C                                                                       00002000
C          COEFFICIENTS OF THE FULL ORDER MODEL                         00002010
C                                                                       00002020
      CALL  SRCOEF( X,K,K,N,MJ,JND,IPR,A,SDX )                          00002030
      WRITE( 6,614 )                                                    00002040
      WRITE( 6,615 )      (A(I),I=1,K)                                  00002050
C                                                                       00002060
C          .....  REGRESSION MODELS WITH K-1 REGRESSORS  .....          00002070
C                                                                       00002080
      DO 10  II=1,K                                                     00002090
      L = K - 1                                                         00002100
      KND(II) = 0                                                       00002110
      CALL  COMPSD( X,D,N,K,L,IND,JND,KND,MND,MJ,SDX )                  00002120
      CALL  SRCOEF( X,L,K,N,MJ,JND,IPR,A,SDX )                          00002130
      DO 15  I=1,L                                                      00002140
      J = K - I                                                         00002150
   15 B(J,II) = A(I)                                                    00002160
      KND(II) = 1                                                       00002170
   10 SD(II) = SDX/DN                                                   00002180
C                                                                       00002190
      CALL  SRTMIN( SD,K,MND )                                          00002200
C                                                                       00002210
C                                                                       00002220
      WRITE( 6,601 )  KM1                                               00002230
      WRITE( 6,607 )                                                    00002240
      DO 20  I=1,K                                                      00002250
      AIC = N*(DLOG(PAI2) + 1.0D0 + DLOG(SD(I))) + 2.0D0*K              00002260
      IF( I .EQ. 1 )  AIC1(K) = AIC                                     00002270
   20 WRITE( 6,600 )  I , SD(I) , AIC , MND(I)                          00002280
      WRITE( 6,611 )                                                    00002290
      DO 25  II=1,K                                                     00002300
      J = MND(II)                                                       00002310
   25 WRITE( 6,610 )  II , (B(I,J),I=1,L)                               00002320
C                                                                       00002330
C          .....  REGRESSION MODELS WITH M REGRESSORS (M=K-2,....,1)  ...00002340
C                                                                       00002350
      DO 200  II=2,KM1                                                  00002360
      M = K - II                                                        00002370
      ICOUNT = 0                                                        00002380
      WRITE( 6,601 )  M                                                 00002390
      DO 30  I=1,KMAX                                                   00002400
      ISD(I) = 0                                                        00002410
   30 SDM(I) = 1.D60                                                    00002420
```

```
      JJ = II                                                           00002430
      JJ1 = JJ + 1                                                      00002440
C                                                                       00002450
C          INITIALIZATION OF KND(I) (I=1,...,K)                         00002460
C                                                                       00002470
      IDEC = 0                                                          00002480
      IBIT = 1                                                          00002490
      DO 40  I=1,JJ                                                     00002500
      IBIT = IBIT*2                                                     00002510
   40 KND(I) = 0                                                        00002520
      DO 50  I=JJ1,K                                                    00002530
      IDEC = IDEC + IBIT                                                00002540
      IBIT = IBIT*2                                                     00002550
   50 KND(I) = 1                                                        00002560
C                                                                       00002570
C     CHECKING OF THE MODELS WITH M REGRESSORS                          00002580
C                                                                       00002590
   60 IG = 0                                                            00002600
      DO 70  I=1,JJ                                                     00002610
   70 IF( KND(I) .EQ. 0 )  IG = IG + 1                                  00002620
      IF( IG .NE. II )   GO TO 90                                       00002630
C                                                                       00002640
C          RESIDUAL VARIANCE OF THE REGRESSION MODEL SPECIFIED BY KND(I)00002650
C                                                                       00002660
      CALL  COMPSD( X,D,N,K,M,IND,JND,KND,MND,MJ,SDX )                  00002670
C                                                                       00002680
      ICOUNT = ICOUNT + 1                                               00002690
      IF( SDM(KMAX) .LT. SDX )   GO TO 80                               00002700
      ISD(KMAX) = IDEC                                                  00002710
      SDM(KMAX) = SDX                                                   00002720
C                                                                       00002730
      CALL  STORE( SDM,KMAX,ISD )                                       00002740
C                                                                       00002750
C  ...  SPECIFICATION OF THE NEXT MODEL  ...                            00002760
C                                                                       00002770
   80 CONTINUE                                                          00002780
   90 IDEC = IDEC - 1                                                   00002790
      DO 100  I=1,K                                                     00002800
      IF( KND(I) .EQ. 0 )  GO TO 100                                    00002810
      KND(I) = 0                                                        00002820
      GO TO 110                                                         00002830
  100 KND(I) = 1                                                        00002840
      GO TO 120                                                         00002850
  110 CONTINUE                                                          00002860
C  .........................................                            00002870
      IF( KND(JJ1) .NE. 0 )   GO TO 60                                  00002880
      JJ = JJ + 1                                                       00002890
      JJ1 = JJ + 1                                                      00002900
C                                                                       00002910
      IF( SDM(KMAX) .LT. SD(JJ) )   GO TO 120                           00002920
      GO TO  60                                                         00002930
C                                                                       00002940
C  ...  CHECKING COMPLETED  ...                                         00002950
C                                                                       00002960
  120 AIC1(M+1) = DN*(DLOG(PAI2) + 1.0D0 + DLOG(SDM(1))) + 2.0D0*(M+1)  00002970
C                                                                       00002980
C          PRINT OUT OF BEST KMAX SUBSET REGRESSION MODELS              00002990
C                                                                       00003000
      WRITE( 6,6 )     ICOUNT                                           00003010
      WRITE( 6,608 )                                                    00003020
      LMAX = MIN0( KMAX,ICOUNT )                                        00003030
      DO 140  I=1,LMAX                                                  00003040
      AIC = DN*(DLOG(PAI2) + 1.0D0 + DLOG(SDM(I))) + 2.0D0*M            00003050
      J = ISD(I)                                                        00003060
      CALL  BINARY( J,K,KND )                                           00003070
      J = M                                                             00003080
      DO 130  I1=1,K                                                    00003090
      IF( KND(I1) .EQ. 0 )  GO TO 130                                   00003100
      LND(J) = MND(I1)                                                  00003110
      J = J - 1                                                         00003120
  130 CONTINUE                                                          00003130
  140 WRITE( 6,609 )  I , SDM(I) , AIC , (LND(J),J=1,M)                 00003140
C                                                                       00003150
C          COEFFICIENTS OF SUBSET REGRESSION MODELS                     00003160
C                                                                       00003170
      WRITE( 6,611 )                                                    00003180
      DO 150  I=1,LMAX                                                  00003190
      J = ISD(I)                                                        00003200
      CALL  BINARY( J,K,KND )                                           00003210
      CALL  COMPSD( X,D,N,K,M,IND,JND,KND,MND,MJ,SDX )                  00003220
      CALL  SRCOEF( X,M,K,N,MJ,JND,IPR,A,SDX )                          00003230
  150 WRITE( 6,610 )  I , (A(J),J=1,M)                                  00003240
C                                                                       00003250
  200 CONTINUE                                                          00003260
C                                                                       00003270
C  ...  MODEL WITH NO REGRESSORS  ...                                   00003280
C                                                                       00003290
      M = 0                                                             00003300
C     CALL  AICCOM( X,N,0,K,MJ,OSD,AIC )                                00003310
      OSD = 0.0D0                                                       00003320
      DO 205 I=1,K+1                                                    00003330
  205 OSD = OSD + X(I,K+1)**2                                           00003340
      OSD = OSD/N                                                       00003350
      AIC = N*(DLOG(PAI2) + 1.0D0 + DLOG(OSD)) + 2.0D0*1                00003360
      AIC1(1) = AIC                                                     00003370
      WRITE( 6,601 )  M                                                 00003380
      WRITE( 6,604 )  OSD , AIC                                         00003390
C                                                                       00003400
      WRITE( 6,612 )                                                    00003410
      DO 210  I=1,K1                                                    00003420
```

```
      J = I - 1                                                         00003430
      II = BICOEF( K,J ) + 0.5D0                                        00003440
  210 WRITE( 6,602 )  J , AIC1(I) , II                                  00003450
      AICMIN = AIC1(1)                                                  00003460
      IMIN = 0                                                          00003470
      DO 220  I=2,K1                                                    00003480
      IF( AICMIN .LE. AIC1(I) )  GO TO 220                              00003490
      IMIN = I - 1                                                      00003500
      AICMIN = AIC1(I)                                                  00003510
  220 CONTINUE                                                          00003520
      WRITE( 6,603 )  AICMIN , IMIN                                     00003530
      WRITE( 6,613 )                                                    00003540
C                                                                       00003550
      RETURN                                                            00003560
C                                                                       00003570
    6 FORMAT( 1H0,'NUMBER OF SUBSETS CHECKED  =',I7 )                   00003580
  600 FORMAT( 1H ,I5,3X,D15.8,F18.3,I15 )                               00003590
  601 FORMAT( 1H0,130(1H-),/,1X,'NUMBER OF REGRESSORS  =',I3 )          00003600
  602 FORMAT( 1H ,10X,I5,F16.3,15X,I20 )                                00003610
  603 FORMAT( 1H0,'MINIMUM OF MAIC =',F15.3,5X,'ATTAINED AT  M =',I3 )  00003620
  604 FORMAT( 1H0,'OSD =',D15.8,5X,'AIC =',F18.3 )                      00003630
  605 FORMAT( 1H+,40X,'.....  FULL ORDER MODEL  .....' )                00003640
  606 FORMAT( //1H0,26(1H-),/,1H ,'SUBSET REGRESSION ANALYSIS',/,1H ,26(00003650
     11H-) )                                                            00003660
  607 FORMAT( 1H0,4X,'I',10X,'SD',18X,'AIC',9X,'DELETED REGRESSOR' )    00003670
  608 FORMAT( 1H0,4X,'I',10X,'SD',18X,'AIC',15X,'JND(J) J-TH REGRESSOR')00003680
  609 FORMAT( 1H ,I5,3X,D15.8,F18.3,10X,(/,1H+,51X,25I3) )              00003690
  610 FORMAT( 1H ,I5,3X,(/,1H+,8X,10F12.6) )                            00003700
  611 FORMAT( 1H0,'<<< SUBSET REGRESSION COEFFICIENTS (A(JND(J));J=1,...00003710
     1,M >>>' )                                                         00003720
  612 FORMAT( 1H0,130(1H-),////,10X,'<< MAIC(M) = MINIMUM AIC AT EACH OR00003730
     1DER M >>',/,1H0,14X,'M',9X,'MAIC(M)',10X,'NUMBER OF POSSIBLE COMBI00003740
     2NATION OF REGRESSORS' )                                           00003750
  613 FORMAT( //1H ,2(/,26X,35(1H-),3X,'WARNING',3X,35(1H-)),/,1H0,25X, 00003760
     1'THE SIMPLE MINDED CHOICE OF THE REGRESSORS WITH THE MINIMUM OF MA00003770
     2IC MAY NOT PRODUCE',/,1H0,25X,81HA GOOD RESULT.  RATHER IT IS REC00003780
     3MMENDED TO OBSERVE THE GENERAL BEHAVIOR OF AIC'S,/,1H ,2(/,26X,35(00003790
     41H-),3X,'WARNING',3X,35(1H-)) )                                   00003800
  614 FORMAT( 1H0,'<<< REGRESSION COEFFICIENTS OF FULL ORDER MODEL >>>')00003810
  615 FORMAT( 1H ,(/,1H+,8X,10F12.6) )                                  00003820
      END                                                               00003830
```

```
      SUBROUTINE  SRCOEF( X,M,K,N,MJ,JND,IPR,A,SD )                     00003840
C                                                                       00003850
C     SUBSET REGRESSION COEFFICIENTS AND RESIDUAL VARIANCE COMPUTATION. 00003860
C                                                                       00003870
C       INPUTS:                                                         00003880
C          X:      TRIANGULAR MATRIX                                    00003890
C          M:      NUMBER OF REGRESSORS                                 00003900
C          K:      HEIGHEST ORDER OF THE MODELS                         00003910
C          N:      DATA LENGTH                                          00003920
C         JND(I):    (I=1,...,M)  SPECIFICATION OF I-TH REGRESSOR       00003930
C       OUTPUTS:                                                        00003940
C          A:      REGRESSION COEFFICIENTS                              00003950
C          SD:     INNOVATION VARIANCE                                  00003960
C                                                                       00003970
      IMPLICIT  REAL * 8(A-H,O-Z)                                       00003980
      DIMENSION  X(MJ,1) , A(1) , JND(1)                                00003990
      K1 = K + 1                                                        00004000
      M1 = M + 1                                                        00004010
C                                                                       00004020
C           REGRESSION COEFFICIENTS COMPUTATION                         00004030
C                                                                       00004040
      L = JND(M)                                                        00004050
      A(M) = X(M,K1)/X(M,L)                                             00004060
      MM1 = M - 1                                                       00004070
      IF( MM1 .EQ. 0 )  GO TO 60                                        00004080
      DO 10  II=1,MM1                                                   00004090
      I = M - II                                                        00004100
      SUM = X(I,K1)                                                     00004110
      I1 = I + 1                                                        00004120
      DO 20  J=I1,M                                                     00004130
      L = JND(J)                                                        00004140
   20 SUM = SUM - A(J)*X(I,L)                                           00004150
      L = JND(I)                                                        00004160
   10 A(I) = SUM/X(I,L)                                                 00004170
C                                                                       00004180
C           RESIDUAL VARIANCE AND AIC COMPUTATION                       00004190
C                                                                       00004200
   60 CONTINUE                                                          00004210
      SD = 0.0D00                                                       00004220
      DO 30  I=M1,K1                                                    00004230
   30 SD = SD + X(I,K1)*X(I,K1)                                         00004240
      OSD = SD/N                                                        00004250
      AIC = N*DLOG( OSD ) + 2.0D00*M                                    00004260
C                                                                       00004270
C           REGRESSION COEFFICIENTS AND RESIDUAL VARIANCE PRINT OUT     00004280
C                                                                       00004290
      IF( IPR .LT. 2 )  RETURN                                          00004300
      WRITE( 6,5 )                                                      00004310
      WRITE( 6,6 )                                                      00004320
      DO 40  I=1,M                                                      00004330
      L = JND(I)                                                        00004340
   40 WRITE( 6,7 )  L , A(I)                                            00004350
      WRITE( 6,8 )  OSD , M , AIC                                       00004360
C                                                                       00004370
      RETURN                                                            00004380
    5 FORMAT( 1H0,10X,'SUBSET REGRESSION COEFFICIENTS' )                00004390
    6 FORMAT( 1H0,14X,1HI,12X,'A(I)' )                                  00004400
```

```
    7 FORMAT( 1H ,10X,I5,5F20.10 )                                     00004410
    8 FORMAT( 1H0,10X,'SD  = RESIDUAL VARIANCE    =',D19.12,/,11X,'M   =00004420
     1 NUMBER OF PARAMETERS =',I3,/,11X,'AIC =  N*LOG(SD) + 2*M      =', 00004430
     2F15.3 )                                                           00004440
      END                                                               00004450


      SUBROUTINE  SRTMIN( X,N,IX )                                      00004460
C                                                                       00004470
C       THIS SUBROUTINE ARRANGES X(I) (I=1,N) IN ORDER OF INCREASING    00004480
C       MAGNITUDE OF X(I)                                               00004490
C                                                                       00004500
C       INPUTS:                                                         00004510
C          X:    VECTOR                                                 00004520
C          N:    DIMENSION OF THE VECTOR                                00004530
C       OUTPUTS:                                                        00004540
C          X:    ARRANGED VECTOR                                        00004550
C          IND:  INDEX OF ARRANGED VECTOR                               00004560
C                                                                       00004570
      IMPLICIT  REAL * 8  ( A-H , O-Z )                                 00004580
      DIMENSION  X(1) , IX(1)                                           00004590
C                                                                       00004600
      NM1 = N - 1                                                       00004610
      DO 30  I=1,N                                                      00004620
   30 IX(I) = I                                                         00004630
      DO 20  II=1,NM1                                                   00004640
      XMIN = X(II)                                                      00004650
      MIN = II                                                          00004660
      DO 10  I=II,N                                                     00004670
      IF( XMIN .LT. X(I) )  GO TO 10                                    00004680
      XMIN = X(I)                                                       00004690
      MIN = I                                                           00004700
   10 CONTINUE                                                          00004710
      IF( XMIN .EQ. X(II) )  GO TO 20                                   00004720
      XT = X(II)                                                        00004730
      X(II) = X(MIN)                                                    00004740
      X(MIN) = XT                                                       00004750
      IT = IX(II)                                                       00004760
      IX(II) = IX(MIN)                                                  00004770
      IX(MIN) = IT                                                      00004780
   20 CONTINUE                                                          00004790
C                                                                       00004800
      RETURN                                                            00004810
      END                                                               00004820


      SUBROUTINE  COMPSD( X,D,N,K,L,IND,JND,KND,MND,MJ,SD )             00004830
C                                                                       00004840
C     THIS SUBROUTINE COMPUTES RESIDUAL VARIANCE OF A SUBSET REGRESSION 00004850
C                                                                       00004860
C       INPUTS:                                                         00004870
C          X:        (K+1)*(K+1)  MATRIX                                00004880
C          D:        WORKING AREA                                       00004890
C          N:        DATA LENGTH                                        00004900
C          K:        NUMBER OR VARIABLES                                00004910
C          L:        NUMBER OR REGRESSORS OF THE MODEL                  00004920
C          IND:      SPECIFICATION OF THE PRESENT FROM OF X             00004930
C                    IND(J)=I  MEANS VARIABLE J IS THE I-TH REGRESSOR   00004940
C          IND:      SPECIFICATION OF THE PRESENT FROM OF X             00004950
C                    JND(I)=J  MEANS I-TH REGRESSOR IS VARIABLE J       00004960
C          KND(I):   =0      VARIABLE MND(I) IS NOT USED AS A REGRESSOR 00004970
C                    =1      VARIABLE MND(I) IS USED AS A REGRESSOR     00004980
C          MND:      SPECIFICATION OF VARIABLE                          00004990
C          MJ:       ABSOLUTE DIMENSION OF X                            00005000
C                                                                       00005010
C       OUTPUTS:                                                        00005020
C          X:        TRANSFORMED MATRIX                                 00005030
C          IND:      SPECIFICATION OF TRANSFORMED X                     00005040
C                    IND(J)=I  MEANS VARIABLE J IS THE I-TH REGRESSOR   00005050
C          JND:      SPECIFICATION OF TRANSFORMED X                     00005060
C                    JND(I)=J  MEANS I-TH REGRESSOR IS VARIABLE J       00005070
C          SD:       INNOVATION VARIANCE OF THE MODEL                   00005080
C                                                                       00005090
      IMPLICIT  REAL*8(A-H,O-Z)                                         00005100
      DIMENSION  X(MJ,1), D(20), IND(20), JND(20), MND(20), KND(20)     00005110
      DIMENSION  LND(20)                                                00005120
C                                                                       00005130
      K1 = K + 1                                                        00005140
      LND(K1) = K1                                                      00005150
      J = K1                                                            00005160
      DO 10  I=1,K                                                      00005170
      II = MND(I)                                                       00005180
      IF( KND(I) .NE. 0 )  GO TO  10                                    00005190
      J = J - 1                                                         00005200
      LND(J) = II                                                       00005210
   10 CONTINUE                                                          00005220
      DO 50  I=1,K                                                      00005230
      II = MND(I)                                                       00005240
      IF( KND(I) .EQ. 0 )  GO TO 50                                     00005250
      J = J - 1                                                         00005260
      LND(J) = II                                                       00005270
   50 CONTINUE                                                          00005280
      DO 20  I=1,K                                                      00005290
      M = I                                                             00005300
   20 IF( JND(I) .NE. LND(I) )  GO TO 30                                00005310
   30 CONTINUE                                                          00005320
```

```
      CALL   HUSHL1( X,D,MJ,K1,K,M,IND,LND )                             00005330
C                                                                        00005340
      L1 = L + 1                                                         00005350
      SUM = 0.D0                                                         00005360
      DO 40  I=L1,K1                                                     00005370
   40 SUM = SUM + X(I,K1)**2                                             00005380
      SD = SUM/DFLOAT(N)                                                 00005390
C                                                                        00005400
      DO 60  I=1,K1                                                      00005410
   60 JND(I) = LND(I)                                                    00005420
      DO 70  I=1,K1                                                      00005430
      J = JND(I)                                                         00005440
   70 IND(J) = I                                                         00005450
      RETURN                                                             00005460
      END                                                                00005470


      SUBROUTINE  BINARY( M,K,MB )                                       00005480
C                                                                        00005490
C       DECIMAL TO BINARY CONVERSION                                     00005500
C                                                                        00005510
C       INPUTS:                                                          00005520
C          M:     NUMBER IN DECIMAL REPRESENTATION                       00005530
C          K:     NUMBER OF BITS USED FOR THE REPRESENTATION             00005540
C                                                                        00005550
C       OUTPUT:                                                          00005560
C          MB:    NUMBER IN BINARY REPRESENTATION                        00005570
C                                                                        00005580
      DIMENSION  MB(1)                                                   00005590
C                                                                        00005600
      N = M                                                              00005610
      DO 10  I=1,K                                                       00005620
        L = N / 2                                                        00005630
        MB(I) = N - L*2                                                  00005640
   10 N = L                                                              00005650
      RETURN                                                             00005660
C                                                                        00005670
      END                                                                00005680


      REAL FUNCTION  BICOEF * 8( K,J )                                   00005690
C                                                                        00005700
C     THIS FUNCTION RETURNS BINOMIAL COEFFICIENTS                        00005710
C                                                                        00005720
C          F(K,J) = K!/(J!*(K-J)!)                                       00005730
C                                                                        00005740
C       INPUTS:                                                          00005750
C          K:     NUMBER OF OBJECTS                                      00005760
C          J:     NUMBER OF OBJECTS TAKEN                                00005770
C                                                                        00005780
C       OUTPUT:                                                          00005790
C          F:     NUMBER OF COMBINATIONS OF SELECTING J OBJECTS FROM     00005800
C                 A SET OF K OBJECTS                                     00005810
C                                                                        00005820
      IMPLICIT REAL*8(A-H,O-Z)                                           00005830
C                                                                        00005840
      KMJ = K-J                                                          00005850
      SUM = 0.D0                                                         00005860
      DO 10  I=1,K                                                       00005870
      DI = I                                                             00005880
   10 SUM = SUM + DLOG( DI )                                             00005890
C                                                                        00005900
      IF( J .EQ. 0 )  GO TO 30                                           00005910
      DO 20  I=1,J                                                       00005920
      DI = I                                                             00005930
   20 SUM = SUM - DLOG( DI )                                             00005940
C                                                                        00005950
   30 IF( KMJ .EQ. 0 )  GO TO 50                                         00005960
      DO 40  I=1,KMJ                                                     00005970
      DI = I                                                             00005980
   40 SUM = SUM - DLOG( DI )                                             00005990
C                                                                        00006000
   50 BICOEF = DEXP( SUM )                                               00006010
      RETURN                                                             00006020
C                                                                        00006030
      END                                                                00006040


      SUBROUTINE  HUSHL1( X,D,MJ1,K,L,M,IND,JND )                        00006050
C                                                                        00006060
C     THIS SUBROUTINE PERFORMS THE HOUSEHOLDER TRANSFORMATION OF THE MAT00006070
C                                                                        00006080
C       INPUTS:                                                          00006090
C          X:   ORIGINAL (K+1)*(K+1) MATRIX                              00006100
C          N:   NUMBER OF ROWS OF X,  NOT GREATER THAN MJ1               00006110
C          K:   NUMBER OF COLUMNS OF X                                   00006120
C          L:   END POSITION OF THE HOUSEHOLDER TRANSFORMATION           00006130
C          M:   STARTING POSITION OF THE HOUSEHOLDER TRANSFORMATION      00006140
C        IND:     SPECIFICATION OF THE PRESENT FORM OF X                 00006150
C                 IND(J) = I:     VARIABLE I IS THE J-TH REGRESSOR       00006160
C        JND:     SPECIFICATION OF THE REQUIRED FORM OF X                00006170
C                 JND(I) = J:     THE I-TH REGRESSOR IS VARIABLE J       00006180
C                                                                        00006190
C       OUTPUTS:                                                         00006200
C          X:   TRANSFORMED MATRIX                                       00006210
C                                                                        00006220
C                                                                        00006230
      IMPLICIT  REAL * 8  ( A-H , O-Z )                                  00006240
      DIMENSION  X(MJ1,1) , D(20) , IND(20) , JND(20)                    00006250
C                                                                        00006260
```

```
      TOL = 1.0D-60                                                     00006270
C                                                                       00006280
      NN = 0                                                            00006290
      DO  100       II=M,L                                              00006300
      JJ = JND(II)                                                      00006310
      NN = MAX0( NN,IND(JJ) )                                           00006320
      H = 0.0D00                                                        00006330
      DO  10        I=II,NN                                             00006340
      D(I) = X(I,JJ)                                                    00006350
   10 H = H + D(I)*D(I)                                                 00006360
      IF( H .GT. TOL )      GO TO 20                                    00006370
      G = 0.0D00                                                        00006380
      GO TO 100                                                         00006390
   20 G = DSQRT( H )                                                    00006400
      F = X(II,JJ)                                                      00006410
      IF( F .GE. 0.0D00 )      G = -G                                   00006420
      D(II) = F - G                                                     00006430
      H = H - F * G                                                     00006440
C                                                                       00006450
C     ( I - D*D'/H ) * X                                                00006460
C                                                                       00006470
      II1 = II + 1                                                      00006480
      IF( II .EQ. NN )    GO TO 35                                      00006490
      DO  30        I=II1,NN                                            00006500
   30 X(I,JJ) = 0.D0                                                    00006510
   35 CONTINUE                                                          00006520
      IF( II .EQ. K )      GO TO 100                                    00006530
      DO  60        J1=II1,K                                            00006540
      J = JND(J1)                                                       00006550
      S = 0.0D00                                                        00006560
      DO  40        I=II,NN                                             00006570
   40 S = S + D(I)*X(I,J)                                               00006580
      S = S / H                                                         00006590
      DO  50        I=II,NN                                             00006600
   50 X(I,J) = X(I,J) - D(I)*S                                          00006610
   60 CONTINUE                                                          00006620
  100 X(II,JJ) = G                                                      00006630
      RETURN                                                            00006640
      END                                                               00006650


      SUBROUTINE  STORE( X,K,IND )                                      00006660
C                                                                       00006670
C     THIS SUBROUTINE ALLOCATES THE NEW DATA X(K) TO AN APPROPRIATE     00006680
C     POSITION WITHIN THE OUTPUT VECTOR (X(I),I=1,K) WITH X(I)'S        00006690
C     ORDERED IN THE ORDER OF INCREASING MAGUNITUDE.  EACH X(I) IS      00006700
C     ACCOMPANIED BY A LABEL IND(I).                                    00006710
C                                                                       00006720
C       INPUTS:                                                         00006730
C          X(I) (I=1,K-1):      ORIGINAL ORDERED DATA                   00006740
C          X(K):         NEW DATA                                       00006750
C          K:            NUMBER OF DATA                                 00006760
C          IND(I):       LABEL ATTACHED TO X(I)                         00006770
C       OUTPUTS:                                                        00006780
C          X(I) (I=1,K):        ORDERED DATA                            00006790
C          IND(I) (I=1,K):      LABEL ATTACHED TO X(I)                  00006800
C                                                                       00006810
      REAL*8  X(1)                                                      00006820
      DIMENSION  IND(1)                                                 00006830
C                                                                       00006840
      XX = X(K)                                                         00006850
      II = IND(K)                                                       00006860
      KM1 = K - 1                                                       00006870
      J = K                                                             00006880
   20 J = J - 1                                                         00006890
      IF( J .EQ. 0 )  GO TO 40                                          00006900
      IF( XX .LT. X(J) )    GO TO 20                                    00006910
      IF( J .EQ. K-1 )    RETURN                                        00006920
   40 J1 = J + 1                                                        00006930
      DO 10  L=J1,K                                                     00006940
      LL = K - L + J1                                                   00006950
      IND(LL) = IND(LL-1)                                               00006960
   10 X(LL) = X(LL-1)                                                   00006970
      X(J1) = XX                                                        00006980
      IND(J1) = II                                                      00006990
      RETURN                                                            00007000
C                                                                       00007010
      END                                                               00007020
```

```
PROGRAM SUBREG  ...  SUBSET REGRESSION MODEL.

ORIGINAL DATA      MEAN LOWEST TEMPERETURE
N (DATA LENGTH) =   20
READING FORMAT =(4F10.0)
```

	REGRESSAND	REGRESSORS		
I	Z(I)	Y(I,1)	Y(I,2)	Y(I,3)
1	-8.00000	45.42000	141.68000	2.80000
2	-13.60000	43.77000	142.37000	111.90000
3	-9.50000	43.05000	141.33000	17.20000
4	-5.40000	40.82000	140.78000	3.00000
5	-6.70000	39.70000	141.17000	155.20000
6	-3.20000	38.27000	140.90000	38.90000
7	-0.10000	36.55000	136.65000	26.10000
8	-5.50000	36.67000	138.20000	418.20000

```
  9    -7.60000     36.15000    137.25000    560.20000
 10   -10.00000     36.33000    138.55000    999.10000
 11    -0.90000     35.17000    136.97000     51.10000
 12    -4.70000     35.52000    137.83000    481.80000
 13    -0.40000     35.68000    139.77000      5.30000
 14     0.50000     35.48000    134.23000      7.10000
 15    -0.60000     35.02000    135.73000     41.40000
 16     0.20000     34.37000    132.43000     29.30000
 17     1.50000     33.58000    130.38000      2.50000
 18     2.00000     31.57000    130.55000      4.30000
 19     0.10000     33.55000    133.53000      1.90000
 20    13.50000     26.23000    127.68000     34.90000

--------------------------
SUBSET REGRESSION ANALYSIS
--------------------------
--------------------------------------------------------------------------------
NUMBER OF REGRESSORS  =  4               .....  FULL ORDER MODEL  .....

OSD = 0.24620075D+01      AIC =             84.777

<<< REGRESSION COEFFICIENTS OF FULL ORDER MODEL >>>
           38.324695   -1.171541    0.023061   -0.009830

--------------------------------------------------------------------------------
NUMBER OF REGRESSORS  =  3

    I          SD                     AIC         DELETED REGRESSOR
    1    0.24636078D+01              82.790               3
    2    0.28561525D+01              85.747               1
    3    0.70055188D+01             103.692               2
    4    0.77365018D+01             105.677               4

<<< SUBSET REGRESSION COEFFICIENTS (A(JND(J));J=1,...,M >>>
    1      40.742078   -1.151646   -0.009759
    2      -1.459949    0.381307   -0.010902
    3     147.083474   -1.088758   -0.006374
    4      94.258110   -0.693462   -0.524227

--------------------------------------------------------------------------------
NUMBER OF REGRESSORS  =  2

NUMBER OF SUBSETS CHECKED  =       6

    I          SD                     AIC                JND(J) J-TH REGRESSOR
    1    0.87339516D+01             104.102              2  1
    2    0.95950289D+01             105.982              1  3
    3    0.10571989D+02             107.922              2  3
    4    0.24597733D+02             124.811              4  2
    5    0.26267426D+02             126.124              4  3

<<< SUBSET REGRESSION COEFFICIENTS (A(JND(J));J=1,...,M >>>
    1      -1.144547   39.021938
    2     155.828891   -1.159606
    3      -1.384557    0.349068
    4      -0.008304   -0.060604
    5      -0.009194   -0.012239

--------------------------------------------------------------------------------
NUMBER OF REGRESSORS  =  1

NUMBER OF SUBSETS CHECKED  =       4

    I          SD                     AIC                JND(J) J-TH REGRESSOR
    1    0.28351852D+02             125.651              4
    2    0.29167099D+02             126.218              2
    3    0.31799545D+02             127.947              3
    4    0.32660599D+02             128.481              1

<<< SUBSET REGRESSION COEFFICIENTS (A(JND(J));J=1,...,M >>>
    1      -0.012061
    2      -0.093973
    3      -0.022370
    4      -2.920000

--------------------------------------------------------------------------------
NUMBER OF REGRESSORS  =  0

OSD = 0.41187000D+02      AIC =            133.120

--------------------------------------------------------------------------------
```

III-4 Fitting Variance Analysis Models (VARMOD)

VARMOD is a FORTRAN program to fit analysis of variance models to a given set of data and to calculate *AIC* for each model. To use this program, prepare data with the following (card image) format:

record number	contents	FORMAT
1	number of explanatory variables	(I5)
2	number of categories for each explanatory variables	(16I5)
3	first data*	(F10.0, 14I5)
4 to n	second to the last data*	

(*) Punch values of objective variable, the first explanatory variable, the second explanatory variable, ..., and then the k-th explanatory variable.

[*Note*] Subroutine NORMC is a prototype version of a procedure to check whether a given set of data is distributed as normal or not. This routine fits two models, Normal distribution model and Histogram model, to the data and computes *AIC* values. If *AIC* of the normal distribution model is larger than that of the histogram model, it is quite probable that the data is not normally distributed. We applied this program to five different types of data to see the performance. The results are summarized in the following table.

'Fig. 7.5' and 'Fig. 7.6' in the table refer to distributions defined by density functions,

$$f(x) = \frac{12}{15625}(-x^5+5x^4+65x^3+35x^2-280x+176)$$

on $-4 \leqq x \leqq 1$ and

$$f(x) = \frac{1}{4819500810000}(3072915625x^5 - 24687487500x^4 - 14625037500x^3 + 213749964000x^2 + 373680172800x + 174182565888)$$

on $-6/5 \leq x \leq 24/5$, respectively. The graphs of these two density functions are shown in Figures 7.5 and 7.6 of Chapter 7. The triangular distribution is defined by the density function

$$f(x) = \begin{cases} \dfrac{x+a}{a^2} & -a \leq x \leq 0 \\ \dfrac{x-a}{a^2} & 0 < x \leq a \end{cases}$$

Count of cases when the given set of data is judged to be normal by NORMC (out of 100 experiments)

Sample size (n)	True distribution				
	Normal	Fig. 7.5	Fig. 7.6	Uniform	Triangular
10	100	100	100	100	100
20	88	84	76	65	87
50	91	93	89	74	90
100	96	87	80	3	91
200	98	89	74	0	91
500	100	72	38	0	69
1000	100	68	0	0	30

It is observed from this table that;

1. If a set of data is judged to be not normally distributed by this routine, we may conclude that the data *is not* normally distributed.

2. Even if a set of data is judged to be normal, the real distribution is not necessarily normal. However, if the

judgement is made on a large set of data (1000 samples, say) then we can think that the real distribution is approximately normally distributed.

As a sample input the data of [Example 9.1] are given. Compare the sample output with the results of hand calculation given in the section 9.3.

```
C86-03-11-14:33:02 AICPVM00 PAIR       :VARMOD                            00000005
C83-03-03-13:57:15 AICPVM00 FORT       :VARMOD                            00000010
      PROGRAM  VARMOD                                                     00000020
C                                                                         00000030
C                                                                         00000040
C     THIS PROGRAM COMPUTES AIC VALUES OF VARIANCE ANALYSIS MODELS.       00000050
C     IT REQUIRES FOLLOWING INPUTS:                                       00000060
C      NFCTR:  NUMBER OF INDEPENDENT VARIABLE                             00000070
C      NLEVEL(1),....,NLEVEL(NFCTR): NUMBER OF LEVELS OF EACH INDEPENDENT00000080
C                                   VARIABLE                              00000090
C      RDATA(I),IDATA(1,I),....,IDATA(NFCTR,I): RESULT OF I-TH EXPERIMENT00000100
C      WRHER    I=1,2,...                                                 00000110
C               RDATA(I) = VALUE OF DEPENDENT VARIABLE                    00000120
C               IDATA(1,I),....,IDATA(NFCTR,I) = VALUES OF INDEPENDENT    00000130
C                                                VARIABLES                00000140
C                                                                         00000150
C                                                                         00000160
      IMPLICIT  REAL * 8  ( A-H , O-Z )                                   00000170
      DIMENSION  NLEVEL(5) , RDATA(500) , IDATA(5,500) , CFL(5,10)        00000180
      DIMENSION  CFLFL(5,10,5,10) , SUMFL(5,10)                           00000190
      DIMENSION  A(100,101) , B(100)                                      00000200
      DIMENSION  SMFLFL(5,10,5,10)                                        00000210
      DIMENSION  CNTRBN(5,10) , MARK(5)                                   00000220
      DIMENSION  LIST(10)                                                 00000230
      DIMENSION MARKL(5,20),CONTL(5,10,20),AICL(20),NPARL(20),ERRVL(20),00000240
     *          GMEANL(20)                                                00000250
C                                                                         00000260
C                                                                         00000270
C     INITIALIZATION                                                      00000280
C                                                                         00000290
C   MAXIMUM OF THE NUMBER OF DATA                                         00000300
      MAXN=500                                                            00000310
C   MAXIMUM OF THE NUMBER OF FACTORS                                      00000320
      MAXF=5                                                              00000330
C   MAXIMUM OF THE NUMBER OF CATEGORIES                                   00000340
      MAXL=10                                                             00000350
C   WORKING AREA SIZE                                                     00000360
      MAXA=100                                                            00000370
C   INPUT DEVICE                                                          00000380
      MI=5                                                                00000390
C   ERROR MESSAGE CONTROL                                                 00000400
      CALL ERRSET( 203,300,-1,1,1,0,3)                                    00000410
C                                                                         00000420
C   DATA INPUT AND WORKING AREA CHECK                                     00000430
C                                                                         00000440
      READ( 5,100 )      NFCTR                                            00000450
      READ( 5,100 )      (NLEVEL(I),I=1,NFCTR)                            00000460
      NDATA = 0                                                           00000470
      I=0                                                                 00000480
   10 I=I+1                                                               00000490
      READ( MI,101,END=15 )      RDATA(I),(IDATA(J,I),J=1,NFCTR)          00000500
      NDATA=NDATA+1                                                       00000510
      GO TO 10                                                            00000520
   15 CONTINUE                                                            00000530
C                                                                         00000540
      WRITE( 6,600 )      NFCTR                                           00000550
      IF(NFCTR .LE. MAXF) GO TO 1                                         00000560
      WRITE(6,2)                                                          00000570
    2 FORMAT(1H ,'NFCTR .GT. MAXF')                                       00000580
      STOP                                                                00000590
    1 WRITE( 6,606 )                                                      00000600
      DO  18    I=1,NFCTR                                                 00000610
      WRITE( 6,601 )      I , NLEVEL(I)                                   00000620
      IF(NLEVEL(I) .LE. MAXL) GO TO 18                                    00000630
      WRITE(6,3)                                                          00000640
      STOP                                                                00000650
    3 FORMAT(1H ,'NLEVEL .GT. MAXL')                                      00000660
   18 CONTINUE                                                            00000670
      WRITE( 6,602 )      NDATA                                           00000680
      IF(NDATA .LE. MAXN) GO TO 4                                         00000690
      WRITE(6,5)                                                          00000700
      STOP                                                                00000710
    5 FORMAT(1H ,'NDATA .GT. MAXN')                                       00000720
    4 CONTINUE                                                            00000730
      WRITE( 6,604 )                                                      00000740
      DO  20    I=1,NDATA                                                 00000750
   20 WRITE( 6,603 )      RDATA(I) , (IDATA(J,I),J=1,NFCTR)               00000760
C                                                                         00000770
C                                                                         00000780
C     PRE-PROCESSING OF THE DATA                                          00000790
C                                                                         00000800
      CALL  PROLG( RDATA,IDATA,NDATA,NLEVEL,NFCTR,SUM,PSUM,SUMFL,SMFLFL,00000810
     1 CFL,CFLFL,MAXF,MAXL )                                              00000820
C                                                                         00000830
C                                                                         00000840
C     *************                                                       00000850
C     * MAIN LOOP *                                                       00000860
C     *************                                                       00000870
C                                                                         00000880
      IIEND = 2**NFCTR                                                    00000890
      DO  80    IIP1 = 1,IIEND                                            00000900
C                                                                         00000910
C     EQUATION BUILDING                                                   00000920
C                                                                         00000930
      II=IIP1 - 1                                                         00000940
      CALL  MAKBIT( MARK,II,NFCTR )                                       00000950
      CALL  ABMAKE( SUM,SUMFL,CFL,CFLFL,NLEVEL,NFCTR,NDATA,A,B,PAR,LDIM,00000960
     1MAXF,MAXL,MAXA,LIST,LFCTR,MARK ,IG)                                 00000970
C                                                                         00000980
C     COMPUTATION OF AIC VALUE                                            00000990
```

```
C                                                                       00001000
      AIC=1.D50                                                         00001010
      IF(IG .EQ. 1) GO TO 30                                            00001020
      IF(PAR .GT. 0.5D0*NDATA) GO TO 30                                 00001030
      CALL  AICCMP( A,B,PSUM,PAR,MARK,NLEVEL,NDATA,NFCTR,GMEAN,CNTRBN,  00001040
     1 RINT,0,AIC,LDIM,MAXA,MAXF,LIST,LFCTR,ERRVAR )                    00001050
C                                                                       00001060
C     RESULT ARRANGEMENT                                                00001070
C                                                                       00001080
   30 NPAR=PAR                                                          00001090
      III=IIP1                                                          00001100
      IF(III .GT. 21) III=21                                            00001110
      II1=III-1                                                         00001120
      IF(II1 .EQ. 0) GO TO 46                                           00001130
      DO 40 I=1,II1                                                     00001140
      III=III-1                                                         00001150
      IF(AIC .GE. AICL(III)) GO TO 50                                   00001160
      IF(III .EQ. 20) GO TO 40                                          00001170
      AICL(III+1)=AICL(III)                                             00001180
      NPARL(III+1)=NPARL(III)                                           00001190
      ERRVL(III+1)=ERRVL(III)                                           00001200
      DO 45 K=1,NFCTR                                                   00001210
      MARKL(K,III+1)=MARKL(K,III)                                       00001220
      JE=NLEVEL(K)                                                      00001230
      DO 45 J=1,JE                                                      00001240
   45 CONTL(K,J,III+1)=CONTL(K,J,III)                                   00001250
      GMEANL(III+1)=GMEANL(III)                                         00001260
   40 CONTINUE                                                          00001270
   46 III=III-1                                                         00001280
   50 IF(III .EQ. 20) GO TO 55                                          00001290
      AICL(III+1)=AIC                                                   00001300
      NPARL(III+1)=NPAR                                                 00001310
      ERRVL(III+1)=ERRVAR                                               00001320
      DO 47 K=1,NFCTR                                                   00001330
      MARKL(K,III+1)=MARK(K)                                            00001340
      JE=NLEVEL(K)                                                      00001350
      DO 47 J=1,JE                                                      00001360
   47 CONTL(K,J,III+1)=CNTRBN(K,J)                                      00001370
      GMEANL(III+1)=GMEAN                                               00001380
   55 CONTINUE                                                          00001390
   80 CONTINUE                                                          00001400
C                                                                       00001410
C     ********************                                              00001420
C     * END OF MAIN LOOP *                                              00001430
C     ********************                                              00001440
C                                                                       00001450
C     RESULT OUTPUT                                                     00001460
C                                                                       00001470
      WRITE(6,608)                                                      00001480
      II1=II1+1                                                         00001490
      IF(II1 .GT. 20) II1=20                                            00001500
      DO 90 I=1,II1                                                     00001510
      WRITE(6,102) AICL(I),NPARL(I),(MARKL(J,I),J=1,NFCTR)              00001520
      IF(AICL(I) .GT. 1.D40) GO TO 90                                   00001530
      WRITE(6,605) ERRVL(I)                                             00001540
      WRITE( 6,103 )      GMEANL(I)                                     00001550
      DO  60   K=1,NFCTR                                                00001560
      JE = NLEVEL(K)                                                    00001570
   60 WRITE( 6,104 )      K , (CONTL(K,J,I),J=1,JE)                     00001580
   90 CONTINUE                                                          00001590
C                                                                       00001600
C     NORMALITY CHECK                                                   00001610
C                                                                       00001620
      WRITE(6,607)                                                      00001630
      CALL CHECK(RDATA,IDATA,GMEANL,CONTL,NDATA,NFCTR)                  00001640
C                                                                       00001650
  100 FORMAT( 16I5 )                                                    00001660
  101 FORMAT( F10.0,14I5 )                                              00001670
  102 FORMAT(1H0,                                                       00001680
     *         1X,'AIC =',F10.1,5X,'NO. OF FREE PARAMETER =',I4,5X,     00001690
     1  'MODEL =',20I1 )                                                00001700
  103 FORMAT( 5X,'GRAND MEAN =',D13.5 ,/,5X,                            00001710
     *        'CONTRIBUTION FROM EACH FACTOR')                          00001720
  104 FORMAT( 5X,I5,5X,7(D13.5,3X) )                                    00001730
  105 FORMAT( ///,5X,'MINIMUM AIC AT II =',I5,5X,D13.5 )                00001740
  600 FORMAT( 1H1,'**      PROGRAM  VARMOD      **',/,1H0,'NUMBER OF FACTO00001750
     1R =',I5 )                                                         00001760
  601 FORMAT( 1H0,'FACTOR',I5,'  =',I5 )                                00001770
  602 FORMAT( 1H0,'NUMBER OF DATA =',I5 )                               00001780
  603 FORMAT( 1H ,8X,F10.1,10X,20I5 )                                   00001790
  604 FORMAT( 1H0,'ORIGINAL DATA',/,                                    00001800
     *        1H0,'DEPENDENT VARIABLE',5X,'INDEPENDENT VARIABLES' )     00001810
  605 FORMAT( 5X,'RESIDUAL VARIANCE =',D13.5 )                          00001820
  606 FORMAT( //1H0,15X,'NUMBER OF LEVEL' )                             00001830
  607 FORMAT(1H0,/,1H0,'** ANALYSIS OF THE RESIDUALS OF MAICE MODEL **')00001840
  608 FORMAT(1H0,/,1H0,'***** RESULTS *****')                           00001850
      STOP                                                              00001860
      END                                                               00001870


C                                                                       00001880
      SUBROUTINE  PROLG( RDATA,IDATA,NDATA,NLEVEL,NFCTR,SUM,PSUM,SUMFL, 00001890
     1 SMFLFL,CFL,CFLFL,MAXF,MAXL )                                     00001900
      IMPLICIT  REAL * 8  ( A-H , O-Z )                                 00001910
      DIMENSION  RDATA(1) , IDATA(MAXF,1) , NLEVEL(1) , SUMFL(MAXF,1)   00001920
      DIMENSION  CFL(MAXF,1) , CFLFL(MAXF,MAXL,MAXF,1)                  00001930
      DIMENSION  SMFLFL(MAXF,MAXL,MAXF,1)                               00001940
C                                                                       00001950
C                                                                       00001960
      PSUM = 0.D0                                                       00001970
```

```
      SUM = 0.D0                                                        00001980
      DO  10   J=1,NDATA                                                00001990
      PSUM = PSUM + RDATA(J)*RDATA(J)                                   00002000
   10 SUM = SUM + RDATA(J)                                              00002010
C                                                                       00002020
      DO  30   IF=1,NFCTR                                               00002030
      ILE = NLEVEL(IF)                                                  00002040
      DO  30   IL=1,ILE                                                 00002050
      S = 0.D0                                                          00002060
      S2 = 0.D0                                                         00002070
      DO  20   J=1,NDATA                                                00002080
      J1 = IDATA(IF,J)                                                  00002090
      IF( IL .NE. J1 )     GO TO 20                                     00002100
      S2 = S2 + 1.D0                                                    00002110
      S = S + RDATA(J)                                                  00002120
   20 CONTINUE                                                          00002130
      CFL(IF,IL) = S2                                                   00002140
   30 SUMFL(IF,IL) = S                                                  00002150
C                                                                       00002160
C                                                                       00002170
C                                                                       00002180
      DO  70   IF1=1,NFCTR                                              00002190
      ILE1 = NLEVEL(IF1)                                                00002200
      DO  70   IL1=1,ILE1                                               00002210
      DO  70   IF2=1,NFCTR                                              00002220
      ILE2 = NLEVEL(IF2)                                                00002230
      DO  70   IL2=1,ILE2                                               00002240
      S = 0.D0                                                          00002250
      S2 = 0.D0                                                         00002260
      DO  60   J=1,NDATA                                                00002270
      J1 = IDATA(IF1,J)                                                 00002280
      J2 = IDATA(IF2,J)                                                 00002290
      IF( J1 .NE. IL1  .OR.  J2 .NE. IL2 )     GO TO 60                 00002300
      S = S + 1.D0                                                      00002310
      S2 = S2 + RDATA(J)                                                00002320
   60 CONTINUE                                                          00002330
      SMFLFL(IF1,IL1,IF2,IL2) = S2                                      00002340
   70 CFLFL(IF1,IL1,IF2,IL2) = S                                        00002350
C                                                                       00002360
      RETURN                                                            00002370
      END                                                               00002380


      SUBROUTINE  MAKBIT( MARK,II,NFCTR )                               00002390
      DIMENSION  MARK(1)                                                00002400
C                                                                       00002410
      DO  10   I=1,NFCTR                                                00002420
   10 MARK(I) = 0                                                       00002430
C                                                                       00002440
      JB = II                                                           00002450
      DO  20   I=1,NFCTR                                                00002460
      JJ = JB / 2                                                       00002470
      IB = JB - JJ*2                                                    00002480
      MARK(I) = IB                                                      00002490
      JB = JJ                                                           00002500
      IF( JB .EQ. 0 )     RETURN                                        00002510
   20 CONTINUE                                                          00002520
      RETURN                                                            00002530
      END                                                               00002540


      SUBROUTINE  ABMAKE( SUM,SUMFL,CFL,CFLFL,NLEVEL,NFCTR,NDATA,A,B,   00002550
     1  PAR,LDIM,MAXF,MAXL,MAXA,LIST,LFCTR,MARK ,IG)                    00002560
      IMPLICIT  REAL * 8  ( A-H , O-Z )                                 00002570
      DIMENSION  MARK(1)                                                00002580
      DIMENSION  SUMFL(MAXF,1) , CFL(MAXF,1) , CFLFL(MAXF,MAXL,MAXF,1)  00002590
      DIMENSION  NLEVEL(1) , A(MAXA,1) , B(1)                           00002600
      DIMENSION  LIST(1)                                                00002610
C                                                                       00002620
      IG=0                                                              00002630
      LDIM = 0                                                          00002640
      LFCTR = 0                                                         00002650
      DO  1   I=1,NFCTR                                                 00002660
      IF( MARK(I) .EQ. 0 )     GO TO 1                                  00002670
      LFCTR = LFCTR +1                                                  00002680
      LIST(LFCTR) = I                                                   00002690
      LDIM = LDIM + NLEVEL(I)                                           00002700
    1 CONTINUE                                                          00002710
      NPAR = LDIM - LFCTR + 2                                           00002720
      LDIM = LDIM + LFCTR + 1                                           00002730
      IF(LDIM .GT. MAXA) IG=1                                           00002740
      IF(IG .EQ. 1) RETURN                                              00002750
      PAR = NPAR                                                        00002760
      DO  7   I=1,LDIM                                                  00002770
      DO  5   J=1,LDIM                                                  00002780
    5 A(I,J) = 0.D0                                                     00002790
    7 B(I) = 0.D0                                                       00002800
C                                                                       00002810
C     -----  B  -----                                                   00002820
      B(1) = SUM                                                        00002830
      K = 1                                                             00002840
      IF( LFCTR .EQ. 0 )     GO TO 16                                   00002850
      DO  15   IL=1,LFCTR                                               00002860
      I = LIST(IL)                                                      00002870
      JE = NLEVEL(I)                                                    00002880
      DO  10   J=1,JE                                                   00002890
      K = K + 1                                                         00002900
   10 B(K) = SUMFL(I,J)                                                 00002910
   15 K = K + 1                                                         00002920
   16 CONTINUE                                                          00002930
```

```
C                                                                       00002940
C     -----  A  -----                                                   00002950
      A(1,1) = NDATA                                                    00002960
      K = 1                                                             00002970
      KK = 1                                                            00002980
      IF( LFCTR .EQ. 0 )     RETURN                                     00002990
      DO  25   IL=1,LFCTR                                               00003000
      II = LIST(IL)                                                     00003010
      IE = NLEVEL(II)                                                   00003020
      KK = KK + IE + 1                                                  00003030
      DO  20   I=1,IE                                                   00003040
      K = K + 1                                                         00003050
      A(KK,K) = 1.D0                                                    00003060
      A(K,KK) = 1.D0                                                    00003070
      A(K,K) = CFL(II,I)                                                00003080
      A(K,1) = CFL(II,I)                                                00003090
   20 A(1,K) = A(K,1)                                                   00003100
   25 K = K + 1                                                         00003110
C                                                                       00003120
      IF( LFCTR .EQ. 1 )     RETURN                                     00003130
      IORGN = 1                                                         00003140
      LFCTR1 = LFCTR - 1                                                00003150
      DO  60   IL=1,LFCTR1                                              00003160
      II = LIST(IL)                                                     00003170
      IE = NLEVEL(II)                                                   00003180
      JORGN = IORGN + IE + 1                                            00003190
      ILP1 = IL + 1                                                     00003200
      DO  50   JL=ILP1,LFCTR                                            00003210
      JJ = LIST(JL)                                                     00003220
      JE = NLEVEL(JJ)                                                   00003230
      DO  40   I=1,IE                                                   00003240
      I1 = IORGN + I                                                    00003250
      DO  30   J=1,JE                                                   00003260
      J1 = JORGN + J                                                    00003270
      A(I1,J1) = CFLFL(II,I,JJ,J)                                       00003280
   30 A(J1,I1) = A(I1,J1)                                               00003290
   40 CONTINUE                                                          00003300
      JORGN = JORGN + JE + 1                                            00003310
   50 CONTINUE                                                          00003320
      IORGN = IORGN + IE + 1                                            00003330
   60 CONTINUE                                                          00003340
      RETURN                                                            00003350
      END                                                               00003360


      SUBROUTINE  AICCMP( PARTA,PARTB,PSUM,PAR,MARK,NLEVEL,NDATA,NFCTR, 00003370
     1 GMEAN,CNTRBN,RINT,ICNT,AIC,LDIM,MAXA,MAXF,LIST,LFCTR,ERRVAR )    00003380
      IMPLICIT  REAL * 8  ( A-H , O-Z )                                 00003390
      DIMENSION  PARTA(MAXA,1) , PARTB(1)                               00003400
      DIMENSION  CNTRBN(MAXF,1)                                         00003410
      DIMENSION  MARK(1) , NLEVEL(1)                                    00003420
      DIMENSION  LIST(1)                                                00003430
      DIMENSION  RINT(10,1)                                             00003440
      DIMENSION  WORK1(100) , WORK(100)                                 00003450
      DATA AICO /2.837877067D0/                                         00003460
C                                                                       00003470
      LDIMP1=LDIM+1                                                     00003480
      DO  10   I=1,LDIM                                                 00003490
   10 PARTA(I,LDIMP1) = PARTB(I)                                        00003500
C                                                                       00003510
      CALL  HUSHLD(PARTA,WORK1,MAXA,LDIM,LDIMP1)                        00003520
      CALL  SOLVE( PARTA,LDIM,WORK,MAXA )                               00003530
C                                                                       00003540
      AIC = 0.D0                                                        00003550
      DO  20   I=1,LDIM                                                 00003560
   20 AIC = AIC + WORK(I)*PARTB(I)                                      00003570
C                                                                       00003580
      AIC = PSUM - AIC                                                  00003590
      ERRVAR = AIC / NDATA                                              00003600
      DATA = NDATA                                                      00003610
      AIC = DATA*(AICO + DLOG(ERRVAR)) + 2.D0*PAR                       00003620
C                                                                       00003630
C                                                                       00003640
      DO  50   I=1,NFCTR                                                00003650
      IF( MARK(I) .NE. 0 )     GO TO 50                                 00003660
      JE = NLEVEL(I)                                                    00003670
      DO  40   J=1,JE                                                   00003680
   40 CNTRBN(I,J) = 0.D0                                                00003690
   50 CONTINUE                                                          00003700
C                                                                       00003710
      GMEAN = WORK(1)                                                   00003720
      JJ = 1                                                            00003730
      IF( LFCTR .EQ. 0 )     GO TO 200                                  00003740
      DO  100   K=1,LFCTR                                               00003750
      KK = LIST(K)                                                      00003760
      JE = NLEVEL(KK)                                                   00003770
   70 DO  80   J=1,JE                                                   00003780
      J1 = JJ + J                                                       00003790
   80 CNTRBN(KK,J) = WORK(J1)                                           00003800
      JJ = JJ + JE + 1                                                  00003810
  100 CONTINUE                                                          00003820
C                                                                       00003830
  200 CONTINUE                                                          00003840
      IF( ICNT .EQ. 0 )     RETURN                                      00003850
      IE = NLEVEL(1)                                                    00003860
      JE = NLEVEL(2)                                                    00003870
      DO  210   I=1,IE                                                  00003880
      DO  210   J=1,JE                                                  00003890
  210 RINT(I,J) = 0.D0                                                  00003900
      IF( MARK(3) .EQ. 0 )     RETURN                                   00003910
```

```
      DO  220   I=1,IE                                                  00003920
      DO  220   J=1,JE                                                  00003930
      JJ = JJ + 1                                                       00003940
  220 RINT(I,J) = WORK(JJ)                                              00003950
C                                                                       00003960
C                                                                       00003970
C                                                                       00003980
      RETURN                                                            00003990
      E N D                                                             00004000
C                                                                       00004010
C                                                                       00004020
C                                                                       00004030
      SUBROUTINE HUSHLD(X,D,MJ,N,K)                                     00004040
C                                                                       00004050
C     THIS SUBROUTINE TRANSFORMS MATRIX X TO AN UPPER TRIANGULAR        00004060
C     FORM BY HOUSEHOLDER TRANSFORMATION.                               00004070
C                                                                       00004080
C     INPUTS:                                                           00004090
C        X:      N*K DATA MATRIX                                        00004100
C        D:      WORKING AREA                                           00004110
C        MJ:     DIMENSION OF THE FIRST ARRAY                           00004120
C        N:      NUMBER OF RAWS OF X, NOT GREATER THAN MJ.              00004130
C        K:      NUBER OF COLUMNS OF X                                  00004140
C     OUTPUTS:                                                          00004150
C        X:      SQUARE ROOT OF DATA COVARIANCE MATRIX (UPPER TRIANGULAR 00004160
C                MATRIX)                                                00004170
C                                                                       00004180
      IMPLICIT REAL*8(A-H,O-Z)                                          00004190
      DIMENSION X(MJ,1), D(1)                                           00004200
C                                                                       00004210
          TOL = 1.0D-30                                                 00004220
C                                                                       00004230
      DO 100 II=1,K                                                     00004240
      H=0.0D0                                                           00004250
      IF(II .GT. N) GO TO 11                                            00004260
      DO 10 I=II,N                                                      00004270
      D(I)=X(I,II)                                                      00004280
   10 H=H+D(I)**2                                                       00004290
   11 CONTINUE                                                          00004300
      IF(H.GT.TOL) GO TO 20                                             00004310
      G=0.0D0                                                           00004320
      GO TO 100                                                         00004330
   20 G=DSQRT(H)                                                        00004340
      F=X(II,II)                                                        00004350
      IF(F.GE.0.0D0) G=-G                                               00004360
      D(II)=F-G                                                         00004370
      H=H-F*G                                                           00004380
C                                                                       00004390
      IF(II.EQ.K) GO TO 100                                             00004400
      DO 60 J=II+1,K                                                    00004410
      S=0.0D0                                                           00004420
      DO 40 I=II,N                                                      00004430
   40 S=S+D(I)*X(I,J)                                                   00004440
      S=S/H                                                             00004450
      DO 50 I=II,N                                                      00004460
   50 X(I,J)=X(I,J)-D(I)*S                                              00004470
   60 CONTINUE                                                          00004480
  100 X(II,II)=G                                                        00004490
C                                                                       00004500
      RETURN                                                            00004510
      END                                                               00004520


      SUBROUTINE   SOLVE( UL,NN,X,MM )                                  00004530
      IMPLICIT  REAL * 8  ( A-H , O-Z )                                 00004540
      DIMENSION  UL(MM,1) , X(1)                                        00004550
      N = NN                                                            00004560
      NP1 = N + 1                                                       00004570
C                                                                       00004580
      DO  10    K=1,NN                                                  00004590
   10 X(K)=UL(K,NP1)                                                    00004600
      DO  4    IBACK=1,N                                                00004610
      IP = NP1 - IBACK                                                  00004620
      IP1 = IP - 1                                                      00004630
      X(IP)=X(IP)/UL(IP,IP)                                             00004640
      IF(IP1 .LE. 0) GO TO 4                                            00004650
      DO  3    J=1,IP1                                                  00004660
    3 X(J)=X(J) - X(IP)*UL(J,IP)                                        00004670
    4 CONTINUE                                                          00004680
      RETURN                                                            00004690
      END                                                               00004700


      SUBROUTINE SORT(X,N)                                              00004710
      DIMENSION X(1)                                                    00004720
      DO 1000 K=2,N                                                     00004730
      RX=X(K)                                                           00004740
      I=K-1                                                             00004750
      ITEM=I                                                            00004760
      DO 20 J=1,I                                                       00004770
      IF(RX .GE. X(ITEM)) GO TO 10                                      00004780
      X(ITEM+1)=X(ITEM)                                                 00004790
      ITEM=ITEM-1                                                       00004800
      IF(ITEM .LT. 1) GO TO 10                                          00004810
   20 CONTINUE                                                          00004820
   10 X(ITEM+1)=RX                                                      00004830
 1000 CONTINUE                                                          00004840
 1001 CONTINUE                                                          00004850
      RETURN                                                            00004860
      END                                                               00004870
```

```
      SUBROUTINE NORMC(X,N)                                             00004880
C                                                                       00004890
C    THIS ROUTINE CHEKS THE NORMALITY                                   00004900
C    OF GIVEN DATA X(I) I=1,...,N.                                      00004910
C    THE DICISION IS MADE ON THE COMPARISON OF AIC'S OF                 00004920
C    NORMAL MODEL AND HISTOGRAM MODEL FITTED BY THE PROGRAM.            00004930
C    WHEN N IS GREATER THAN 10 AND GIVEN SET OF DATA IS JUDGED TO BE    00004940
C    NOT-NORMAL, IT IS RESONABLE TO THINK THAT THE DISTRIBUTION         00004950
C    OF THE DATA IS NOT NORMAL.                                         00004960
C    IF N IS LARGER THAN 1000 AND THE SET OF DATA IS JUDGED TO BE NORMAL,00004970
C    THE TRUE DISTRIBUTION OF THE DATA HAS TO BE                        00004980
C    VERY CLOSE TO THE NORMAL DISTRIBUTION.                             00004990
C                                                                       00005000
C                                                                       00005010
C        PERCENTAGE   OF   CORRECT  JUDGEMENT                           00005020
C              (SIMULATIONAL RESULT)                                    00005030
C                          REAL DISTRIBUTION                            00005040
C        NO. OF DATA (N)     NORMAL    TRIANGLE                         00005050
C               10            100          0                            00005060
C               20             88         13                            00005070
C               50             91         10                            00005080
C              100             96          9                            00005090
C              200             98          9                            00005100
C              500            100         31                            00005110
C             1000            100         70                            00005120
C                                                                       00005130
      IMPLICIT REAL*8 (A-H,O-Z)                                         00005140
      REAL*4 X(1)                                                       00005150
      DIMENSION T(200),  Q(2), NC(200), H(200)                          00005160
      DATA DLP /2.837877067D0/                                          00005170
C                                                                       00005180
C     INITIALIZATION                                                    00005190
C                                                                       00005200
      WRITE(6,1)                                                        00005210
    1 FORMAT(1H0,'**** NORMALITY CHECK ROUTINE ****')                   00005220
      CALL SORT(X,N)                                                    00005230
C                                                                       00005240
C     NORMAL MODEL FIT                                                  00005250
C                                                                       00005260
      SUM=0.D0                                                          00005270
      SUM2=0.D0                                                         00005280
      DO 10 I=1,N                                                       00005290
      SUM=SUM+X(I)                                                      00005300
   10 SUM2=SUM2+X(I)**2                                                 00005310
      AN=N                                                              00005320
      Q(1)=SUM/AN                                                       00005330
      VAR=SUM2/AN                                                       00005340
      Q(2)=VAR-Q(1)**2                                                  00005350
      AICN=DLP+DLOG(Q(2))                                               00005360
      AICN=AICN*N + 4.D0                                                00005370
      WRITE(6,2) AICN,Q(1),Q(2)                                         00005380
      Q(2)=DSQRT(Q(2))                                                  00005390
    2 FORMAT(1H0,'AIC OF THE NORMAL MODEL =',F10.1,                     00005400
     *      /,1H ,'      ESTIMATED MEAN     =',D10.3,                   00005410
     *      /,1H ,'      ESTIMATED VARIANCE =',D10.3)                   00005420
C                                                                       00005430
C     CATEGORIZATION                                                    00005440
C                                                                       00005450
      ANNNN=2.D0*DSQRT(AN) - 1.D0                                       00005460
      Q2=Q(2)*7.5D0/ANNNN                                               00005470
      TEM=Q(1)-0.5D0*Q2                                                 00005480
      IF(TEM .LT. X(1)) GO TO 100                                       00005490
      DO 20 I=1,100                                                     00005500
      TEM=TEM-Q2                                                        00005510
      IF(TEM .LT. X(1)) GO TO 100                                       00005520
   20 CONTINUE                                                          00005530
C                                                                       00005540
C     FREQUENCY COUNT                                                   00005550
C                                                                       00005560
  100 NN=0                                                              00005570
      T(1)=TEM                                                          00005580
      DO 30 I=1,N                                                       00005590
   33 IF(X(I) .GE. TEM) GO TO 35                                        00005600
      NC(NN)=NC(NN)+1                                                   00005610
      GO TO 30                                                          00005620
   35 NN=NN+1                                                           00005630
      NC(NN)=0                                                          00005640
      TEM=TEM+Q2                                                        00005650
      T(NN+1)=TEM                                                       00005660
      GO TO 33                                                          00005670
   30 CONTINUE                                                          00005680
C                                                                       00005690
C     HISTOGRAM MODEL FIT                                               00005700
C                                                                       00005710
      CALL POOL(H,NC,Q2,NN,N,AICH)                                      00005720
C                                                                       00005730
C     RESULT OUTPUT                                                     00005740
C                                                                       00005750
      WRITE(6,3) AICH                                                   00005760
      ITEM=(NN-1)/8 + 1                                                 00005770
      II=-7                                                             00005780
      DO 900 I=1,ITEM                                                   00005790
      II=II+8                                                           00005800
      JJ=II+7                                                           00005810
      IF(JJ .GT. NN) JJ=NN                                              00005820
      JJP1=JJ+1                                                         00005830
      WRITE(6,4) (T(K),K=II,JJP1)                                       00005840
      WRITE(6,5) (NC(K),K=II,JJ)                                        00005850
      WRITE(6,6) (H(K),K=II,JJ)                                         00005860
  900 CONTINUE                                                          00005870
```

```
    3 FORMAT(1H0,'AIC OF THE HISTOGRAM     =',F10.1)                    00005880
    4 FORMAT(1H ,'     CLASSIFICATION  =',9D11.3)                       00005890
    5 FORMAT(1H ,'     FREQUENCY       =      ',8I11)                   00005900
    6 FORMAT(1H ,'     DENSITY         =      ',8D11.3,/,1H0)           00005910
      RETURN                                                            00005920
      END                                                               00005930


      SUBROUTINE POOL(H,NX,DD,M,NN,AIC)                                 00005940
      IMPLICIT REAL*8 (A-H,O-Z)                                         00005950
      DIMENSION N(200),  H(1),NX(1),NP(200),D(200)                      00005960
      DTEM=NN*DD                                                        00005970
      DO 50 I=1,M                                                       00005980
      N(I)=NX(I)                                                        00005990
      D(I)=DTEM                                                         00006000
      NP(I)=1                                                           00006010
   50 H(I)=N(I)/DTEM                                                    00006020
      ICNT=0                                                            00006030
      NPARAM=M+1                                                        00006040
      II0=1                                                             00006050
    1 ICHK=0                                                            00006060
      DO 10 I=1,M                                                       00006070
      IF(I .EQ. 1) II=II0                                               00006080
      IF(I .GT. 1) II=II+NEXT                                           00006090
      NEXT=NP(II)                                                       00006100
      IF(II .GT. M)  GO TO 11                                           00006110
      IF(N(II) .GE. 5)  GO TO 10                                        00006120
      I1 = 0                                                            00006130
      IF(II .GT. 1) I1=II-NP(II-1)                                      00006140
      IF(ICNT .GE. 1 .AND. I1 .LT. 1) I1=M+II0-NP(M)                    00006150
      I2=II+NP(II)                                                      00006160
      IF(ICNT .GE.1 .AND.I2 .GT. M) I2=II0                              00006170
      H1=1.D50                                                          00006180
      H2=H1                                                             00006190
      IF(I1 .GE. 1)  H1=H(I1)                                           00006200
      IF(I2 .LE. M)  H2=H(I2)                                           00006210
      IF(H1 .GT. 1.D45 .AND. H2 .GT. 1.D45) GO TO 11                    00006220
    2 CONTINUE                                                          00006230
      IF(I1 .EQ. II) GO TO 11                                           00006240
      ICHK=1                                                            00006250
      J1=I1                                                             00006260
      J2=II                                                             00006270
      IF(H1 .LE.H2) GO TO 5                                             00006280
      J1=II                                                             00006290
      J2=I2                                                             00006300
      NEXT=NP(II)+NP(I2)                                                00006310
    5 IF(J1 .GT. J2) II0=II0 + NP(J2)                                   00006320
      NP0 =NP(J1) + NP(J2)                                              00006330
      NPARAM = NPARAM -1                                                00006340
      N0  = N(J1) +  N(J2)                                              00006350
      D0  = D(J1) + D(J2)                                               00006360
      H0=N0/D0                                                          00006370
      JJ=J1-1                                                           00006380
      DO 4 J=1,NP0                                                      00006390
      JJ=JJ+1                                                           00006400
      IF(JJ .GT. M) JJ=1                                                00006410
      NP(JJ) = NP0                                                      00006420
      N(JJ) = N0                                                        00006430
      D(JJ) = D0                                                        00006440
    4 H(JJ) = H0                                                        00006450
      IF(NP0 .EQ. M) GO TO 12                                           00006460
   10 CONTINUE                                                          00006470
   11 CONTINUE                                                          00006480
      IF(ICNT .EQ. 0) ICNT = 1                                          00006490
      IF(ICHK .EQ. 1)  GO TO 1                                          00006500
   12 CONTINUE                                                          00006510
      AIC = 0.D0                                                        00006520
      DO 20 I=1,M                                                       00006530
   20 AIC = AIC - NX(I)*DLOG(H(I))                                      00006540
      AIC =(AIC + NPARAM)*2.D0                                          00006550
      RETURN                                                            00006560
      END                                                               00006570


      SUBROUTINE CHECK(RDATA,IDATA,GMEAN,CNTRBN,NDATA,NFCTR)            00006580
      IMPLICIT REAL*8 (A-H,O-Z)                                         00006590
      REAL*4 W                                                          00006600
      DIMENSION RDATA(1),IDATA(5,1),CNTRBN(5,10,1),W(500),GMEAN(1)      00006610
      DO 10 I=1,NDATA                                                   00006620
      SUM=RDATA(I)-GMEAN(1)                                             00006630
      DO 5 J=1,NFCTR                                                    00006640
      JJ=IDATA(J,I)                                                     00006650
    5 SUM=SUM-CNTRBN(J,JJ,1)                                            00006660
   10 W(I)=SUM                                                          00006670
      CALL NORMC(W,NDATA)                                               00006680
      RETURN                                                            00006690
      END                                                               00006700
```

Sample input

```
2
4    4
   268     1     1
   233     1     2
   254     1     3
   281     1     4
   240     2     1
   249     2     2
   231     2     3
   314     2     4
   256     3     1
   250     3     2
   280     3     3
   291     3     4
   265     4     1
   250     4     2
   248     4     3
   271     4     4
```

Sample output

```
**      PROGRAM  VARMOD      **

NUMBER OF FACTOR =     2


                NUMBER OF LEVEL

FACTOR     1  =     4

FACTOR     2  =     4

NUMBER OF DATA =    16

ORIGINAL DATA

DEPENDENT VARIABLE      INDEPENDENT VARIABLES
             268.0                  1    1
             233.0                  1    2
             254.0                  1    3
             281.0                  1    4
             240.0                  2    1
             249.0                  2    2
             231.0                  2    3
             314.0                  2    4
             256.0                  3    1
             250.0                  3    2
             280.0                  3    3
             291.0                  3    4
             265.0                  4    1
             250.0                  4    2
             248.0                  4    3
             271.0                  4    4


***** RESULTS *****

 AIC =       138.8       NO. OF FREE PARAMETER =    5       MODEL =01
    RESIDUAL VARIANCE =  0.18370D+03
    GRAND MEAN =  0.26131D+03
    CONTRIBUTION FROM EACH FACTOR
         1        0.0              0.0              0.0              0.0
         2     -0.40625D+01     -0.15812D+02     -0.80625D+01     0.27937D+02

 AIC =       142.9       NO. OF FREE PARAMETER =    8       MODEL =11
    RESIDUAL VARIANCE =  0.16266D+03
    GRAND MEAN =  0.26131D+03
    CONTRIBUTION FROM EACH FACTOR
         1     -0.23125D+01     -0.28125D+01      0.79375D+01    -0.28125D+01
         2     -0.40625D+01     -0.15812D+02     -0.80625D+01     0.27937D+02

 AIC =       147.6       NO. OF FREE PARAMETER =    2       MODEL =00
    RESIDUAL VARIANCE =  0.46171D+03
    GRAND MEAN =  0.26131D+03
    CONTRIBUTION FROM EACH FACTOR
         1        0.0              0.0              0.0              0.0
         2        0.0              0.0              0.0              0.0

 AIC =       152.8       NO. OF FREE PARAMETER =    5       MODEL =10
    RESIDUAL VARIANCE =  0.44067D+03
    GRAND MEAN =  0.26131D+03
    CONTRIBUTION FROM EACH FACTOR
         1     -0.23125D+01     -0.28125D+01      0.79375D+01    -0.28125D+01
         2        0.0              0.0              0.0              0.0


** ANALYSIS OF THE RESIDUALS OF MAICE MODEL **

**** NORMALITY CHECK ROUTINE ****

AIC OF THE NORMAL MODEL =       132.8
     ESTIMATED MEAN      = 0.310D-05
     ESTIMATED VARIANCE  = 0.184D+03

AIC OF THE HISTOGRAM     =       138.5
     CLASSIFICATION   = -0.363D+02 -0.218D+02 -0.726D+01  0.726D+01  0.218D+02  0.363D+02
     FREQUENCY        =            1          4          7          2          2
     DENSITY          =      0.968D-02  0.968D-02  0.301D-01  0.968D-02  0.968D-02
```

III-5 Generation of Normal Random Numbers (NRAND)

The normal random number table is not so popular as the random digit table. For the convienience of readers, we introduce here a program to convert a random number table to a normal random number table. The subroutine SMM in the program is provided by Prof. Shimizu of the Institute of Statistical Mathematics [Shimizu 1976, *Central limiting theorem*, Kyoiku Syuppan, Tokyo, in Japanese].

The table of random numbers used as the sample input was generated using the method by Niki [1980, Machine generation of random numbers, Proceedings of the Institute of Statistical Mathematics, Vol. 27, No. 1, pp. 115-131, in Japanese].

The sample input and output given in III-5.1 and III-5.2 can be used as a random number table and normal random number table, respectively.

```
C83-03-03-13:57:54 AICPRN00 FORTLE                                      00000005
C82-5-17-16-56 RANDOM NUMBER TABLE                                      00000010
C81-10-1-14-03 RANDOM DIGIT TABLE                                       00000020
C                                                                       00000030
C     THIS PROGRAM CONVERTS UNIFORM RANDOM NUMBERS INTO NORMAL          00000040
C     RANDOM NUMBERS.                                                   00000050
C     SUBROUTINES SMM IS PROVIDED BY DR. SHIMIZU OF THE INSTITUTE       00000060
C     OF STATISTICAL MATHEMATICS.                                       00000070
C                                                                       00000080
      PROGRAM NRAND                                                     00000090
      DIMENSION G(10000),U(24000)                                       00000100
C                                                                       00000110
C   **********************************                                  00000120
C   *  UNIFORM RANDOM NUMBER LOADING  *                                 00000130
C   **********************************                                  00000140
C                                                                       00000150
      N3 = 0                                                            00000160
      DO 100 K=1,24000,10                                               00000170
      K1 = K+9                                                          00000180
      READ(5,555,END=150) (U(I), I=K,K1)                                00000190
      N3 = N3+10                                                        00000200
      IF(MOD(K1,100) .EQ. 0) READ(5,555,END=150) DUMMY                  00000210
  100 CONTINUE                                                          00000220
  555 FORMAT(10F7.7)                                                    00000230
C                                                                       00000240
C   ****************************************                            00000250
C   *  CONVERSION TO NORMAL RANDOM NUMBERS  *                           00000260
C   ****************************************                            00000270
C                                                                       00000280
  150 CONTINUE                                                          00000290
      CALL SMM(N3,N,G,U)                                                00000300
      WRITE(6,603)                                                      00000310
  603 FORMAT(1H1,'NORMAL RANDOM NUMBERS',/,1H )                         00000320
      DO 200 K=1,N,10                                                   00000330
      K1=K+9                                                            00000340
      WRITE(6,6) (G(I),I=K,K1)                                          00000350
    6 FORMAT(1H ,10F10.5)                                               00000360
      IF(MOD(K1,100) .EQ. 0) WRITE(6,601)                               00000370
  200 CONTINUE                                                          00000380
      STOP                                                              00000390
  601 FORMAT(1H0)                                                       00000400
      END                                                               00000410


      SUBROUTINE  SMM(N3,N,G,URAN)                                      00000420
      DIMENSION G(1), URAN(1)                                           00000430
      DATA A1,A2,A3/-0.01441012,-0.06794095,-0.06185342/                00000440
      DATA B1,B2,B3/-0.9855899,-1.062783,-0.2152481/                    00000450
      DATA C2,C3/0.4923018,0.07289628/                                  00000460
      DATA Z1,Z2,Z3,Z4/0.02710381,0.2372523,0.7627477,0.9728962/        00000470
      DATA DE,DE1,DE2,DE3/0.466,0.4666459,0.4909798,0.4984253/          00000480
      DATA T1,T2,T3,T4,T5/-2.486797,-1.675003,-1.185167,-2.759952,      00000490
     + -31.89943/                                                       00000500
      DATA S0,S1,S2/2.370334,4.434955,34.38623/                         00000510
      DATA AL1,AL2,AL3,AL4/0.1568,2.02080,2.9528,8.719028/              00000520
      KU = 1                                                            00000530
      J = 0                                                             00000540
  500 CONTINUE                                                          00000550
      J = J+1                                                           00000560
      U = URAN(KU) - 0.5                                                00000570
      KU = KU + 1                                                       00000580
      IF(KU .GT. N3) GO TO 900                                          00000590
      UA=ABS(U)                                                         00000600
      IF(UA-DE ) 1,1,101                                                00000610
    1 CONTINUE                                                          00000620
      V = URAN( KU )                                                    00000630
      KU = KU + 1                                                       00000640
      IF(KU .GT. N3) GO TO 900                                          00000650
      IF(V-Z4) 2,2,305                                                  00000660
    2 IF(V-Z3) 3,3,304                                                  00000670
    3 IF(V-Z2) 4,4,303                                                  00000680
    4 IF(V-Z1) 301,301,302                                              00000690
  101 IF(UA-DE1)  102,102,103                                           00000700
  103 IF(UA-DE2)  104,104,105                                           00000710
  105 IF(UA-DE3)  106,106,107                                           00000720
  102 CONTINUE                                                          00000730
      U1 = URAN(KU)                                                     00000740
      KU = KU + 1                                                       00000750
      IF(KU .GT. N3) GO TO 900                                          00000760
      X=AL1*U1                                                          00000770
      U2 = URAN(KU)                                                     00000780
      KU = KU + 1                                                       00000790
      IF(KU .GT. N3) GO TO 900                                          00000800
      IF(EXP(-0.5*X*X)+B1+A1*U2) 102,102,310                            00000810
  104 CONTINUE                                                          00000820
      U1 = URAN(KU)                                                     00000830
      KU = KU + 1                                                       00000840
      IF(KU .GT. N3) GO TO 900                                          00000850
      X=4.0*DE*U1+AL1                                                   00000860
      U2 = URAN(KU)                                                     00000870
      KU = KU + 1                                                       00000880
      IF(KU .GT. N3) GO TO 900                                          00000890
      IF(EXP(-0.5*X*X)+C2*X+B2+A2*U2) 104,104,310                       00000900
  106 CONTINUE                                                          00000910
      U1 = URAN(KU)                                                     00000920
      KU = KU + 1                                                       00000930
      IF(KU .GT. N3) GO TO 900                                          00000940
      X=2.0*DE*U1+AL2                                                   00000950
      U2 = URAN(KU)                                                     00000960
      KU = KU + 1                                                       00000970
```

```
      IF(KU .GT. N3) GO TO 900                                  00000980
      IF(EXP(-0.5*X*X)+C3*X+B3+A3*U2) 106,106,310               00000990
  107 CONTINUE                                                  00001000
      U1 = URAN(KU)                                             00001010
      KU = KU + 1                                               00001020
      IF(KU .GT. N3) GO TO 900                                  00001030
      X=AL4-2.0*LOG(U1)                                         00001040
      U2 = URAN(KU)                                             00001050
      KU = KU + 1                                               00001060
      IF(KU .GT. N3) GO TO 900                                  00001070
      IF(AL4-U2*U2*X) 107,107,309                               00001080
  301 G(J)=S2*V+U+T1                                            00001090
      GO TO 500                                                 00001100
  302 G(J)=S1*V+U+T2                                            00001110
      GO TO 500                                                 00001120
  303 G(J)=S0*V+U+T3                                            00001130
      GO TO 500                                                 00001140
  304 G(J)=S1*V+U+T4                                            00001150
      GO TO 500                                                 00001160
  305 G(J)=S2*V+U+T5                                            00001170
      GO TO 500                                                 00001180
  309 X = SQRT(X)                                               00001190
  310 IF(U) 320,320,321                                         00001200
  320 G(J)=-X                                                   00001210
      GO TO 500                                                 00001220
  321 G(J)=X                                                    00001230
      GO TO 500                                                 00001240
  900 CONTINUE                                                  00001250
      J = J-1                                                   00001260
      N = J/100                                                 00001270
      N = N*100                                                 00001280
      RETURN                                                    00001290
      END                                                       00001300
```

III-5.1 Sample input: Random number table

```
793613 395070 878677 810155 171001 762524 565596 018039 945319 182408
583237 838454 693255 308921 865716 654394 385737 332286 195432 057777
984641 647219 251266 253194 980036 696742 075166 285071 408315 289572
607482 259379 541923 314208 104504 802972 844075 391267 714695 259515
644112 477864 713893 231078 020619 445100 799765 292238 619696 466048
921108 910130 392164 329068 484887 197787 938657 631763 724110 388826
300962 745590 266635 126413 194686 776678 453861 466392 346664 998176
815618 881463 432005 601661 363229 339937 128568 976110 493726 855083
881392 156589 723433 667960 781105 772554 706716 535992 617528 374151
745747 809900 560061 775196 518559 264687 589503 423649 878151 165425

324487 999286 553185 796079 750274 787377 730964 264823 888413 663708
320607 859277 333191 264168 298142 843211 834629 204988 100309 973445
617586 672923 958302 225655 090846 518132 881907 783891 831729 304387
314423 863962 836077 040984 756482 624937 916159 875754 907591 834251
278055 414093 614477 669665 583702 100934 785869 225204 356561 530592
421588 198025 659991 978827 325988 372303 026208 591271 709278 395448
286019 263814 826084 469976 710732 524266 919484 876025 925657 343266
239239 794182 804995 665442 911623 800608 475337 883754 544243 146831
484424 900218 241047 579368 797022 214412 248959 290205 572215 275564
054373 516767 118374 197697 238010 908560 035617 389260 144768 204641

775522 473577 329502 161973 648013 957248 800711 180729 872641 834575
521272 873951 685566 700281 989938 673053 421036 332752 802504 239238
746517 848246 487634 304944 693939 294230 766050 816869 792537 424546
350067 352148 640979 154524 545005 452029 650550 594440 369839 163629
834535 231723 334434 601817 677599 128446 746820 914816 028036 170920
155932 120917 557795 266751 606952 763314 457488 807771 218967 867539
723660 145914 796874 104862 233621 611881 538989 456201 765904 254345
559802 993539 304524 058171 034555 917937 731948 894641 735086 154783
941514 041643 271116 228634 503923 278921 804722 622150 494810 183244
227030 412601 018201 656510 633495 210589 275558 680742 177676 474507

015349 503660 290443 044801 997093 620157 274984 804473 730543 645095
605204 775674 506312 110302 409547 203971 679940 743861 352738 485003
446384 480725 226277 560205 278937 464155 078859 714518 475274 743726
969685 023143 995549 224007 279879 946922 854589 044349 446749 022674
949519 213097 412156 275480 286824 327949 910042 306523 942747 854479
350822 690000 540379 674839 377896 991341 277019 247250 651654 105619
968304 261928 831128 678348 111529 808070 172258 494751 235369 276730
206024 811081 480135 329302 460145 084921 047241 978952 268533 270934
227431 086503 764879 139662 840341 383819 553064 110295 221756 557557
475470 495149 272644 487880 130377 140467 525196 706499 252650 334633

560938 559791 075264 471307 683696 439163 655352 470979 269269 134268
047504 031744 090068 243032 319070 294937 999833 316701 264322 071652
293418 672441 480033 665097 336323 400944 536921 323255 124322 741129
393877 482325 280281 447846 015920 194894 131547 871952 674604 585828
435481 630989 021972 391874 188250 605674 983588 967941 468509 432978
722675 555553 014381 116785 895299 395860 617609 735316 964441 434538
157450 972276 917483 695964 979741 909902 578743 688919 527693 255144
253402 613768 940861 829861 014274 534673 383892 224948 615933 779401
508917 560862 478932 904581 656787 527562 955348 985007 695943 154298
565122 683266 549142 843341 627956 165544 945743 329578 203422 342336

757835 047238 649157 347299 916384 690452 829251 545252 325398 533730
123253 863650 931041 785584 697908 463692 082350 099334 538341 621358
297006 418204 467823 146532 542550 558351 512231 073100 189459 304452
728111 597373 237762 114447 259857 830685 946695 808100 471816 645847
029743 196722 406455 563245 400401 318378 276841 603730 033896 871032
812416 078846 196393 153280 641004 905911 564161 852581 017677 199703
487541 225780 022336 922432 588385 321841 414504 348337 990235 153034
907211 796610 496876 082274 247169 409169 675088 159684 890073 272963
284391 968301 594350 824400 896075 284010 569656 501268 478056 763151
540529 090457 193743 451528 180344 721281 229166 603441 634209 965225

023141 513245 732134 541854 031101 403083 590629 764736 744210 351140
341167 236337 262307 435142 195281 082825 848922 048502 695263 088177
232430 932130 038494 491585 491105 458768 243455 456667 518173 250848
928105 525436 897070 577931 288494 997874 665393 804750 206605 532475
350113 273190 446364 000163 812574 263906 681011 682140 697040 185930
229141 091485 947471 169171 617501 523767 354890 053499 122704 820513
236255 779237 545067 282011 804102 968680 729469 049870 422456 680203
119783 613028 475661 935585 500222 691084 127032 662111 042159 357798
164363 930875 157012 296236 193295 192030 639158 518262 379022 955632
918725 768848 217863 308114 832834 689294 228670 677083 977367 264151

236565 961263 731727 145257 449748 085242 897976 593739 900291 727057
320088 522361 138750 115092 468503 176507 709681 295837 376850 940349
716763 044158 464834 566191 324534 474994 379315 005652 852001 003362
011306 237327 505149 793333 397733 860345 319288 220200 832299 483386
120139 012345 759003 760012 609191 226078 986058 076479 389156 615900
725595 210870 738974 834202 518068 232415 663685 195835 641991 223311
323124 781804 772775 504326 755362 753779 161678 285637 114611 984514
246633 581214 016988 091780 085484 573808 774313 792484 391788 087885
294206 168944 329052 220934 129476 748123 805547 172879 322249 257420
432926 510802 988416 754617 709651 677671 444772 758417 108825 653502

469055 195604 723422 519481 287574 404535 344195 655320 820470 217658
682336 870946 846633 008522 925529 712916 382751 789394 746543 177881
448378 532397 421192 566566 172931 649182 061808 615770 860118 655504
323958 376431 944401 182978 775497 914738 306102 705645 583264 881509
744758 857429 663941 328849 024539 132801 217237 234411 345611 767446
530191 728194 838559 833552 077211 425609 064927 218094 450255 210966
242251 956850 393498 214304 226899 888802 003214 496129 562897 164797
312322 044378 642249 307358 137676 557504 585398 487839 585677 087371
809843 602707 833582 067349 934027 150453 550688 553254 011961 019450
309030 678954 120137 593775 060706 188516 639811 371322 160595 191155
```

```
017688 007410 668812 527829 123678 950144 245128 535003 162296 618708
667201 360506 409737 430419 309585 082511 478975 994488 720875 901594
271301 379404 817772 729171 664904 756924 946330 522432 542540 946975
979831 322017 149292 351558 649784 127970 411986 977973 980397 426549
585149 450216 142115 927246 758967 529213 852541 144084 940583 103584
446594 867300 257412 472346 831057 639792 552642 121888 257749 501967
510069 277048 431859 734984 971639 936271 094954 356131 779455 304283
162935 262259 705791 002689 945136 408898 578887 549586 296917 498175
780294 819228 638841 013571 193055 895944 686124 343721 823070 081320
106876 724314 997058 914836 715496 217760 747418 243171 819511 758786

542080 675248 732399 472353 106725 948465 634913 424161 439027 350848
629858 717208 563769 576686 501396 368118 379914 626229 047930 951851
071155 521024 777590 217135 265934 906820 883604 151687 236346 158894
523186 395188 608342 868394 210936 228652 594806 358950 553414 613127
976275 123817 370696 068365 801324 175724 632583 595010 136679 422096
515483 999476 233588 474235 757605 549914 025191 040144 484679 004073
598365 130460 388734 584151 485950 357592 078869 543271 443433 560082
232332 867277 330858 609687 736917 931203 379655 318494 452437 173549
269950 658942 871610 607053 811277 646137 993458 264589 806899 716908
260986 682820 763879 286795 079254 069894 133141 301197 180608 036803

983622 888967 151681 575990 022783 873324 527840 860516 567159 283415
294879 126736 968418 721485 319145 188666 380494 241506 251625 727183
367701 793606 434513 479347 709802 534035 149841 130263 468295 267568
073032 636094 372060 904400 550679 784559 100840 010118 845662 507424
092496 631169 743941 134720 624342 459867 113152 599285 348715 110176
644995 299895 367988 021777 456399 346652 963560 882515 462590 037663
372540 079213 639865 069148 038128 006196 936264 335345 331796 432880
294650 451931 588704 838171 535086 975107 695067 186677 020652 501067
024496 193984 328606 283267 588428 882577 864208 442614 454790 688156
481111 919533 733547 712856 353876 738940 366926 740250 163690 379701

416240 043198 656013 986283 407295 220778 037224 753281 529491 112241
949154 825975 020863 458138 699312 236623 603955 716217 781041 960457
750034 605943 198269 313701 415662 946483 391733 959792 354377 978113
496618 933232 374672 043054 857902 403904 851752 103485 097232 806801
869972 483356 339738 032977 200184 995513 930799 081702 898411 600511
372864 554627 389102 596251 730864 679241 205212 651387 800458 772676
400728 559259 941588 072938 975317 560900 605993 800786 686707 759638
968113 271278 830928 026193 964979 614099 189300 730002 767411 459006
994833 801956 488056 525791 679596 566680 663933 820746 641045 811879
875700 736117 159935 757979 333386 270070 517493 722851 308159 985668

374073 261675 414300 315335 515136 993604 149019 110975 728326 289614
923458 040733 442963 238012 712174 154004 672076 969645 693938 990331
571527 471195 336346 720721 777263 522819 108488 527818 937165 328922
512607 561115 769173 629919 141216 805482 135253 224500 920464 955710
189101 376031 216710 660412 755186 019306 384563 446433 247714 128996
875093 758621 750596 610378 727581 018505 171667 451215 420584 114383
371482 345420 787766 520225 908500 097740 948570 486235 787100 623653
539383 237216 176888 224588 058878 177519 373056 146816 515267 915375
263962 289856 017532 393889 837130 098143 667102 658860 262733 166441
344096 997052 892362 694053 761076 876239 753555 111588 469811 385065

469616 210573 150487 189945 614075 182748 433520 226561 952182 811645
639799 919521 197407 609526 231977 975347 429521 439931 566617 478495
529073 211763 327228 975232 576458 485586 237661 704118 816712 938160
715380 945047 963414 662268 375866 706465 520227 132898 367158 601249
242832 862422 264461 916101 582522 205579 255155 574484 157463 556869
803395 372374 047615 108354 299539 699755 656925 684877 131273 951434
050283 679056 142972 551317 647950 482986 923143 355244 068932 477087
916194 081811 503997 697871 863739 668229 347618 345865 274503 811760
730093 410208 081980 739428 374826 442667 961328 949365 123462 978147
319324 228311 858275 320601 023721 795421 347548 873794 585707 700167

445142 328396 799835 821899 846802 012059 051105 462376 281619 044680
863760 158934 859033 763029 206326 378090 961124 898592 497902 146320
277913 877755 882878 625345 985451 264475 038629 241796 194384 187573
308031 475753 294310 850411 727200 044400 396268 397339 003597 442594
111228 145991 842122 677229 309641 157347 226295 291572 947980 192440
474246 582542 140556 997923 543331 141623 990070 350971 139541 698796
923691 133794 305865 232161 748597 001122 615279 177721 223621 204053
063302 753186 106900 088214 627811 337781 207312 245611 695493 974012
629582 022726 550094 524084 483323 477123 814810 426543 771117 826102
237560 216865 774835 578803 576771 737712 917110 889297 214848 251967

542059 155980 456906 823206 198241 166012 545598 310149 863911 058939
303285 647353 630500 419102 394067 294806 457554 233988 239002 682240
808224 790746 698025 895654 268697 613823 115769 317452 288342 659006
380781 085310 028413 417084 640751 632137 602464 584279 324778 232959
568243 557071 401442 525757 254693 878551 932445 231364 153389 409700
872432 971691 304177 360373 075811 860220 099439 094783 982042 204098
358814 120639 600876 734084 545967 867627 041810 340792 562961 238683
354005 088499 660368 829615 343053 592383 940886 189273 422686 128237
922672 685499 909600 704279 898352 178951 674835 656426 074979 774187
446208 770442 725965 783181 812334 549871 171859 542015 382210 218123

887254 999001 576344 395670 278267 980185 095396 453504 594528 739897
643903 100531 311485 912350 646564 389606 312326 014924 451878 245556
654529 631265 112974 043425 836800 725160 053951 118216 184674 669921
604079 104220 577110 453578 558326 687685 529959 525871 498815 687034
702614 293944 432920 691342 550515 111991 373583 738382 817041 819769
316366 353192 477098 325525 625855 872061 206334 375818 723985 573113
012888 123348 731273 511909 057452 585798 687248 198469 600841 872996
202862 299788 946568 143020 672413 766316 076011 622787 364985 216117
354339 382633 311858 395301 845306 972000 538652 022768 484520 641624
648033 893129 186169 542543 030352 309054 378379 256405 946248 208343
```

912245	944034	615962	484687	627820	425264	810774	299740	171790	173330
414080	524241	555422	586475	978901	687104	798413	364611	127313	898219
442002	476028	152091	071226	849700	544612	426052	951770	954030	092011
749126	356109	323769	718364	867136	601396	665324	316560	413527	871869
117976	421495	334681	861656	421722	996059	615271	767495	391660	784251
070866	448427	202834	224051	811757	422871	404996	312838	891414	171629
249308	556486	681211	269130	920733	282696	494070	943078	348395	078369
550843	622213	661378	167238	897596	706778	295974	233525	203086	659603
153826	021382	710506	648473	548286	432637	317745	203604	075800	730056
238491	227696	643165	550635	163824	831023	063803	876814	845954	653505
161487	853445	752161	690994	793198	016121	084504	094310	149934	948539
062586	430223	318523	164865	452269	709888	237801	118702	788308	121607
989768	605461	356702	653540	206035	699855	828461	836390	667752	950967
424577	653669	037670	953108	065711	997980	602635	681267	535257	523307
422556	706512	717251	162845	219062	576588	577929	151202	048833	986894
747885	407581	491720	400234	102505	973208	976163	551970	994559	662419
580459	724560	305617	182116	725216	846862	793548	817734	745425	517459
146686	324914	236789	013531	577859	954041	382016	944374	201817	732712
728869	857139	309903	398148	423219	078386	152724	325571	313539	602273
961612	531341	226549	059929	344391	935417	186238	122020	945508	184697
295856	655463	590299	391329	681104	148027	691588	876186	545534	149116
905739	158398	359060	897347	007891	117720	029145	439430	311822	906621
601572	763357	878953	080138	919022	199391	950536	796628	148692	018408
043984	194043	117243	044977	149146	355764	172703	596356	865656	761924
205195	435779	052445	693292	429970	227382	636630	214083	404261	969434
392227	074831	897283	379438	341701	182319	176040	612629	495963	002752
448229	970554	942158	188056	710355	579598	477099	658274	147355	024099
533654	078334	126914	063003	866041	128437	380591	779320	511653	122787
735344	594258	475978	780650	479508	251341	052865	771931	202429	818907
331790	162742	142546	964430	631593	923753	896723	218910	806613	883485
873172	842300	399537	424656	032320	108181	999205	309669	177252	032192
653907	171503	411313	317563	899962	843999	632536	146444	304352	838101
249052	107756	766100	908722	777234	318516	762867	476673	787180	996948
624939	904705	639834	091967	651723	638805	798796	822936	698026	493321
059442	723113	381521	787498	086611	026802	980373	651510	819321	904684
595843	388236	955457	243721	201513	681372	103612	446730	859654	554175
509506	292627	534362	790564	844407	872163	608629	250978	940770	761827
147427	925015	164795	031829	224821	315556	113970	543268	767394	036251
716658	631327	685730	676920	719039	682449	554139	044142	004517	265392
233650	251064	757585	038558	787617	568330	166379	784043	739717	594717
022904	240222	122969	777772	642233	045928	469157	261969	117130	628994
209387	235667	134725	430463	332268	921092	518166	090697	487654	833857
836956	450197	306180	264603	113100	714381	684427	981719	958344	495825
316085	699154	732337	359553	521378	873563	318272	050372	497464	203859
061004	186894	098866	466539	112833	383807	901827	151729	112477	361387
126998	393119	906148	789647	599949	571801	137420	722727	799397	523368
910221	744919	444437	857659	052422	424044	478724	032982	481261	731539
716847	502030	596130	772517	178225	834612	141432	166853	491324	495795
320919	446997	963722	505954	055738	994670	158102	478824	690248	274933
150950	378482	024215	555506	763259	250402	345806	759918	729358	099207
550854	429685	305207	937187	991255	889707	552217	115388	901845	479301
495238	964397	350195	992472	906905	437064	950271	721402	682779	056733
078202	859704	946076	936574	494002	385806	331428	421775	238609	307033
077920	198825	138278	059360	409590	808098	783711	251244	110383	233350
323463	528209	381205	608525	137697	397109	117214	867508	719884	040372
332766	833237	854156	280274	485478	193139	494108	116215	201780	492535
954092	699390	822539	916826	143527	584009	318122	323419	859725	254607
442205	885859	401671	941705	182255	066940	699742	724255	053434	088840
408783	337539	931676	060308	032870	037829	933108	270974	099089	817050
472152	089157	809782	146470	613513	077219	647729	688000	754140	551705
713737	457845	026180	753245	852836	888464	674922	615561	860526	959106
330398	241985	669346	005461	158929	491923	814993	693318	927463	437717
586063	723815	194494	268302	737506	705181	190324	043891	813803	376234
857020	467696	033888	952597	801602	263487	647412	721678	035968	355414
252068	778485	639570	876779	735191	185577	696409	900278	491110	750517
805759	847039	069500	933971	475321	149082	865631	822718	398780	244839
400496	694323	769793	495404	266497	969554	106523	186387	233068	861792
801109	560640	355121	262465	071192	223213	497249	940751	240180	552191
809575	274901	659363	357550	708068	586128	975122	013138	632108	214129
856390	415978	852850	695699	820143	240229	882657	497438	916002	049903
237299	920044	660153	658031	258615	231467	762488	446374	711427	694029
623125	488731	984303	614598	313791	084679	629628	096030	650063	369859
920202	026000	628510	663013	557277	673020	154381	521910	865266	837347
462726	301969	406653	218923	127765	735031	714342	981436	165139	650680
156568	916144	038870	351046	223491	746123	823376	511241	877573	944971
901204	585589	607065	658977	861116	476166	944501	477783	197935	764577
408900	987745	135369	912898	257057	394532	861780	931832	210343	290945
420521	691965	306140	269651	391257	180867	320779	882423	684582	308687
110879	400218	752917	447408	559115	602942	751186	562221	795660	733442
148654	165143	646238	894298	143126	809607	727690	267721	751496	450541
859089	630548	865501	752065	705335	528247	676007	178679	261619	078677
022983	747411	702271	818204	504607	078786	741963	382974	470998	193945
025469	376977	599963	345812	949035	169912	859328	562992	383297	362208
238927	532275	100724	310628	925729	072890	266386	025559	828723	585207
501619	237936	919383	447178	423783	011813	081780	413546	198598	306847
921750	020540	082871	094030	421619	389252	139176	010474	422204	396549
094417	148218	508201	535147	255102	191296	490255	230448	219628	750987
627969	106525	679253	700255	609350	797062	565310	528799	616882	835230
335599	364487	673225	646761	726403	530119	012511	864271	227780	342306
092450	572054	993461	388050	052484	546175	076889	579727	565451	469221

III-5.2 Sample output:

```
NORMAL RANDOM NUMBERS

  0.04489   1.21173   0.29327  -1.80091  -0.42071   1.04179  -0.25967   0.73168  -0.51180  -1.72333
  0.62875  -1.14490  -0.43225   0.04434  -0.14125  -1.39732   1.28645   0.40016   0.10111   0.48486
 -2.04671   0.73207   0.07596   1.29117   0.15452  -0.20676   1.10074   0.66298  -0.58296  -0.30756
 -1.18517   0.16731  -0.39707   2.33133   0.77779  -0.43126   1.17397   0.52747  -0.21593  -0.36618
  1.09848  -0.13194   0.88610  -1.15716   0.55262   1.54014   1.34750  -0.42557   0.51664  -1.14366
 -0.39036  -0.07093  -0.87518   1.91872  -0.47670  -0.89391   0.66751   0.46948   1.34218   1.72132
 -0.77483   1.10435   1.44850   0.24215   0.48862  -0.39009  -0.21359   0.85417  -0.88064  -0.03862
 -1.65830  -1.13325  -0.92331  -1.10848  -1.14370   0.38409  -0.43056   0.01281   1.24841   0.79090
  0.38500   0.81380   2.34106  -0.01412   0.63187   0.32356   0.31893   0.26464   0.43167   1.07178
 -0.43085   0.18631  -0.23880   0.30888  -0.21409   0.60480  -0.66072   0.52278   0.21349  -1.13585

 -1.31253  -0.24209   0.02026   0.16254  -0.39612   0.89769   0.42006  -1.03404   0.20430   0.59300
 -0.10391   2.30462   0.84561   1.43969  -0.75346  -1.04880  -0.88991  -0.52011   0.59426  -0.86751
 -0.48014  -0.54934  -0.70628  -1.98450  -0.49306   1.93164  -0.41321   0.85094   0.39446   0.29063
 -0.60410   0.13049  -0.10520  -0.14209  -0.69075  -0.46379  -1.36111   0.15591   1.78428  -2.01250
  1.21948  -1.12373  -1.76037  -0.28041  -0.62003  -0.62099  -0.04856   1.47237   0.30118   0.45480
  2.06693  -0.82208  -1.05493   0.64503  -1.00203  -0.60298  -0.63640  -0.98457   0.26399  -0.26517
 -1.88057   0.65735  -1.56393  -0.79073   0.06495  -1.13279  -0.14182  -0.03603  -0.25609  -1.42166
  0.51467  -0.63933   0.20266  -0.49275   0.03949   0.08656  -1.31026  -1.98672  -1.01903  -0.66700
  3.31943  -0.91802   0.12511  -0.22287  -0.01154  -1.30039  -0.01042  -0.53848  -1.99152  -1.39670
  0.78582  -0.06710  -1.60027  -0.94825   2.02811   1.03010   0.35435  -0.89468   1.75261  -1.04218

  1.14042   0.02307   1.36130  -0.57610   0.30054   0.01092   0.77622   1.50453   1.91966  -0.67029
 -1.20767  -0.21280   0.86782   0.43651  -0.09465   0.69355   1.15512   0.11184  -1.65211   0.32600
 -0.39688  -1.05732   0.18086  -1.33858  -0.77406   0.45892  -1.42967   0.68395   1.27063   0.31752
 -0.52349  -0.17284  -0.71058  -1.42095   1.21411  -1.22516  -0.01250   0.55799  -1.52637  -0.32983
 -1.99297   0.63193  -0.38081   1.99934   0.91652   0.28920  -1.01702   0.32418   0.84716  -0.73810
  1.82708   0.07266   0.60264  -1.23330  -0.42115   0.20485  -0.02565   1.65499  -1.16681   0.11790
  0.84359  -0.52535  -0.82708  -0.87380   0.58780   0.01134  -1.05601  -1.07215  -0.02950  -0.64933
 -0.00026   1.10700   1.24395  -0.42341   2.34710  -2.53483  -0.24705   0.61274  -0.65337  -1.54013
 -0.47727   0.17383  -1.58285   0.50169   0.43218  -0.47164   1.84020  -1.22436   0.34960  -0.11230
  1.36499   0.45315   0.01129  -0.79491   1.03280  -0.82598  -1.13006   0.18244   1.35725   1.06858

 -0.73697   0.78152   0.14841   0.64918   1.01053  -0.47386   0.80777   1.32654  -0.19939  -1.03729
 -0.45957   0.17352  -0.49607   0.95415  -0.53920  -0.34972  -0.31107  -2.81486  -0.59918   0.05093
 -0.06800   0.65146  -1.15881  -1.99481   0.16388  -0.51421   1.17867  -0.62619  -0.64280  -0.54264
  0.53044   0.28303   0.85690  -0.84643   1.56889  -0.06086  -0.32788   0.74790   0.03598  -0.89992
 -0.73626  -1.37985   1.06073  -0.74845  -0.40157   2.09926   0.75156   0.04676  -0.93395   0.08015
  0.22519  -0.48404  -0.46478   1.03411   0.14985   1.36577  -2.07840   0.72871  -0.46894  -0.41910
  1.57237   0.29355   1.23276   1.28746  -0.24174  -0.40434  -0.63154   0.33901   1.18721  -0.99902
 -1.46145  -0.39982  -0.89999   0.20440  -0.95441  -1.98748   0.14521  -0.78006  -0.11845  -1.25706
  0.25992   0.19092   0.41904   1.03965   0.94976  -0.22940  -2.02225  -0.73283  -0.15759  -1.27824
 -0.16520  -1.16664  -1.14067  -0.15399  -0.92022   0.51503  -0.35345  -0.52093  -0.46733   2.27365

  0.86098   0.77390   0.49950   1.48238   0.75704   0.20659  -0.58065   1.98426  -0.03286   0.99446
  0.32821  -0.68346  -0.77503   1.03308  -0.30814   0.66241  -1.08179  -0.23759  -0.51840   0.48885
  1.90201   0.55303  -1.14811   0.25005  -2.61620   0.90657  -0.18431  -0.99128   0.13857   2.46103
  0.76825  -0.22037   0.15865   0.36836   0.07999  -0.41653  -1.33621  -0.90758   0.70965  -0.83768
  1.56561  -0.97513  -0.28615   0.15199  -0.37112  -0.04663  -0.01444   1.78409  -0.68268   0.36225
 -0.01001  -0.97383  -0.04121   3.40990   0.37592  -0.39998   0.05085  -1.46763  -0.09081  -0.58454
 -0.03365   0.67126   0.14594  -0.29425  -0.87175   1.26454   0.94508   2.40797   0.58320  -0.34964
  0.81064  -1.53672  -1.51463  -1.07282   1.46040  -0.61334  -1.62739   0.43931   0.51970  -0.70279
  1.16167  -0.20720  -0.59460  -0.84723  -0.08641   0.13838   0.47665  -0.97643  -0.44489  -1.58354
 -0.16717   0.52452  -0.94322  -0.32921  -0.83358   0.02921  -0.15151  -1.33766  -0.32932  -1.86998

 -0.40709   1.61752  -1.54538  -1.45116  -1.22847  -2.73561   0.04598  -0.32730  -0.31929   1.04600
  1.66589  -0.65203  -1.09079  -0.71228  -0.00713   1.45535  -0.16455   0.14338   0.97312  -0.13351
 -0.07649  -0.70880  -0.31884  -0.08700   0.91600   0.13758  -1.14773   1.35236  -0.59787   0.92015
  1.05312  -0.68975  -0.38621   0.18985   0.11462   0.85519   0.13012  -0.86430   0.41541   0.33052
 -1.68901   2.03268  -0.88186   0.63665   0.00235   0.11725   0.65572   0.06405   0.96729   0.04119
 -0.90994   0.66246  -1.95552   0.23448   0.17024   2.26468  -0.14879   2.15969  -0.27036  -1.07090
 -0.67804  -0.77983   1.71245   2.34824   0.00325   0.35953   0.33135  -0.32557   0.03165   0.15747
  0.57712   0.45354  -1.04410   1.89904  -0.60475   0.09694  -1.56775  -0.24241  -1.35520   0.98811
  0.51223  -1.62290  -0.44397  -1.24714  -0.49492   0.33570  -0.83303   0.41594   0.58020  -0.58358
 -1.00208  -1.32884  -1.15082   1.31496  -0.73415  -0.33974  -0.40354  -0.70310   2.72311  -0.30262

 -0.77151  -1.18212  -0.75045  -0.73670   1.29184   1.45788  -0.04298   1.37103  -0.21286   0.01564
 -0.70677   1.46234   0.04229   0.22149   1.71746   1.64667   0.84804   0.36526  -1.06538   0.10715
  0.80768   1.06738  -0.68075  -0.06829  -0.20774   0.00088  -1.64684   0.27302   0.59514   1.09089
 -0.02530  -0.23539   0.10762   0.08002  -0.48538  -0.89598   0.47302   0.76250  -0.51773   0.61467
  0.01725   0.14950  -0.26107   1.91177   1.35881  -0.84313  -0.06696  -0.76893   1.31747  -0.68174
 -0.60638   0.98308  -0.58264   1.68639  -1.02818   0.91076   0.67999   0.64978  -1.07112  -0.76746
 -0.51180   0.88895  -0.70148  -2.46577  -1.23912   0.62081  -0.27399  -1.01405   1.23587  -0.36861
 -0.96910   2.52718   0.81101   1.53537  -0.82637   0.32142  -3.11294  -1.04642   0.16343  -1.67688
 -0.25670  -0.89568   1.78865  -1.57575   0.10718  -0.07090   0.14069   1.17489  -0.97566   0.46162
  0.64023   1.60115  -0.87307  -0.94118   0.84783  -1.24051  -0.40441  -1.04970   0.15256  -0.06126

 -0.59231  -0.67972   0.17097   1.05519   1.41026   0.03849  -0.81693   0.16524  -1.41588  -0.59104
 -0.17655  -0.55570   1.75395  -1.26337   1.01894   0.72379   0.63089  -1.65521   0.53724  -0.14025
  0.34304  -1.12195  -0.00997  -0.60737  -0.03138  -0.04240   1.50522  -0.49399   0.96029   1.45958
  1.42848   0.09337  -1.18605   0.14668   0.80605   1.12589  -0.86295  -0.23719   0.89310   3.13387
  1.58369  -0.51482   0.66316  -1.08525   1.09776  -0.11511  -2.16129  -0.65124   0.46567  -1.86944
  0.87050  -1.59677   0.08744  -1.10871  -0.03293   0.50320   0.09128   0.44215  -0.28581   0.38646
 -1.12781   0.43863   1.19273  -0.53162  -0.43647   1.23345  -0.58802   0.39729  -0.38672  -1.40830
  0.52964  -0.06251  -0.40232   1.23782   0.05169  -1.07158  -0.19724  -0.62915  -0.56333   0.88424
  1.16085   0.32022   1.34907  -0.21299  -0.73288   1.19302   0.99416   0.71890   0.28766   0.76106
 -1.11338  -0.53033   0.15561   1.84775   0.96808  -1.24576   0.26074  -1.02446   0.57965  -0.13067

  1.92290   0.18252  -0.56162   1.30412   0.49327  -0.38841  -0.77992  -0.47037   0.17611   2.08685
 -0.97851   0.12893  -0.53864  -0.52242  -0.11680  -0.36603  -0.09435   1.41663  -1.47904   0.34053
 -0.77193   0.88773  -0.84336   0.08140  -2.09772   0.56244  -0.11139  -0.95428   0.12111  -0.92669
  0.26319   0.58942   0.69248   0.70981   0.68652   0.70488  -1.63926  -1.67224   1.09671  -0.60281
 -1.12531   0.44977  -1.41076  -0.84737   1.92940  -1.35427  -0.93047   2.30563   1.22106   1.16021
  0.28681  -0.76833  -2.28473   1.54904   1.31032   0.25342   1.27029  -0.43152  -1.40415  -0.76073
  0.05596   0.53590  -1.68267   1.23297  -1.44761  -0.41037   0.16435  -0.16729  -0.83741   1.31748
 -0.96815  -0.56678   1.07880  -2.13051  -0.50662   0.64738   1.40152   0.89601   1.15502  -0.71893
 -0.23879  -1.85829  -0.69274  -0.09890   0.98650  -0.44703   0.01061  -0.73660  -0.58893   1.44370
 -1.45090   0.11151  -1.02472  -0.05699  -2.39620   1.49264  -0.39883   0.39903   0.35226  -2.01077

 -1.29394  -1.76867  -0.73935   0.57689  -1.11879   0.45877   0.67817  -0.60990   0.21639   0.57429
 -1.12146   1.15980   1.46844  -0.30742   1.46488   1.34878  -0.27905  -0.73402  -0.92349   1.38311
 -0.89299   0.76134  -1.44806   1.53629  -0.15294   0.20757   2.66902   1.37731  -1.12730   0.48074
  1.18853   0.18219   0.08829   0.61409  -1.97857   0.61110  -1.03412   0.99930   0.07671  -0.12593
  1.27552   0.62965   1.16330  -0.75934  -0.51913  -0.57587  -0.28347  -1.24684   0.52795   0.60509
  0.65150  -1.42510  -2.26814   0.36162   0.27165  -0.86879   0.85225  -1.60450  -1.38942   0.61491
 -0.52718  -1.39357  -0.61739  -1.34184  -0.46712   0.46415  -0.43857   1.28577  -0.50922  -1.40881
  0.65153   1.94315   0.28815  -0.10057   1.13564  -1.63333  -0.77343  -1.28513  -0.48045  -0.66258
 -0.60027  -0.71608  -0.62635   1.14824   0.27014   0.16536   0.35479   0.99076   0.98816  -0.62762
 -1.55000   0.53008   0.22166   0.76226   0.61974  -1.29358  -0.01864  -0.30472   0.47783   1.85924
```

-0.39209	-0.34324	-0.63709	-0.62355	0.80357	-0.28025	-0.53204	2.62326	0.51348	0.85508
-0.03199	0.10931	2.65715	0.63100	1.83979	-0.27668	-0.35399	-0.71879	-1.21530	-1.77347
0.73352	-0.30592	-1.02972	-0.10967	0.13844	-0.60619	0.70462	-1.27607	0.76818	-0.16667
-0.83296	-1.16549	-0.31592	0.92671	1.62867	-0.15735	-0.60043	-0.22194	1.11100	1.31814
-1.69587	0.73130	-1.72757	-0.47630	-0.97586	-0.66190	0.25104	0.42055	-0.08447	-0.07261
0.79040	0.55833	-1.62872	1.27558	0.80309	1.17200	0.05709	0.14339	-2.16892	0.77322
0.27983	0.61658	-0.85471	0.72385	-1.79002	0.02044	0.28045	-0.04131	-1.29381	-0.73227
0.60931	0.93426	0.15113	0.37920	1.06407	-1.01973	0.37547	0.72816	0.08279	-0.49102
0.84837	-0.55808	0.35536	0.79512	0.44484	-0.70792	-1.11387	1.40948	-0.13611	-0.22399
-0.17829	0.41222	0.55594	0.93838	1.07304	0.89482	1.29991	-1.07279	0.79966	-0.41413

0.35365	0.44753	0.48588	1.93576	-0.32678	-0.10806	-0.04827	1.19096	-0.44457	1.31891
-0.50667	-0.79744	0.18487	2.06279	0.02230	0.95967	-0.81420	0.30688	0.35002	1.80852
0.60408	0.48389	0.30462	0.39184	0.32885	1.97427	0.92408	-0.49294	1.73446	-0.78519
0.37554	-0.73986	-0.98161	0.97433	-0.26889	-0.62564	0.12825	0.30312	0.39867	0.84899
-1.29395	1.35246	0.47374	-0.32289	0.13426	0.66853	0.96298	0.27229	-0.70656	-1.56446
-0.87066	-1.91707	0.11392	1.29479	0.72104	-0.21363	-0.75662	-1.19602	1.15624	-0.98085
-1.09182	0.29418	-2.15681	-0.62315	-0.75924	-1.35876	-1.67511	-0.34089	-2.48746	-0.32301
-1.42325	0.09151	-1.07151	-0.66272	0.31455	-1.07460	0.65392	0.88433	0.13357	1.06114
-0.48561	0.52110	0.29779	-1.76780	-1.42268	2.33400	-1.55419	-1.28783	0.23487	-0.80510
0.40154	0.36551	0.02007	-0.15214	0.44325	-1.30456	0.87457	1.20157	0.06861	1.71042

0.11775	-0.49011	0.03714	-0.43562	-0.04622	0.56319	-1.38682	0.12821	0.63044	-1.62760
-1.45986	-0.01226	0.68752	-1.31087	-0.96262	-0.10695	1.51079	1.93585	-0.29713	-2.29711
-0.17945	-0.97810	-1.92013	-2.62329	0.44309	-1.65254	1.41204	0.36271	1.03395	0.88399
0.91322	-1.83964	-0.19403	1.06598	1.24110	-0.02876	-1.73796	-0.07107	-0.13240	0.24712
-0.23720	0.26908	-0.01752	1.31727	-0.72435	-1.65232	1.12205	-0.45507	0.77914	-0.27365
-0.01825	-0.25116	0.77331	2.14992	-1.17863	0.06559	-0.18618	-0.55106	-0.25185	-0.16969
-1.24625	-1.97622	-1.21859	0.83121	0.40707	-0.97157	0.78039	-0.19385	0.79797	-0.69821
-0.65758	-1.28979	-0.09458	0.51927	-0.11574	-1.78504	-1.06359	-0.88849	0.07937	-0.79658
-1.98302	1.78420	-0.94323	0.57723	-1.52889	1.76487	0.09894	-0.37259	0.43482	-0.84924
1.56139	0.58233	-0.50950	0.04652	-0.02175	-1.41546	1.00813	-0.41856	-1.31523	-1.13825

REFERENCES

The Akaike Information Criterion has a wide range of applications. Since we do not have enough space to describe every topics, we have presented only a few introductory examples in this book. In particular we have not given any explanation of models in time series analysis apart from the AR model. We have also not given any explanation of a new technique developed recently which applies an information criterion to Bayesian models whose parameters are distributed according to a prior distribution. The list of references covers topics not necessarily discussed in the text.

Theory of Information Criterion

Akaike, H. (1973). Information theory and an extention of the maximum likelihood principle, 2nd Inter. Symp. on Information Theory (Petrov, B. N. and Csaki, F. eds.), Akademiai Kiado, Budapest, 267-281.

Akaike, H. (1974). A new look at the statistical model identification, IEEE Trans. Autom. Contr., AC-19, 716-723.

Akaike, H. (1977). On entropy maximization principle, Application of Statistics (Krishnaiah ed.), North-Holland, 27-41.

Akaike, H. (1978). Canonical correlation analysis of time series and the use of an information criterion, System Identification: Advances and Case Studies (R. K. Mehra and D. G. Lainiotics, eds.), Academic Press, 27-96.

Akaike, H. (1980). Likelihood and the Bayes procedure, Bayesian Statistics (Bernardo, J. M., De Groot, M. H., Lindley, D. U. and Smith, A. F. M. eds.), University Press, Valencia, Spain.

Boltzmann, L. (1876). Uber die Beziehung zwischen dem zweiten Hauptsatz der mechanischen Warmetheorie und der Wahrscheinlichkeitrechnung respective den Satzen uber das Warmegleichgewicht. Wiener Berichte, 76, 373-435.

Ogata, Y. (1980). Maximum likelihood estimators of incorrect Markov models for times series and derivation of AIC, Journal of Applied Probability, Vol. 17, No. 1, 59-72.

Shimizu, R. (1978). Entropy maximization principle and selection of the order of an autoregressive Gaussian process, Ann. Inst. Statist. Math, Vol. 30 A, 263-270.
Shibata, R. (1980). Asymptotically efficient selection of the order of the model for estimating parameters of a linear process, Ann. Statist., Vol. 8, 147-167.

General Statistical Theory

Cox, D. R. (1975). Partial likelihood, Biometrika, Vol. 62, No. 2, 269-276.
Kullback, S. (1959). Information theory and statistics, Wiley & Sons, New York.
Kullback, S. and Leibler, R. A. (1951). On information and sufficiency, Ann. Math. Statist., Vol. 22, 79-86.
Rao, C. R. (1965). Linear statistical inference and its applications, John Wily & Sons, New York.
Sanov, I. N. (1961). On the probability of large deviation of random variables, Selected Translations in Math. Stat. and Prob. (Inst. Math. Statist. and American Mathematical Society eds.), Vol. 1, American Mathematical Society, Rhode Island, 213-244.

Computational Algorithm and Program Package

Akaike, H., Arahata, E. and Ozaki, T. (1975). TIMSAC-74, A time series analysis and control program package--(1) & (2), Computer Science Monographs, No. 5 & No. 6, The Institute of Statistical Mathematics, Tokyo.
Akaike, H., Kitagawa, G., Arahata, E. and Tada, F. (1979). TIMSAC-78, Computer Science Monographs, No. 11, The Institute of Statistical Mathematics, Tokyo.
Akaike, H. and Ishiguro, M. (1980). BAYSEA, a Bayesian seasonal adjustment program, Computer Science Monographs, No. 13, The Institute of Statistical Mathematics, Tokyo.
Forsythe, G. E. and Moler, C. B. (1969). Computer solution of linear algebraic systems, Prentice-Hall, Inc.
Golub, G. H. (1965). Numerical methods for solving linear least square problems, Number. Math., 7, 206-216.
Katsura, K. and Sakamoto, Y. (1980). CATDAP, A categorical data analysis program package, Computer Science Monographs, No. 14, The Institute of Statistical Mathematics, Tokyo.

Discrete Distribution Models

Sakamoto, Y. (1977). A model for the optimal pooling of categories of the predictor in a contingency table, Research Memo., No. 119, The Institute of Statistical Mathematics, Tokyo.

Sakamoto, Y. and Akaike, H. (1978). Analysis of cross-classified data by AIC, Ann. Inst. Statist. Math., Vol. 30 B, No. 1, 185-197.

Sakamoto, Y. and Akaike, H. (1978). Robot data screening of cross-classified data by an information criterion, Proc. International Conference on Cybernetics and Society, IEEE, New York, 398-403.

Sakamoto, Y. (1982). Efficient use of Akaike s information criterion for model selection in high dimentional contingency table analysis, Metron, 40, 257-275.

Time Series Models

Akaike, H. (1969). Fitting autoregressive models for prediction, Ann. Inst. Statist. Math., Vol. 21, 243-247.

Akaike, H. (1979). On the likelihood of a time series model, The Statistician, Vol. 27, 217-235.

Haggan, V. and Ozaki, T. (1981). Modeling nonlinear random vibrations using an amplitude-dependent autoregressive time series model, Biometrika, Vol. 68, No. 1, 189-196.

Kitagawa, G. (1977). On a search procedure for the optimal AR-MA order, Ann. Inst. Statist. Math., Vol. 29 B, No. 2, 319-332.

Kitagawa, G. and Akaike, H. (1978). A procedure for the modeling of non-stationary time series, Ann. Inst. Statist. Math., Vol. 30 B, No. 2, 351-363.

Kitagawa, G. (1980). Changing spectrum estimation, J. Sound and Vibration, Vol. 89, No. 4, 433-445.

Kitagawa, G. (1981). A nonstationary time series model and its fitting by a recursive technique, J. Time Series Analysis, Vol. 2, No. 2, 103-116.

Nakamura, H. and Akaike, H. (1979). Use of statistical identification for optimal control of a supercritical thermal power plant, Identification and System Parameter Identification, Pergamon Press, 221-232.

Nakamura, H. and Akaike, H. (1981). Statistical identification for optimal control of supercritical thermal power plants, Automatica, Vol. 17, No. 1, 143-155.

Ohtsu, K., Horigome, M. and Kitagawa, G. (1979). A new ship s auto pilot design through a stochastic model, Automatica, Vol. 15, No. 3, 255-268.
Ooe, M. and Sato, T. (1981). An extended response method for analysis of disturbed earch tides data and rank decision with the AIC, Proc. of the 19th International Symposium on Earth Tides, Schweizerbartische Verlagsbuchhandlung.
Otomo, T., Nakagawa, T. and Akaike, H. (1972). Statistical approach to computer control of cement rotaly kilns, Automatica, Vol. 8, 35-48.
Ozaki, T. and Tong, H. (1975). On the fitting of non-stationary autoregressive models in time series analysis, Proc. of 8th Hawaii Int. Conf. on System Science, Western Periodicals Company, 224-226.
Ozaki, T. and Oda, H. (1977). Non-linear time series model identification by Akaike s information criterion, IFAC Workshop on Information and Systems, 82-90.
Ulrych, T. J. and Ooe, M. (1979). Autoregressive and mixed autoregressive moving average method, Topics in Applied Physics, Vol. 34 (Haykin, S. ed.), Springer-Verlag, 73-125.

Other Models

Ishiguro, M. and Ishiguro, M. (1979). Fitting a Gaussian model to aperture synthesis data by Akaike s information criterion (AIC), Image Formation from Coherence Function in Astoronomy (Schooneveld, C. van ed.), D. Reidel Pub. C.. 277-286
Kitagawa, G. (1979, 1981). On the use of AIC for the detection of outliers, Technometrics, Vol. 21, No. 2, Vol. 23, No. 3, 193-199, 320-321.
Ogata, Y., Nakamura, M. and Omura, Y. (1980). Evaluation of neuronal spiketrain data by new methodologies of wave form discrimination and point process analysis, Proceedings of the 7th International CODATA Conference (Mashiko, Y. ed.), Pergamon Press, Oxford. (to appear)
Ogata, Y. and Akaike, H. (1982). On linear intensity models for mixed doubly stochastic Poisson and self-exciting point processes, J. R. S. S., B, Vol. 44, No. 1, 102-107.
Ozaki, T. (1977) On the order determination of ARIMA models, Appl. Statist., Vol. 26, No. 3, 290-301.
Tong, H. (1975). Determination of the order of a Markov chain by Akaike s information criterion, J. Aappl. Prob., Vol. 12, 488-497

Bayesian Models

Akaike, H. (1980). Likelihood and the Bayes procedure, Bayesian Statistics (J. M. Bernado, etc. eds.), University Press, Valencia, Spain, 141-166.
Akaike, H. (1980). Seasonal adjustment by Bayesian modeling, Journal of Time Series Analysis, Vol. 1, No. 1, 1-13.
Akaike, H. and Ishiguro, M. (1980). Trend estimation with missing observations, Ann. Inst. Statist. Math., Vol. 32 B, 481-488.
Akaike, H. and Ishiguro, M. (1981). Comparative Study of the X-11 and BAYSEA Procedure of Seasonal Adjustment, A paper presented at the ASA-CENSUS-NBER Conference on Applied time Series Analysis of Economic Data, Washington, D. C., October, 13-15.
Ishiguro, M. and Akaike, H. (1981). A Bayesian approach to the trading-day adjustment of monthly data, Time Series Analysis (Anderson, O. D. and Perryman eds.), North-Holland. 213-226
Ishiguro, M., Akaike, H., Ooe, M. and Nakai, S. (1983). A Bayesian Approach to the Analysis of Earth Tides, Proc. 9th International Symp. on Earth Tides, (Kuo J. T. ed.), Schweizerbartische Verlagsbuchhandlung. Stuttgart.
Ishiguro, M. (1984). Computationally efficient implementation of a Bayesian seasonal adjustment procedure, J. Time Series Analysis, Vol. 5, No. 4, 245-253.
Ishiguro, M. and Sakamoto, Y. (1983). A Bayesian approach to binary response curve estimation, Ann. Inst. Statist. Math., 35, B, 115-137.
Ishiguro, M. and Sakamoto, Y. (1984). A Bayesian approach to the probability density estimation, Ann. Inst. Statist. Math., Vol. 36, B, 523-538.
Kitagawa, G. and Akaike, H. (1982). A quasi Bayesian approach to outlier detection, Ann. Inst. Statist. Math., Vol. 34, No. 2, 389-398.
Kitagawa, G. and Gersch, W. (1984). A smoothness priors state space approach to the modeling of time series with trend and seasonality, JASA Vol. 79, No. 386, 378-389.
Kitagawa, G. and Gersch, W. (1985). A smoothness priors long AR model method for spectral estimation, IEEE Trans. on Autom. Control, Vol. 30, No. 1, 57-65.
Sakamoto, Y. and Ishiguro, M. (1985). Bayesian binary regression involving two explanatory variables, Ann. Inst. Statist. Math., Vol. 37, No. 1, 369-387.

2.1 The number of free parameters: binomial distribution 1, Poisson distribution 1, multinomial distribution (with c categories) $c-1$, uniform distribution on [0, 1] 0, normal distribution 2, k dimensional normal distribution $k^2+3k/2$, chi-square distribution 1.

5.1 From (5.5), $AIC(1)=552.51$; replacing 0.1 in (5.6) by 0.5, $AIC(0)=554.52$; hence a "significant difference".

5.2 By the same procedure as the third model in section 5.2.1, for "no difference",

$$AIC(0)$$

$$=(-2)\{(\sum_{i=1}^{2} n(i)) \log \frac{\sum_{i=1}^{2} n(i)}{2n} + n(3) \log \frac{n(3)}{n}\} + 2\times 1$$

$$=636.18 .$$

By putting $c=3$ in (5.24), $AIC(1)=634.07$; hence a " significant difference"

5.3 For the hypothesis that the median is \$20,000, replacing 0.1 in (5.6) by 0.5, $AIC(0)=138.63$; for the alternative $AIC(1)=136.60$; hence the median is not \$20,000.

5.4 For the hypothesis in question,

$$AIC(2)$$

$$=(-2)\sum_{i_2=1}^{c_2}\left[\{n(1, i_2)+n(2, i_2)\} \log \frac{n(1, i_2)+n(2, i_2)}{n(1)+n(2)}\right.$$

$$\left.+n(3, i_2) \log \frac{n(3, i_2)}{n(3)}\right] + 2(c_1-1)(c_2-1) = 4454.05.$$

Hence this is best.

5.5 Considering the initial categorization as *0.00~1.00*,

1.00~2.00, …, the respective table is represented by *MODEL*(2, 2, 2) and *MODEL*(3, 0, 3). From (5.60), $AIC(2, 2, 2)=332.68$; $AIC(3, 0, 3)=360.31$; hence the first table is better.

5.6 From $f(x_1, \cdots, x_n \mid \lambda)= \lambda^n e^{-\lambda \sum_{i=1}^{n} x_i}$, $l(\lambda)=n \log \lambda - \lambda \Sigma x_i$

and then from $\partial l(\lambda)/\partial\lambda=0$, $\hat{\lambda}=1/\bar{x}$;

$$F(x)=\int_0^x \lambda e^{-\lambda z} dz=\left[-e^{-\lambda z}\right]_0^x=1-e^{-\lambda x}.$$

5.7 Denoting the classes by $I_1=(a_{i-1}, a_1)$, $i=1, \cdots, 4$ and using the results of the preceding problem

$$AIC(0)$$
$$=(-2)\sum_{i=1}^{4} n(i) \log \{e^{-1/23.2a_{i-1}}-e^{-1/23.2a_i}\}+2\times 1=229.58 .$$

Using (5.45), $AIC(1)=232.58$; hence accept the exponential distribution.

6.1
$$(-2)\sum_{i_1, i_2, i_3} n(i_1, i_2, i_3) \log \frac{n(i_1, i_2)}{n(i_2)}$$
$$+2c_2(c_1-1)+(-2)\sum_{i_1} n(i_1) \log \frac{n}{n(i_1)} -2(c_1-1)$$
$$=(-2)\sum_{i_1, i_2} n(i_1, i_2) \log \frac{n(i_1, i_2)}{n(i_2)}$$
$$+(-2)\sum_{i_1, i_2} n(i_1, i_2) \log \frac{n}{n(i_1)}+2(c_1-1)(c_2-1)$$
$$=(-2)\sum_{i_1, i_2} n(i_1, i_2) \log \frac{n \cdot n(i_1, i_2)}{n(i_1)n(i_2)} + 2(c_1-1)(c_2-1).$$

6.2 Regarding I_4 in Table 6.4 as J, from (6.38),

$$AIC(I_1;I_4)$$
$$=(-2)\sum_{i_1,i_4} n(i_1,i_4)\log\frac{n\cdot n(i_1,i_4)}{n(i_1)n(i_4)}+2(c_1-1)(c_4-1),$$

which gives -68.77 for the values in Table 6.4; hence I_4 is the third best explanatory variable after (I_2, I_3) and I_2.

6.3 Using (6.38), $AIC(I_1;J)=-30.12$.

6.4 $AIC(I_1;J)=(-2)\{n\log n+\sum_{i_1,j} n(i_1,j)\log n(i_1,j)$
$$-\sum_{i_1} n(i_1)\log n(i_1)-n(j)\log n(j)-(c_1-1)(c_J-1)\}.$$

7.1 $AIC(\mu,\sigma^2)=48.2$, $AIC(\mu_0,\sigma_0^2)=49.2$, $AIC(\mu_0,\sigma^2)=49.3$, $AIC(\mu,\sigma_0^2)=49.4$, $\hat{\mu}=0.3013$, $\hat{\sigma}^2=0.5332$; hence reject the hypothesis of $N(0.0, 0.1)$.

7.2 $\hat{\sigma}_1^2=1.3134$, $\hat{\sigma}_2^2=5.0492$, $AIC=73.9$.

8.1 (i) $\Sigma x_i=2.44$, $\Sigma x_i^2=0.9408$, $\Sigma y_i=54$, $\Sigma x_i y_i=20.54$, $\Sigma y_i^2=452$.

(ii) 1) $MODEL(0)$: $\hat{a}_0=5.4$, $d(0)=16.04$, $AIC(0)=29.75$ 2) $MODEL(1)$: $\hat{a}_0=0.198$, $\hat{a}_1=21.318$, $d(1)=0.342$, $AIC(1)=-4.74$ 3) In this case, it is natural to take $y=a_1x_i+\varepsilon_i$ under the assumption $a_0=0$. For this model, the normal equation $0.9408a_1=20.54$ yields $\hat{a}_1=21.832$, $d=0.356$, $AIC=-6.33$; from the comparison of AIC values, select $y_i=21.83x_i+\varepsilon_i$. (Reader should find $d(2)$ and $AIC(2)$.)

(iii) $d=0.672$, $AIC=-1.97$; this value is greater than that in (ii); hence the subjective opinion is not quite correct.

8. 2 (i)

Order	*AIC*	*d*
0	31. 09	0. 179
1	16. 02	0. 093
2	6. 64	0. 060
3	7. 51	0. 057
4	7. 97	0. 054

Hence *MAICE* is the second order AR model

$$z_t = 1.092 z_{t-1} - 0.588 z_{t-2} + \varepsilon_t, \quad \varepsilon_t \sim N(0, 0.06).$$

(ii) The original series is not symmetrical; AR model which always yields symmetric distribution is not adequate for such a process; taking logarithms of the original series, we have apporoximately symmetrical data and AR model better fits to them.

9. 1 *MODEL(2)* with *AIC(2)=112. 1* gives *MAICE* among models with *4* free parameters or less.

INDEX